Mammalian Cell Technology and Its Applications

Mammalian Cell Technology and Its Applications

Asok Mukhopadhyay
Former Senior Scientist
National Institute of Immunology
New Delhi

I.K. International Publishing House Pvt. Ltd.
NEW DELHI

Published by
I.K. International Publishing House Pvt. Ltd.
S-25, Green Park Extension
Uphaar Cinema Market
New Delhi–110 016 (India)
E-mail: info@ikinternational.com
Website: www.ikbooks.com

ISBN 978-93-84588-45-8

Published by Krishan Makhijani for International Publishing House Pvt. Ltd., S-25, Green Park Extension, Uphaar Cinema Market, New Delhi–110 016 and Printed by Rekha Printers Pvt. Ltd., Okhla Industrial Area, Phase II, New Delhi–110 020.

Dedicated to my
Parents whose blessings from the
Heaven have been always the
source of my inspiration

PREFACE TO THE SECOND EDITION

The applications of animal cell technology for manufacturing valuable therapeutics are steadily growing, though the rate has come down recently. This is mainly due to the technological advancements in healthcare sector. The emerging trend in the past 10 years is inclined towards permanent healing of aliments rather than temporal relief. Now the cell technology has been diversified for regeneration/reconstruction of diseased organs or tissues, looking for an enduring solution. Thus, the second edition has been carefully revised, and accordingly the name of the book is changed to 'Mammalian Cell Technology and Its Applications'. To cope with the recent developments, the fourth section of the book has been thoroughly revised providing detailed aspects on cell based therapeutic applications for human or animal healthcare. The first part of this section covers basics of different 'stem cells' and how they are derived. Next two chapters are dedicated for 'regenerative medicine' where tissue engineering and cell therapy approaches are elucidated. I hope the book will be useful to the students of medicine, biotechnology, regenerative medicine courses and also for any one who desires to explore the fast growing area of life science.

Asok Mukhopadhyay

PREFACE TO THE FIRST EDITION

Animal cell technology has made tremendous progress in human healthcares. Beginning with the discovery of rabies vaccine by Louis Pasteur in 1885, and gradual development in cell culture processes it has become possible for on-time eradicate of many infectious diseases. With the advent of recombinant DNA and hybridization technology it is now possible to manufacture many complex therapeutic proteins using animal cells, which otherwise could not be produced or isolated from natural sources. Another form of medicine where animal cells are directly used is regenerative medicine and tissue engineering. The future of healthcare is relies on the progress on these new endeavors of animal cell technology. The development of animal cell technology in various domains of healthcare is not possible only by expanding our knowledge on basic biology, but it also require the development of instruments/equipments, computational methods, assay procedures, culture media, bioengineering, etc. In other wards, impact of animal cell technology in today's healthcare is the tireless efforts of medical personnel, engineers, and the biologists. The greatest challenge in animal cell technology is speedy development of manufacturing processes of more complex and delicate materials emerging as potential medicines.

An animal cell is a unique factory, where thousands of genes are encoded and transcribed; products are translated, and finally processed to biologically active molecules. It is therefore, important to understand inside of a cell, how cellular functions are coordinated, limitation of cells, reasons for proliferation and cellular death, etc. The very first section of the book deals basic biology aspects to understand cell and how it functions. The second section projected to offer basic cell culture technology among the readers. This section covers preservation of animal cells, cell culture medium, culture environment, good manufacturing practice and equipment, quantitative analysis, etc. In third section, recombinant therapeutic proteins, large-scale cell culture, and scale up processes are primarily discussed. Forth and the final section are made for the advanced studies, where therapeutic applications of cells and tissues are discussed. Embryonic and somatic stem cells, cloning, tissue engineering are the main subjects of the fourth section.

Since, animal cell technology commands proper understanding of cell biology, DNA technology, immunology and bioengineering, the goal of this book has been to amalgamate knowledge from these areas and pass-on to the readers intended to start professional careers in academic or in industrial research. I hope that the book will be of use to students of biotechnology and bioengineering discipline and for anyone else in the related area who feels the need to explore this fascinating unit of our life.

Asok Mukhopadhyay

ACKNOWLEDGEMENTS

I am indebted to all of my teachers and mentors whose coaching, patience and prodding helped to build my career. I am especially thankful to Professor SK Majumder for introducing the power of microbes and their industrial applications during my graduation in Jadavpur University, Calcutta. I am equally grateful to Professors TK Ghose, KB Ramachandran, Purnendu Ghosh for teaching me the engineering aspects of biology in M.Tech course at IIT Delhi. Professors GP Talwar (Talwar Research Foundation, New Delhi), SN Mukhopadhyay (IIT Delhi), and BB Aggarwal (MD Anderson Cancer Centre, Texas, USA) mentored me during doctoral and postdoctoral research and extended ardent support to build my research career, for that I am indebted to them. Two important personalities, with whom I had an opportunity to work with in the beginning of my professional career, are Prof. (Late) NC Nigam (Ex-Director, Thapar Corporate R&D Centre, Patiala) and Dr. HN Asthana (Ex-Chief Technologist, Indian Drugs & Pharmaceutical Limited, Rishikesh). I am grateful to them for making me to comprehend how to conceive two-dimensional bookish knowledge into three-dimensional reality.

The idea for this book was conceived during various stimulating discussions in a workshop held in the Department of Biotechnology and Bioengineering, IIT Delhi in connection with the curriculum development in Biochemical Engineering. I am thankful to the participants of this workshop. I am grateful to my friend, Dr. PK Roychaudhury of the same department for sharing his views, and explaining the need of a textbook for graduate and postgraduate students in various technical institutions. I am also pleased to have the opportunity to thank Professor J Deb, Dr. James Gomes of IIT Delhi, and Dr. Akhil C Banerjee of NII for reviewing a few chapters of this book and providing valuable suggestions.

Nobody can be an expert of each and every domain of biotechnology. To write this book, I took help of published literature that appeared in the scientific journals, edited books, reports through the Internet, and also from the trade literature. In many cases, I was able to manage copyright permission to reproduce figures and tables in this book. I am wholeheartedly grateful to all of them for granting me the permission to use them. Despite my sincere efforts, I was unable to reach many of them for copyright permission. However, for the interest of students I have had no option but to use related photographs, tables, and figures for this book. I have duly acknowledged all of them at the places where these have been used. I am indebted to all of them to support me for the completion of this book.

I extend thanks to my friends and colleagues those who have directly and indirectly helped me to make this project a success.

My special thanks to Dr Rajarshi Pal, Associate Professor, Manipal Institute of Regenerative Medicine, Bengaluru, India for reviewing and giving valuable suggestions to improve the contents of this book in the second edition.

Last but not the least, my wife Devi to whom I am indebted for encouraging and supporting me throughout the period of this project.

Asok Mukhopadhyay

CONTENTS

Preface to the Second Edition *vii*
Preface to the First Edition *ix*
Acknowledgements *xii*

Section 1: Introduction to Cell Biology

1. Organization of Animal Cells and Special Properties **3**
1.1 Organization of Mammalian Cells 3
1.2 Comparative Physical Properties of Mammalian and Microbial Cells 11
1.3 Glucose Metabolism in Normal and Tumor Cells 13
1.4 Folding and Secretion of Proteins in Mammalian Cells 16
1.5 Post-translational Modification of Proteins in Mammalian Cells 18
1.6 Summary 21

2. Proliferation and Death of Animal Cells: Molecular Mechanism **23**
2.1 Cell Growth 23
2.2 Growth Regulation 29
2.3 Hayflick's Phenomenon 33
2.4 Cell Signaling 35
2.5 Stress-induced Cell Dearth 39
2.6 Summary 45

Section II: Basic Cell Culture Technology

3. Sources of Animal Cells, Preservation and Control of Cell Lines/Tissues **49**
3.1 Sources of Cells 49
3.2 Selection of Cells for Manufacturing Biological 52
3.3 Generation of Cell Banks 55
3.4 Cryopreservation of Animal Cells and Tissues 59
3.5 Summary 63

4. Design of Cell Culture Media **65**
4.1 Selection of Cell Culture Medium 65
4.2 Basic Constituents of Media 66

4.3 Other Important Defined Medium Supplements 68
4.4 Serum and Its Role in Animal Cell Culture 69
4.5 Physicochemical Properties of Cell Culture Media 72
4.6 Development of Serum-free Media 74
4.7 Design Approaches to Optimize SFM 77
4.8 Metabolism and Catabolic Control 78
4.9 Media Preparation and Testing 79
4.10 Summary 81

5. Good Manufacturing Practices, Aseptic Techniques, and Cell Culture Equipment 83
5.1 Laboratory Design 83
5.2 Aseptic Process 88
5.3 Biosafety Cabinet 93
5.4 Risk Assessment 98
5.5 Sterile Filtration 99
5.6 Ultra-pure Water for Cell Culture 105
5.7 Basic Equipment 107
5.8 Culture Vessels 109
5.9 Summary 110

6. Environmental Factors and Cell Culture Methods 113
6.1 Animal Cells and the Culture Environment 113
6.2 Environmental Parameters 115
6.3 Suspension Versus Adhered Cell Culture 116
6.4 Adhesion of Cells to Culture Surface 117
6.5 Solid Microcarriers 118
6.6 Hydrodynamic Cell Death 124
6.7 Oxygen Supply in Cell Culture Medium 126
6.8 Types and Operation of Cell Culture Bioreactor 130
6.9 Cell Retention Systems in Perfusion Bioreactor 136
6.10 Summary 140

7. Separation of Viable Cells 143
7.1 Physical Properties 143
7.2 Biological Properties 146
7.3 Summary 153

Section III: Therapeutic Proteins and Bioengineering Aspects

8. Stable Gene Expression in Mammalian Cells and Monoclonal Antibody 157
8.1 Pathway for Recombinant Protein Production in Mammalian Cells 157
8.2 Selection and Gene Amplification 161
8.3 Monoclonal Antibodies (Mab) 165
8.4 Development of Human Mab 172
8.5 Summary 176

9. Large-scale Animal Cell Culture for the Production of Therapeutic Proteins **179**
9.1 Development of Large-scale Animal Cell Culture Process 179
9.2 Animal Cell Culture-based Products and Market Potential 181
9.3 Cell-line Dependent Variation of Product Quality in Large-scale Manufacturing Processes 182
9.4 Production of Monoclonal Antibodies (Mab) 185
9.5 Fluidized-bed Bioreactor 194
9.6 Stirred Tank Bioreactor 197
9.7 Summary 201

10. Scale-Up Studies in Animal Cell Culture Bioreactors **203**
10.1 Definition of Scale-up and Biological Constrains 203
10.2 Problems in Cell Culture Process and Corresponding Developments 206
10.3 Density Scale-up Versus Volume Scale-up 206
10.4 Bioreactor Scale-up 209
10.5 Summary 217

Section IV: Advanced Cell Culture Technology

11. Embryonic and Adult Stem Cells **221**
11.1 Beginning of Life 221
11.2 Stem Cells and Classification 226
11.3 Pluripotent Stem Cells: Derivation and Characterization 228
11.4 Adult Stem Cells 240
11.5 Plasticity in Adult Stem Cells 249
11.6 Induced Pluripotent Stem (iPS) Cells 252
11.7 Stem Cell Niche 255
11.8 Summary 259

12. Regenerative Medicine: Tissue Engineering **264**
12.1 The Biology of Autonomous Regeneration 264
12.2 Cell and Tissue Therapy in Regenerative Medicine 267
12.3 Tissue Engineering 268
12.4 Bioreactors in Tissue Engineering 282
12.5 Current Approaches in Tissue Engineering 287
12.6 Organ Culture 288
12.7 Bioartificial Organs 294
12.8 Summary 296

13. Regenerative Medicine: Cellular Therapy **300**
13.1 Cloning 301
13.2 Gene Therapy 308
13.3 Adult Stem Cells Therapy 318
13.4 Pluripotent Stem Cells Therapy 323
13.5 Summary 325

14. Present Trend and Future Perspective **328**
14.1 Shift of Goal in Engineering and Production Technologies 329
14.2 Paradigm Shift in Trend of Future Products 331

Section V: Analytical Techniques and Cell Isolation

15. Quantitative Analysis **337**
15.1 Enumeration of Cell Number 337
15.2 Analysis of Culture Medium 340
15.3 Study of Growth and Metabolic Parameters 342
15.4 Assay of Proteins/Antigens 345
15.5 Plating Efficiency 349
15.6 Cytotoxicity Assay 350
15.7 Summary 354

16. Establishing Primary Cultures and Cryopreservation of Cells/Tissues **356**
16.1 Establishing Primary Cultures 356
16.2 Freezing and Reconstituting Cells and Tissues 363

Glossary **367**

Index **377**

Section I

Introduction to Cell Biology

1

Organization of Animal Cells and Special Properties

The most important discovery in the evolution of life is the origin of cells. According to *endosymbiotic* theory, eukaryotic cell seems to have evolved from a *symbiotic community* of prokaryotic cells. In 1838, Theodor Schleiden and Jacob Schwann proposed the **cell theory**, which stated that "every living organism in plant and animal kingdoms is composed of one or more cells and the new cells arise only by the division of preexisting cells". Hence, each cell forms the basic structural and functional unit of all living multicellular organism. In human, as many as 10^{14} such unit of cells are present. Since mammals evolved after many selection processes, the individual cell acquired special properties with respect to its location in the body and functions it needs to perform. This chapter has been devoted to describe various organelles in animal cells and their functions. Also, some of the special properties of animal cells that make them indespensible for manufacturing therapeutically useful products have been discussed.

1.1 ORGANIZATION OF MAMMALIAN CELLS

Transmission electron micrograph of an animal cell (hepatocyte), and sketches of an animal and a bacterial cell structure with different organelles are shown in Figures 1.1 and 1.2, respectively. The obvious distinguishing feature of animal cells when compared with prokaryotes is that they are ordered in structure. In animal cells, besides nucleus, all major organelles and vesicles, e.g., endoplasmic reticulum, Golgi apparatus, mitochondria, lysosomes, peroxisomes, etc., are membrane-bound.

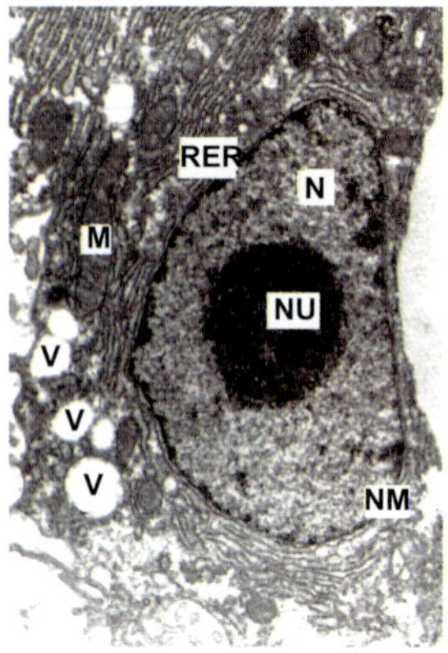

Figure 1.1 Transmission electron micrograph of an animal cell (hepatocyte). Ultrastructure showing nucleolus (NU), nucleus (N), nuclear membrane (NM), rough endoplasmic reticulum (RER), mitochondria (M), vacuoles (V) at 9750 × magnification. [*Source*: Vicentini et al. 2005, *Int. J. Morphol.*, 23, 211].

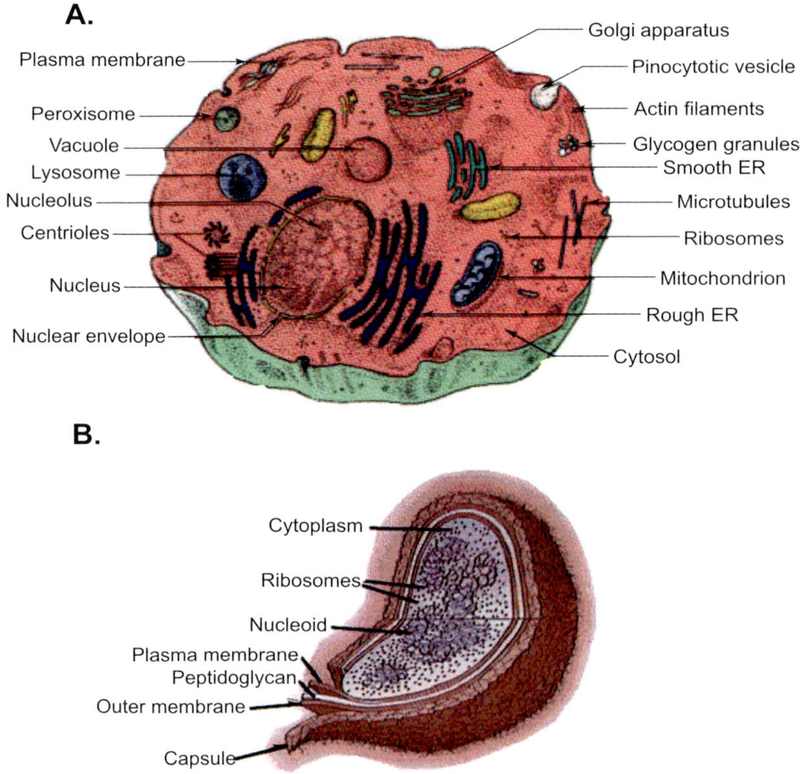

Figure 1.2 **A.** Structure of an animal cell. Various organelles are shown in the picture (Reproduced by permission of Dr. John W. Kimball, MA, USA). **B.** Structure of a bacterium. Different organelles and membrane layers are shown [*Source*: www.emc.maricopa.edu/.../Bio BookDiversity_2.html]

1.1.1 Plasma membrane and the cell surface

Animal cells possess plasma membrane, which envelopes the cytoplasm and physically separates its intracellular components from the surrounding extracellular environment. The plasma membrane is mainly composed of lipid bilayers and proteins (Figure 1.3). Plasma membrane performs many functions, of which selective movement of solutes from surrounding environment to inside of the cells and vice-versa is most important. The plasma membrane is asymmetric in nature, i.e., the phospholipid found in the inner and outer leaflet of the bilayer are distinct. As a result, an electrical gradient is established across the membrane, with the inside being relatively negative as compared with outside of the membrane. Similarly, a concentration gradients is established across the membrane due to lower intracellular ionic concentration than outside. By combining electrical and ionic concentrations gradients, each mammalian cell establishes an *electrochemical gradient* across the plasma membrane. A solute whose gradient is favourable can enter the cell by *passive transport*. For example, Na^+ ion can easily enter a cell, as its intracellular concentration is 10-times lower than its extracellular concentration. Transport, which follows concentration gradient, utilizes specific membrane proteins and does not require an expenditure of cellular energy. This type of transport is termed as *facilitated diffusion*. In case favourable concentration gradient is not present, a solute requires *active transport* in which energy is used to pump solute against concentration gradient. Glucose and other sugars, certain ions, and amino acids utilize the advantage of active transport to move across the membrane. Thus, active transport works against electrochemical gradient utilizing metabolic energy (ATP). It is sensitive to metabolic poisons, requires active membrane proteins, and is specific to certain substances or closely related group of substances. Cell consumes a major part of total generated energy for active transport of solutes. Active and passive transport allow movement of ions and small molecules between a cell and its extracellular surroundings. Larger molecules, such as proteins, nucleic acids, and even molecular aggregates as large as whole cells are internalized in mammalian cells by absorptive mechanisms known as *pinocytosis* and *phagocytosis*. These two absorptive mechanisms for internalization of large molecules are collectively termed as *endocytosis*, and the reverse process is known as *exocytosis*. The outer surface of the plasma membrane makes direct contact with the surrounding environment. Cell receives external, chemical and physical stimuli through these contacts. The membrane surface also anchors the cell in its environment and provides protection against mechanical injury and attack by other living organisms. Glycolipids and glycoproteins on the cell surface function as the receptors of external stimuli, help in cell-cell recognition, and enable adhesion to other cells or a solid support. A list of a few groups of cell surface receptors of lymphocytes is shown in Figure 1.4. Receptors are grouped according to the functions and the agents/ligands they recognize. A brief account of cell surface receptors is given in Chapter 2. Students are also recommended to read the references given at the end of this chapter.

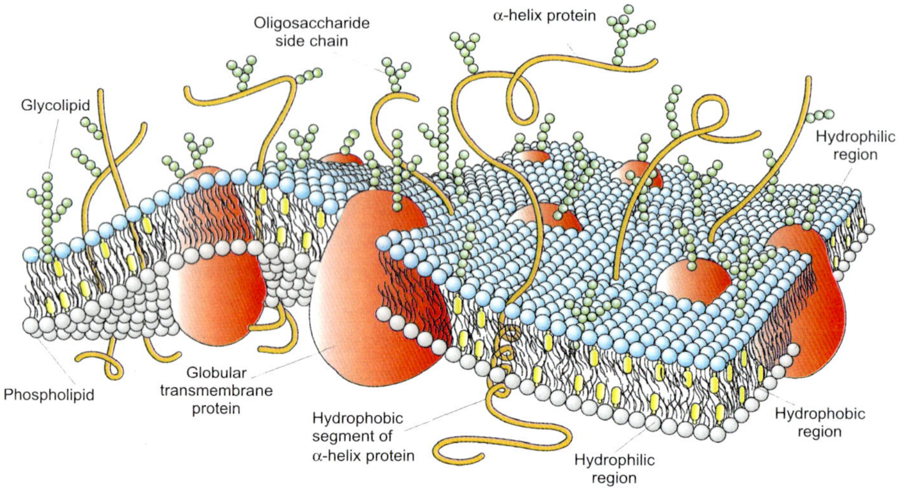

Figure 1.3 Fluid mosaic plasma membrane model of an animal cell [*Source*: Wikipedia free encyclopedia]

Another very important property attributed to plasma membrane and cell surface is the formation of *cell junctions*. Cell junctions hold many tissues of multicellular organisms together and provide a means of communication between some specialized cell types. On the basis of functions, three types of junctions are normally seen:

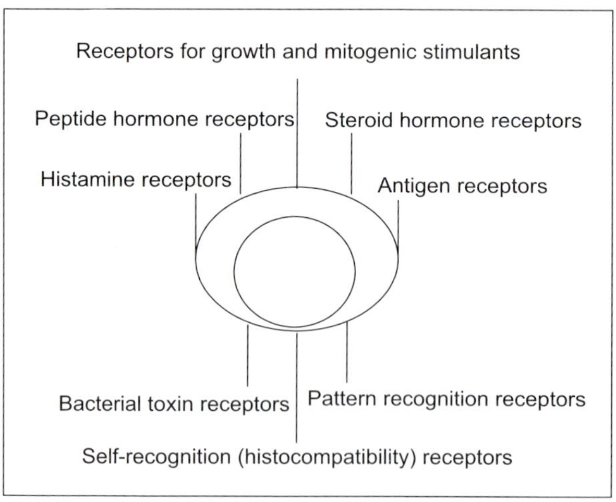

Figure 1.4 Cell surface receptors of lymphocyte (a partial list).

(1) ***Adhesive junctions*** hold cells together in their fixed positions in the tissue. The most primary type of adhesive junctions found in all cell types are known as *desmosomes*, in which 15–20 nm thick layer of plaques of fibrous matter is formed under the plasma membrane in both cells forming the

junction. Desmosomes are also easily formed in the cultured cells. In addition to desmosomes, epithelial cells are adhered together by less dense plaques known as *adherens junctions*.

(2) ***Tight junctions*** join edges of cells together and prevent the cavity and extracellular fluids from mixing, thus diffusion of molecular size particles between cell layers is taken care. The outermost lipid bilayers of the two cells in tight junctions are fused together.

(3) ***Gap junctions*** provide passages by which cells communicate between each other through exchanging ions and small molecules.

1.1.2 The cytoplasm

Cytoplasm accounts for approximately 55% of the total cell volume. In cytoplasm, intermediary metabolism by which cell degrades many compounds and synthesizes the precursors for its own macromolecules are take place. About one-fifth of the weight of the cytoplasm is due to gelatinous protein. Besides, the metabolic pathways that convert several important compounds into simpler forms, many storage compounds like triglycerides and glycogen are also present in the *cytosol*. The cytosol refers only to the fluid, and not the organelles inside the cytoplasm. It also contains dissolved constituents that help to breakdown waste products, and enables movement of materials around the cell through a process called *cytoplasmic streaming*. In addition to organic matter, it also contains many ionic substances.

Cytosol is the place where signal transduction pathways are activated in response to the environmental cues. It also contains *cytoskeleton*, which is considered the 'bone and muscle' of the mammalian cells. Cytoskeleton contains three kinds of filaments: actin filaments, microtubules, and intermediate filaments (Figure 1.5). In animal cells, the cytoskeleton structure is changed to maintain the shape. It also involved in movement of cells and provides a general framework for supporting many enzymatic reactions in the cytosol. It also plays an important role in both intracellular transport and cell division.

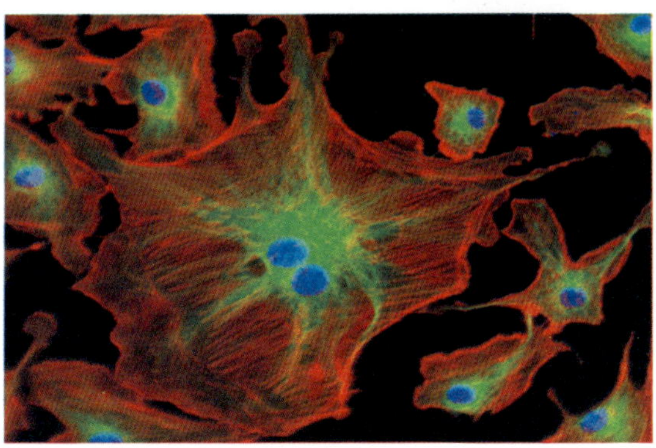

Figure 1.5 The animal cell cytoskeleton. Actin filaments are shown in red, microtubules in green, and the nuclei are in blue. [*Source*: Wikipedia free encyclopedia]

1.1.2.1 The mitochondria

It is a double membrane organelles that constantly fuse and divide. The inner membrane is folded into *cristae* to maximize its surface. The innermost compartment of the mitochondria is called the *mitochondrial matrix*, which contains several copies (2–10) of the mitochondrial genome, 70S ribosomes, tRNAs and various proteins and enzymes required for mitochondrial function (Figure 1.6). Mitochondria carry out one of the most important chemical reactions of the cell, i.e., oxidizing carbohydrates to release energy for cellular activities, Thus termed as 'powerhouse' of the cell.

Mitochondria can self-replicate and maintain various shapes, and sizes in the cytoplasm of all mammalian cells. Under the light microscope, mitochondria appear as spherical, granular, or fibrillar bodies, roughly 1–2 μm length and 0.5 μm in diameter. Depending upon the cell types, number of mitochondria in each cell may vary from several hundred to a thousand. As mentioned earlier, the mitochondria have some of their own DNA and ribosomes, and thus can synthesize corresponding proteins. The nucleus encodes the remaining proteins that are transcribed and translated in the cytoplasm, and finally transported in the mitochondria. Most of the lipids required for the synthesis of various constituents of mitochondria are imported from cytoplasm.

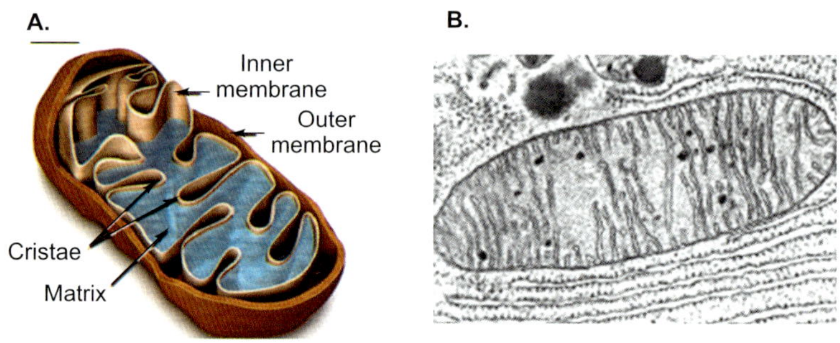

Figure 1.6 Animal cell mitochondria. **A.** Mitochondria structural features. **B.** Ultrastructure of animal cell mitochondria. [*Source*: Don Fawcett-Keith Porter/ Photo Researchers Inc.]

1.1.2.2 The endoplasmic reticulum and Golgi apparatus

Eukaryotic cytoplasm contains large number of ribosomes that are either freely suspended in the cytosol or attached to the membrane of endoplasmic reticulum (ER). The ER membrane studded with ribosomes on the cytoplasmic side is termed as *rough endoplasmic reticulum* (RER), because the ribosomes covering their surfaces give them a ragged or rough appearance in the electron microscope. The *smooth ER* (SER) does not contain any attached ribosomes. RER, is especially abundant in cells specialized for the synthesis and secretion of proteins, such as, β-cells of pancreas (produce insulin), plasma cells (secrete antibody), hepatocytes (synthesize albumin) etc. The *ribosome* is a large complex of RNAs and proteins, and is primarily responsible for translating the genetic instructions carried by mRNA into protein. As proteins are the functional molecules of cells, its synthesis is extremely important. Thus depending upon protein synthesis load, hundreds or even thousands of ribosomes can be seen throughout the cytoplasm of animal cells. In most cells, ribosomes may also be attached to the outermost membrane of the nuclear envelope on the side facing the surrounding cytoplasm (Figure 1.7). SER does not involve in protein synthesis, but important for lipid metabolism. It is present in large number in many specialized cells, where lipid metabolism is carried out. Lipoprotein are principally synthesized in hepatocytes, thus all enzymes required for their biosynthesis are located in the membrame of SER. SER of hepatocytes also contains many detoxify enzymes. In general, all complex cellular macromolecules, synthesized for transportation to the Golgi apparatus, plasma membrane, lysosomes, or cell membrane are synthesized/processed in RER.

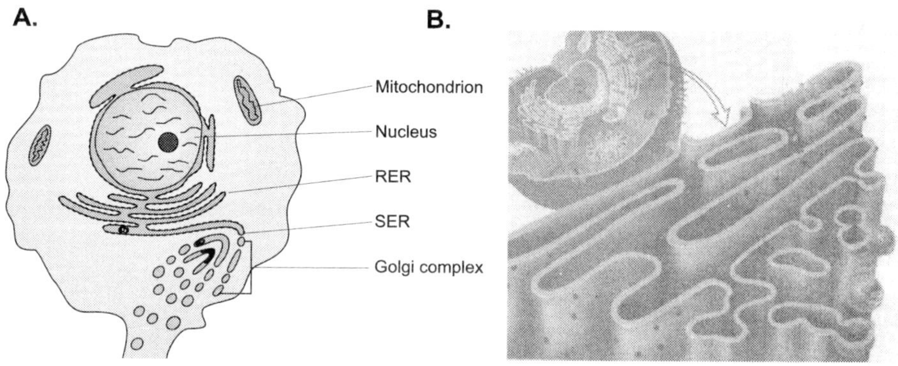

Figure 1.7 Endoplasmic reticulum and Golgi complex. **A.** Diagrammatic presentation of rough endoplasmic reticulum (RER), smooth endoplasmic reticulum (SER), and Golgi complex in pancreatic Acinar cell [*Source*: Health Sciences Library] **B.** Diagram showing flattened continous membrane of RER, outer surface of which is studded with ribosomes.

The proteins synthesized in the RER are passed on to the *Golgi apparatus* for exporting and/or modification. The Golgi apparatus is usually located near the nucleus. It is composed of numerous sets of flattened, disc-shaped membrane-bounded, smooth-surfaced cisternae. Cisternae structurly resembles a stack of plates called Golgi stack of about 1 μm in diameter. The number of Golgi stacks per cell varies from 1 to 100, depending upon the cell type. In some cells, the Golgi apparatus can even account a large fraction of the cell volume. For example, intestinal epithelium as well as those cells, which extensively synthesize glycoproteins. The Golgi apparatus has two distinct faces: *cis* and *trans*. Proteins from ER enter Golgi stacks through the *cis*-side, and after modification exit from the *trans*-side for multiple destinations. In mammalian cells, the *trans*-side is closest to the plasma membrane, where the secretory vesicles are often seen budding from the cisternae.

1.1.2.3 The lysosomes and peroxisomes

Lysosomes are roughly spherical vesicle-like bodies bounded by a single membrane. These are approximately 0.5 μm in diameter. Lysosomes contain about 40 different enzymes that are capable of digesting all major classes of macromolecules that enter the cell. The lysosomal enzymes catalyze the hydrolytic reactions in acidic pH. In most of the cells, lysosomes arise by the fusion of small vesicles budding off from the margins of the Golgi apparatus, the same manner by which secretory vesicles are formed. In some cells, lysosomes are formed without involving Golgi complex. Lysosomes function in many ways: the materials supposed to be digested by lysosomes are brought into the cytoplasm by endocytosis. The vesicle, in which endocytic materials are trapped, fuses with one or more lysosomes, and the lysosomal enzymes digest the endocytic materials. Second process is autolysis, the breakdown of components of the organelles that penetrate the lysosomal membrane and are digested inside. The third process is the release of lysosomal enzymes in the cytoplasm—a process accompanied by the death of cells.

Peroxisomes are membrane bound organelles, similar in size and density to lysosomes and found in all mammalian cells (Figure 1.2), particularly prominent in liver, kidney and brain cells. In liver cells, at least three oxidative enzymes are present in the peroxisomes: D-amino acid oxidase, urate oxidase, and catalase. The major functions of peroxisomes are: (i) detoxification via peroxidative reation, and (ii) catalyzing the breakdown of long chain fatty acids to acetyl-CoA.

1.1.2.4 The nucleus

Nucleus is found in all mammalian cells, except erythrocytes (primitive erythrocyte contain nucleus). The nucleus is spheroid in shape and separated from the cytoplasm by a double membrane called the *nuclear envelope*. The nuclear membrane isolates and protects DNA from structural damage. The outer and the inner membranes are fused at regular intervals to from nuclear pore through which exchange of materials between cytoplasm and nucleus are taken place. The nucleus contains the chromosomal DNA in which genetic information of the cell are stored. The instructions for the synthesin of cell's proteins and its replication are present in DNA. The nucleus functions as the control center for all cellular activities. The nucleus is composed of three main parts, the *nucleolus*, the *nuclear membrane*, and the *chromatin* (Figure 1.8). The nucleolus contains ribosomes, RNA, DNA, and proteins. The main function of the nucleolus (pl. nucleoli) is the production and assembly of ribosome components. Nucleoli are surrounded by a matrix called *nucleoplasm*, which is a gel-like substance composed of nucleotide triphosphates, enzymes, proteins, and transcription factors. The ribosomal DNA, present in the nucleoli, is a fundamental component since it serves as the template for the transcription of ribosomal RNA for inclusion into new ribosomes. Most plant and animal cells have one or more nucleoli, but erythrocytes do not have any. As the production and maturation of ribosomes occur in the nucleuli, a large number of them are found inside the nucleoli.

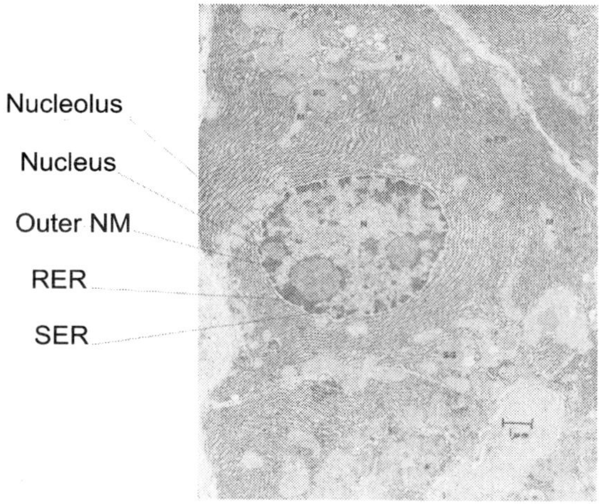

Figure 1.8 Ultrastructure of an animal cell showing nucleolus (NL), nuclear membrane (NM), and attached rough endoplasmic reticulum (RER). [*Source*: www.tqnyc.org/NYC040844/animalscells.htm.]

Most of the space inside the nucleus is occupied by masses of very fine, irregularly coiled and folded fibers collectively called chromatin, which is a complex of DNA and protein. There are two types of chromatin found in the nucleus: *euchromatin* and *heterochromatin*. Euchromatin is the least compact form of DNA, where genes that are frequently expressed by the cell are present. Besides cell's genome, the nucleus contains certain proteins whose interplay is supposed to regulate the expression of genes. Two types of proteins are associated with DNA—*histone* and *nonhistone* proteins. Histone proteins maintain the structure of the fibrous chromatin.

1.2 COMPARATIVE PHYSICAL PROPERTIES OF MAMMALIAN AND MICROBIAL CELLS

Mammalian cells are 10-100 times larger than microbial cells in volume, and thus have a higher total dry weight per cell. Table 1.1 shows comparative cellular dry weights of microbial and most commonly used mammalian cells. Due to much larger size, surface area-to-volume ratio (S/V) in mammalian cells is lower than bacteria and yeast (Table 1.1).

Table 1.1 Comparative dry weight and surface area-to-volume ratio

Organism	*Classification*	*Dry weight (pg/cell)*	*Surface area (S, μm^2)*	*Volume (V, μm^3)*	*S/V (μm^{-1})*
E. coli	Bateria	0.4	3	1	3
S. cerevisiae	Yeast	10	100	50	2
CHO (Chinese hamster ovary)	Mammalian cells	375	700	1750	0.40
MRC-5 (Human diploid fibroblast)	Mammalian cells	280	550	1600	0.34
WI-38 (Human embryonic diploid fibroblast)	Mammalian cells	300	560	1450	0.38
Hybridoma (Mouse x mouse)	Mammalian cells	360	1300	4200	0.30
BHK-21 (Baby hamster kidney)	Mammalian cells	400	670	2000	0.33
HeLa (Human cervical cancer)	Mammalian cells	400	800	2400	0.33
FS-4 (Human foreskin fibroblast)	Mammalian cells	430	540	1750	0.30
Human Egg	Mammalian cells	–	26000	370000	0.07

Contrary to S/V ratio specific glucose uptake and respiration rates of mammalian cells are much higher than bacteria and yeast, as shown in Table 1.2. It may be seen that there is a wide dispersion in the *specific glucose uptake* and *respiration rates* among the mammalian cells. However, despite fundamental differences in cellular organization between *E. coli* (prokaryote) and *S. cerevisiae* (eukaryote), the specific glucose uptake and respiration rates of both are comparable.

Table 1.2 Comparison of specific glucose uptake and respiration rate

Organism	*Nature*	*Sp. glucose uptake (nmol/10^6 cells.h)*	*Sp. respiration rate (μmol O_2/10^6 cells.h)*
E. coli	Bateria	3.6	4.4×10^{-3}
S. cerevisiae	Yeast	6.0	8.0×10^{-2}
WI-38	Mammalian cell	324	0.5
Hybridoma	Mammalian cell	400	0.3
BHK-21	Mammalian cell	300	0.2
HeLa	Mammalian cell	236	0.4
FS-4	Mammalian cell	180	0.05

Besides, the organizational differences between mammalian cells and bacteria, and similarities to that with yeast, the growth rates of mammalian cells are extremely low as compared to both bacteria and yeast. Table 1.3 depicts the average doubling time of mammalian cells, compared to that of bacteria and yeast.

Table 1.3 Population doubling time of various organisms

Organism	*Doubling time (h)*
E. coli	0.4
S. cerevisiae	1.5
CHO	18
Hybridoma	18
CEF (Chick embryo fibroblast)	20
Vero (African green monkey kidney)	24
WI-38	40
FS-4	55
Lymphoblastoma	15

1.3 GLUCOSE METABOLISM IN NORMAL AND TUMOR CELLS

The energy required for the synthesis of macromolecules is generated by the oxidation of carbon substrates. Glucose and glutamine are the major carbon and energy sources used in the culture of mammalian cells. Glucose and glutamine are metabolized in the cytoplasm, whereas energy (ATP) generation occurs in the mitochondria. Both glucose and glutamine have essential anabolic roles, but are complementary each other for the production of other metabolites and energy (Figure 1.9). The study on the regulation of glucose and glutamine metabolism in mammalian cells is important in order to optimize cell culture media composition.

Tumor cells exhibit a high rate of glycolysis than normal (non-tumor) cells. Besides their increased capacity for lactate formation in anoxic conditions, a characteristic property of tumor cells is their high level of lactate production even in the presence of oxygen, known as *aerobic glycolysis*. The inhibition of glycolysis by oxygen is known as *Pasteur effect*, by which a normal animal cell controls the consumption of glucose. In the presence of oxygen, the maximum theoretical yield of ATP per glucose unit is 36-38, whereas under anaerobic condition it is only 2. Thus, aerobic respiration is the most economical biosynthetic process for the generation of energy. Aerobic glycolysis means that the pyruvate production from glucose exceeds the capacity of the tricarboxylic acid (TCA) cycle and respiration chain, and hence a major fraction of pyruvate is converted into lactic acid. Due to the high rate of aerobic glycolysis, tumor cells utilize a small fraction of glucose for energy generation.

The key enzymes in the glycolytic pathway are hexokinase (HK), phosphofructokinase (PFK), and pyruvate kinase (PK). In normal cells, inhibition of glycolytic pathway by oxygen is controlled by metabolic modulation of these enzymes. As shown in Figure 1.10A, the regulation of glycolysis in normal cells by oxygen results in allosteric inhibition of PFK by ATP, which leads to a decrease

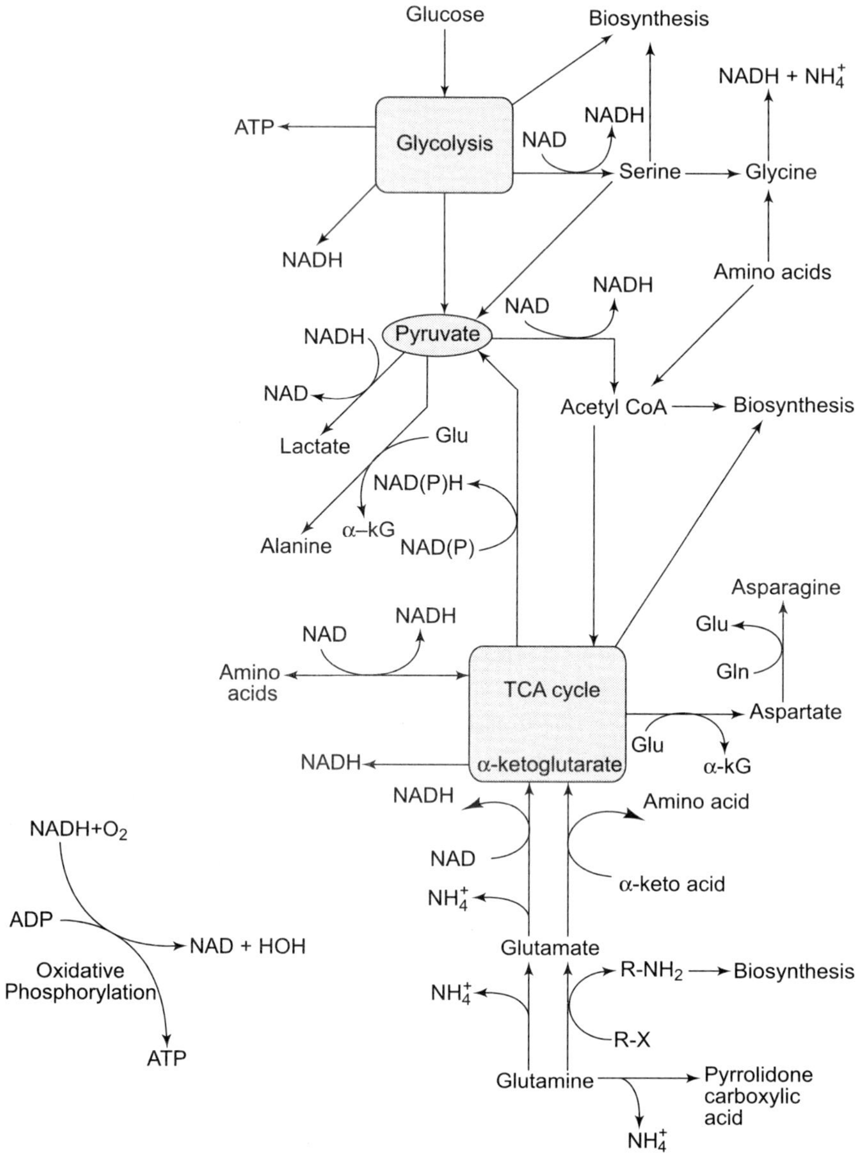

Figure 1.9 Biosynthetic pathways for metabolism of glucose and glutamine by hybridoma. [Adapted from Ref. 3]

in fructose-1,6-bisphosphate (F-1,6-P) level. The steady-state level of F-1, 6-P is either converted into pyruvate or to fructose-6-phosphate (F-6-P) by the action of fructose-1,6-bisphosphatase. Low level of fructose-1,6-bisphosphate efficiently overcomes the inhibition of PFK by ATP. Again, accumulation of glucose-6-phosphate (G-6-P), due to inhibition of PFK, in turn blocks its own synthesis by inhibiting HK. In aerobic glycolysis, mitochondrial ATP production limits consumption of glucose in normal cells by inhibiting PFK. Under anoxic condition, mitochondrial ATP production is dropped, PFK is deinhibited and fructose-1,6-bisphosphate is produced. Accumulation of fructose-1,6-bisphosphate results in the formation of more pyruvate beyond the capacity of mitochondria. Thus, lactate is produced as a by-product in anoxic condition.

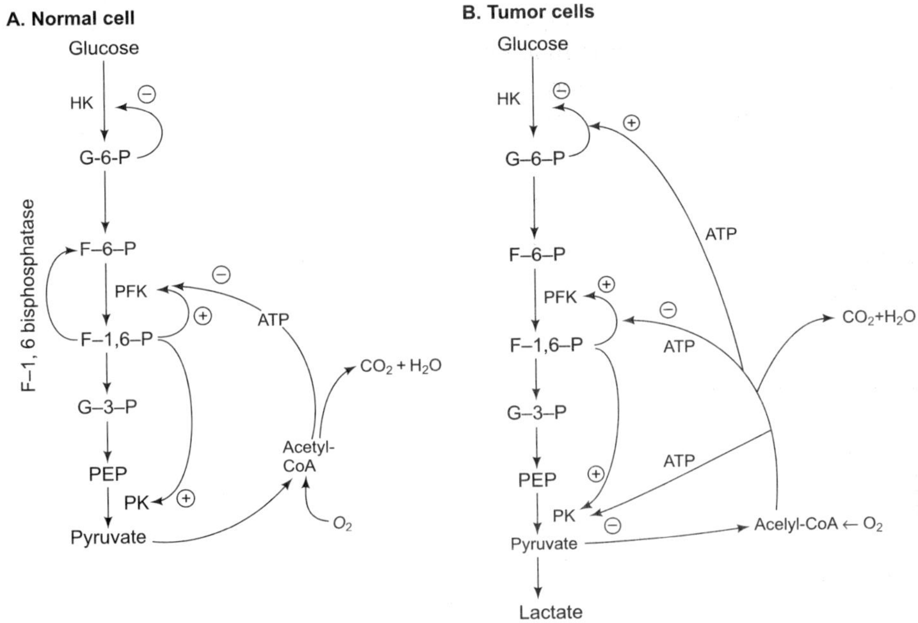

Figure 1.10 Regulation of glucose metabolism by normal and tumor cells.

High aerobic glycolysis in tumor cells is explained by the unusual regulatory mechanism following induction of isoenzymes (Figure 1.10B). Although isoenzymes in normal and tumor cells catalyze the same reaction, they are different in terms of structure and response to metabolites. In tumor cells, mitochondrial HK (type II) is not inhibited by glucose-6-phosphate unless its concentration is appreciable high. Again, tumor cells contain PK (type M_2) that has a lower affinity to phosphoenolpyruvate (PFP), and strongly inhibited by alanine. Fructose-1,6-bisphosphate once formed is slowly converted to pyruvate. In tumor cell, the level of fructose-1,6-bisphosphatase is extremely low as compared to the normal cells. Thus, fructose-1,6-bisphosphate is accumulated in the cells, which ultimately overcomes inhibition of PK (type M_2) by alanine. The accumulation of fructose-1, 6-bisphosphate also reverses ATP inhibition of PFK, thereby allowing mitochondrial control of glycolysis. Overall, in tumor cells, the aerobic glycolysis is maintained at a very high rate, and the total available glycolytic capacity is utilized even in the presence of high level of oxygen. As a consequence, glycolytic intermediates are accumulated in the tumor cells that are later convert into lactate.

1.4 FOLDING AND SECRETION OF PROTEINS IN MAMMALIAN CELLS

In eukaryotic cells protein folding occurs by overlapping of translation, translocation, proteolytic cleavage, glycosylation, and assembly in case of multiunit proteins. In order to become functionally active the nascent polypeptide chain undergo above processing steps in sequence. Improper folding of protein results in a variety of pathological states, including cancer and neurodegenerative diseases. Most of the mammalian proteins of therapeutic value are complex due to the presence of multiple disulfide bonds, high glycosylation, and multimeric forms (homo/heterodimer). Correct folding of these proteins in native conformation is possible in eukaryotic (mammalian) expression systems. The cytosol of animal cells is oxidizing in nature, and moreover oxidative folding occurs in

Table 1.4 Mammalian heat-shock proteins and their functions

Major family	*Location*	*Cellular function*
Hsp60 or chaperones	Mitochondria	Binds to partially folded peptides and assists their correct folding. Helps in assembling multimeric units
TCP-1	Cytoplasm	Same as above
Hsp70 (inducible)	Cytoplasm/nucleus	Binds to extended polypeptides, prevents aggregation of unfolded peptides, dissociates some oligomers, and binds with ATP
Hsc70 (cognate)	Cytoplasm/peroxisome	Same as above
Grp78/BiP	ER	Same as above
mHsp70/Grp75	Mitochondria	Involved in the regulation of HSF1 activity, and the repression of heat shock protein gene transcription
Hsp90 (α and β)	Cytoplasm	Binds to other proteins, regulate protein activity, prevent aggregation of refolded peptide, and correct assembly and folding of newly synthesized protein
Grp94/gp96/Hsp100	ER	Involved in maintaining the HSF1 monomeric state in non-stressful conditions
Hsp110 Apg-1 (mouse)	Cytoplasm	Protein refolding

Abbreviations: TCP-1, tailless complex polypeptide; Grp, glucose-regulated protein; BiP, immunoglobulin heavy chain binding protein; mHsp70, mitochondrial Hsp70; HSF1, heat shock factor 1; Apg-1, protein kinase essential for autophagy

Source: A. Graham Pockley, Expert Reviews in Molecular Medicine, Cambridge University Press, Cambridge, CB2 1SB, UK.

the ER with the help of specific enzymes. Inappropriate environment leads to the misfolding of protein, and its accumulation adversely affects the viability of the cells. Cellular machinery efficiently recognize proteins that are unable to properly fold, and thus target them for degradation. A class of proteins collectively known as *molecular chaperones* bind transiently and reversibly to non-native polypeptides, prevent nonproductive inter- and intra-molecular contacts, discourage aggregation, and ultimately promote folding into its native state. Chaperones are also called heat shock proteins, as their expression increases upon heat treatment of the cells. Major heat shock proteins found in mammalian cells and their functions are shown in Table 1.4.

In animal cells, proteins synthesize in the ribosome are destined to various organelles for specific functions. Here, I will discuss protein secretion in animal cells by ER-Golgi secretory pathways to reach to the cell plasma membrane and cell surface (secreted protein). In order to be in the secretory pathway, the growing polypeptide chain should have signal sequence (15-30 amino acids) at the amino terminal that is recognized by the *signal recognition particle* (SRP). As a sorting machine, SRP and *signal recognition particle receptor* (SR) must interact with different partners, such as signal sequences, the ribosomes, the membrane *translocon*, and with each other at the proper stages during protein translocation (Figure 1.11). The ribosome and the growing polypeptide chain are then delivered to the ER lumen or its membrane via the interaction of SRP with SR. There, SRP releases polypeptide chain to the translocation machinery, the 'translocon', where the nascent protein finishes its synthesis and is either integrated into the membrane or delivered into the secretory pathway. The signal sequence is cleaved in the lumen of ER. The chaperones present in the lumen of the ER bind the newly formed polypeptide chain to assist its folding into native tertiary conformation. At this point, protein is glycosylated with the help of various protein glycosylating enzymes present in ER (see below). Often glycosylation is essential for proper folding of proteins in their native conformations. Reshuffling of disulfide bonds by the enzyme protein disulfide isomerase (PDI) may occur in the ER.

Secretory proteins transported from ER, through the Golgi apparatus, to the plasma membrane are encapsulated in vesicles. A protein is incorporated into a vesicle at one membrane surface, and is released from the vesicle at the next membrane surface. Thus, mammalian protein trafficking from the site of synthesis to the plasma membrane is conducted by protective events of coated vesicles mediated transportation. The biological activity of proteins is protected from accidental exposure to the adverse conditions through their vesicle-mediated transportation.

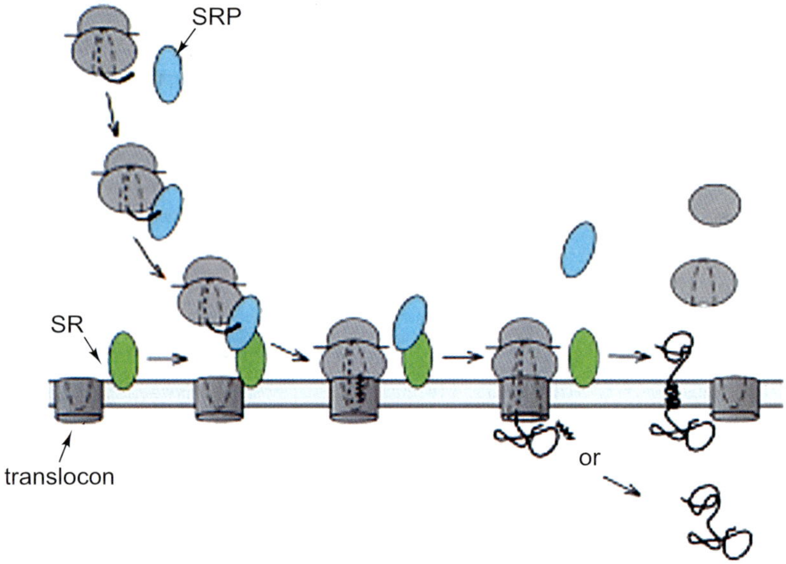

Figure 1.11 Protein translocation pathway from ribosome to the lumen of ER. The ribosome and the polypeptide chain are delivered to the ER or plasma membrane via the interaction of SRP and SR. The polypeptide is released in the translocation machinery (translocon), through which it is delivered to the ER lumen [Reproduced by permission of Dr. Shu-ou Shan, Division of Chemistry and Chemical Engineering, California Institute of Technology, CA, USA].

1.5 POST-TRANSLATIONAL MODIFICATION OF PROTEINS IN MAMMALIAN CELLS

Post-translational modification of protein primarily means its glycosylations followed by or in association with folding to native conformation. Protein glycosylations are of three types: a) N-glycosylation, b) O-glycosylation, and c) glucosylphosphatidylinositol (GPI) anchor.

For therapeutic proteins, N- and O-glycosylations are most important. GPI anchors are found mostly on cell surface membrane proteins, which will not be discussed here. N-linked glycans are characterized by β(1,2)-glycosidic linkage between a N-acetylglucosamine (Gn) residue and the δ amide N of an asparagine (Asn) residue. Depending upon the nature of branching of carbohydrates, the N-linked glycans fall into three main groups, originating from the same biosynthetic pathway. Different branching of carbohydrates in N-linked glycans are shown in Figure 1.12.

Glycosylation begins in the RER and proceeds as glycoproteins migrate through the Golgi complex to their final destination. Asn-glycosylation is a co-translational event and occurs as the polypeptide is beginning to get transferred

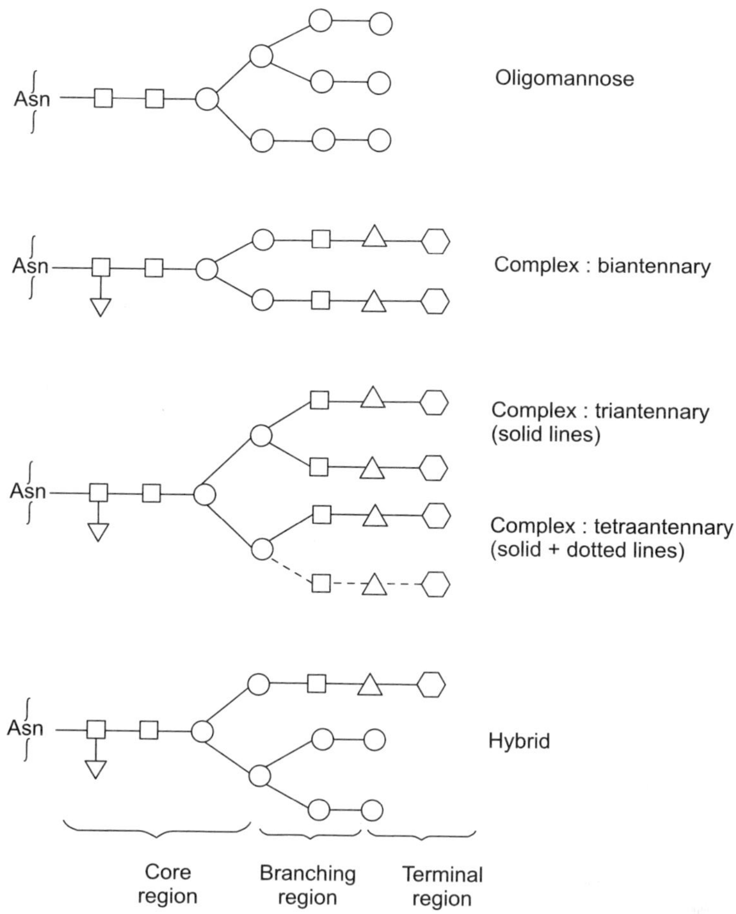

Figure 1.12 The basic forms of common *N*-linked glycans. The *N*-linked glycans posses three regions of oligosaccharides—core, branching, and terminal. All glycans have common core of two *N*-acetylglucosamine residues and three mannose residues. *N*-linked glycans are primarily depend on the branching pattern and the type of monosaccharides present in the branching and terminal regions. The branches are distributed over the two terminating core mannose residues. The complex type *N*-linked glycans can have two to four antennae. In complex and hybrid type *N*- glycans, the antennae usually terminate in sialic acid or galactose. Sialylation adds the greatest degree of microheterogeneity, however variable core fucosylation also adds heterogeneity to the glycan. □-GlcNAc (Gn); Ñ-Fuc; O-man; D-Gal; ⬡-NueAc (SA). [Adapted from Ref. 1]

into the ER, while still in an unfolded state. The initial steps in this pathway appear to be generally conserved, involve the synthesis of a dolichol linked precursor glycan in the ER membrane and its transfer to an Asn residue of the growing peptide chain (Bhatia and Mukhopadhyay, 1999). The transfer of glycan occurs when 12-14 amino acids C-terminal to a sequon have been translocated into the ER lumen. The glycan that is transferred to the polypeptide is a substrate for different enzymes, α-glucosidases (α-Gs) and glycosyl transferases

(Glyc Ts) to give mature high mannose (Man) or complex-type chains (Figure 1.13). The glycan is processed in terms of removal of three glucose (Glc) residues by stepwise action of α-G I and α-G II. The Glc trimming affects the exit of glycoprotein from the ER. A quality control procedure ensures that proteins, which are incompletely folded or incorrectly oligomerized, not transported out

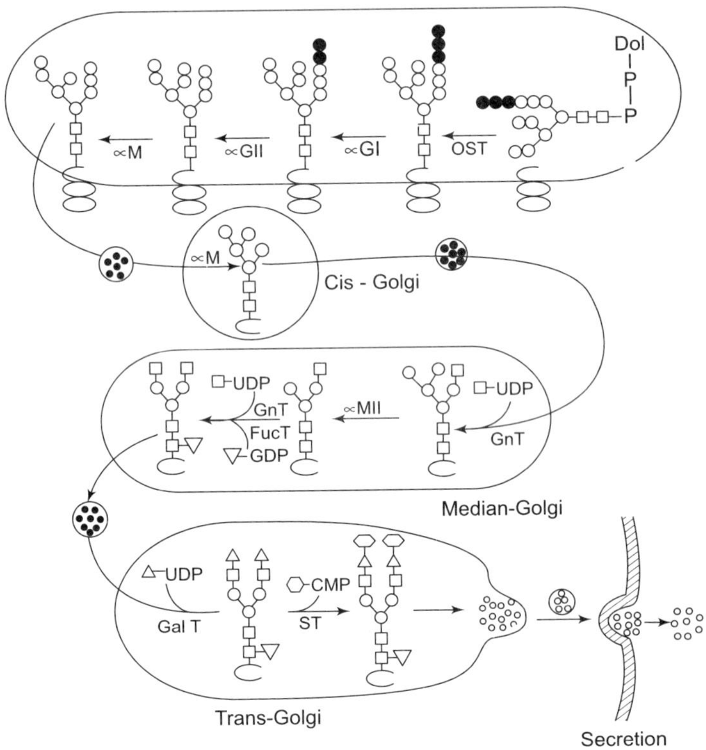

Figure 1.13 Biosynthesis pathway of Asn-linked oligosaccharide. Biosynthesis starts in the rough ER and proceeds through the *cis-*, *medial-*, and *trans-*Golgi apparatus, before glycoprotein is secreted. Some of the enzymes involve are: oligosaccharyl transferase (OST); α-glucosidase I and II (αG I and II); ER-mannosidase (aM); a-mannosidase I and II (aM I and II); N-acetylglucosaminyl transferase I and II (GnT I and II); fucosyl transferase (FucT); galactosyl transferase (GalT); sialyl transferase (ST). □-Gn; O-Man; •Glc; ∇-Fuc; Δ-Gal; -⬡-SA. [Adapted from Ref. 1]

of the ER so long as they have not acquired 'export competent conformation'(Bhatia and Mukhoupadhay). At this stage, polypeptide folds in ER with the help of membrane bound proteins, *calnexin* and *calreticulin*. Before leaving ER, α-mannosidase (α-M) efficiently removes a single terminal Man residue to generate Man_8Gn_2 in the ER. The action of ER α-M alters the conformation of the glycoproteins, making them more susceptible to the Golgi α-M I. The concerted efforts of ER and Golgi α-Ms may be required for fine-tuning of the mechanism to produce surface glycoproteins with particular assortments of high

Man type chains. In Golgi, Man chains are processed in two steps by the action of α-M I, active in *cis*-Golgi and α-M II, active in *medial*-Golgi, resulting in the removal of additional five Man to yield Man_3Gn_2 (Bhatia and Mukhopadhyay, 1999). The processing of Man chains is important for the formation of complex-type glycans. In *medial*-Golgi, N-acetylglucosaminyltransferase I (GnT I) inserts Gn residue to α-1, 3-linked Man followed by the cleavage of two terminal Man by the action of α-M II. Addition of another Gn to the α1, 6-linked Man by GnT-II takes place in the *medial*-Golgi. In mammalian cells, occasionally fucose (Fuc) residue is also incorporated at this stage in the *medial*-Golgi by the action of α-fucosyl transferase (α-FT). The termination of glycan chain occurs in the *trans*-Golgi by the incorporation of galactose (Gal) and sialic acid (SA) with the help of galactosyl transferase (Gal T) and sialyltransferase (ST), respectively. Thus, a complex-biantennary N-glycan is synthesized before secretion of the protein. Diversity of structural features in N-linked glycans is found from cell to cell, where N-acetylgalactosamine (GalNAc), sulfation, and O-acetylation of SAs have been observed (Bhatia and Mukhopadhyay, 1999).

The O-linked glycosylation event takes place in the Golgi after N-glycosylation, folding, and oligomerization of protein. The O-linked glycosylation is limited to the residues present in protein surface. In O-linked glycoproteins, an α-glycosidic linkage between a GalNAc and the hydroxyl group of serine (Ser) or threonine (Thr) residue of peptide chains are involved (Figure 1.14). The O-linked glycosylation is most common in mucine type of proteins.

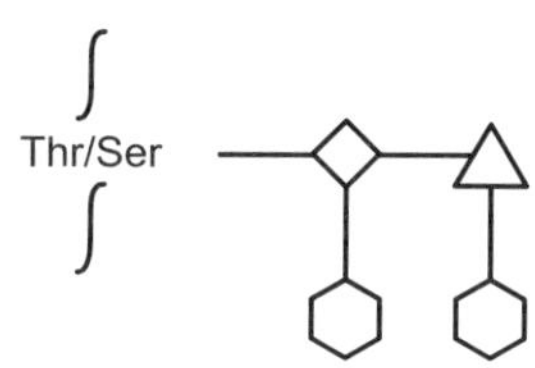

Figure 1.14 The mucine type O-linked glycans. ◇–GalNAc; Δ-Gal; ⬡ –SA. [Adapted from Ref. 1]

1.6 SUMMARY

1. In animal cells (eukaryotes), organelles are membrane-bound, which is not in bacteria (prokaryotes).
2. Besides holding cellular contents, plasma membrane plays an important role in communicating external environment to the cells through the surface receptors.
3. Specific glucose and oxygen consumption rates in animal cells are several folds higher than that in the case of bacteria and yeast.
4. Tumor cells consume more glucose and produce more lactate even in the presence of oxygen.
5. Posttranslational modification of proteins occurs in ER and Golgi complex of mammalian cells.

ABBREVIATIONS

GnT I	N-acetylglucosaminyltransferase I
GalNAc	N-acetylgalactosamine
ATP	Adenosine triphosphate
Asn	Asparagine
Fuc	Fucose
α-FT	α-fucosyl transferase
GPI	Glucosylphosphatidylinositol
GlycTs	Glycosyl transferases
α-Gs	α-glucosidases
Glc	Glucose
Gal	Galactose
Gal T	Galactosyl transferase
Hsp	Heat shock protein
HK	Hexokinase
Man	Mannose
α-M	Mannosidase
PFK	Phosphofructokinase
PK	Pyruvate kinase
RER	Rough endoplasmic reticulum
SER	Smooth ER
SRP	Signal recognition particle
SA	Sialic acid
ST	Sialyltransferase
Ser	Serine
TCA	Tricarboxylic acid cycle
Thr	Threonine
tRNA	transfer RNA

REFERENCES

1. Bhatia PK and Mukhopadhyay A (1999) In: *Adv. Biochem. Engg. Biotechnol.* (T. Scheper, ed.) vol. 64, p. 155. Springer, Berlin.
2. Eigenbrodt E, Fister P and Reinacher M (1985) In: *Regulation of carbohydrate metabolism* (R Beitner, ed) vol. 2, p. 141. CRC Press, Inc. Boca Raton, Florida.
3. Miller WW, Wilke CR and Blanch HW (1989) *Biotechnol. Bioengg.* 33, 477.

SUGGESTION FOR FURTHER READING

1. Wolfe SL (1981) *Biology of the cell.* 2nd edition. Wadsworth Publishing Company, Belmont, California.
2. Alberts B, Bray D, Lewis J, *et al.* (1989) *Molecular Biology of The Cell,* 2nd edition, Garland Publishing Co, New York.

2

Proliferation and Death of Animal Cells: Molecular Mechanism

Proliferation and death are two opposing phenomena in all living organisms. The synthesis of nucleic acids, proteins, carbohydrates, lipids, and other biological molecules accomplish growth at the cellular level. Following a period of growth, cells reproduce by division. In the pattern of growth and division that increases body cell numbers in mammals, the daughter cells are exact genetic duplication of the parent cell. In mammals, unwanted cells are removed during *embryogenesis*, *metamorphosis*, and tissue turnover by the process of appoptosis. Similarly, the proliferation and death phenomenon can be observed in cultured cells. Despite being outside the body, each mammalian cell responds to a stimulant almost similarly as it is believed to respond *in vivo*. The events for the growth of mammalian cells have been described in this chapter, which primarily include cell cycle, molecular basis of proliferation, growth factor signaling, etc. Apoptotic cell death, its induction in cell culture and possible mechanisms of apoptotic death are also included here.

2.1 CELL GROWTH

In multicellular organisms, growth normally refers to an increase in size and/or weight. However, in an isolated cell (derived from a multicellular organism), growth implies proliferation, which means the increase in cell number. It is usual that the individual cell also increases in cell mass before it undergoes a division. In mammals, about 200 different specialized cell types are present. Despite complicated patterns of interdependency, individual cell is responsible for its own maintenance, and in principle cells grow and divide independently of their neighbors. An adult human has about 10^{14} cells derived from single cell (fertilized oocyte), indicating extensive multiplication potential of animal cells. As exceptions, central and peripheral nerve cells, epithelium of lung etc. do not multiply. On the contrary, bone marrow is the most proliferating tissue in the body. Even under normal situations, bone marrow cells proliferate to maintain tissue homeostasis. Muscle and skin epithelial cells can multiply, but only in tissue injury state. It is now known that complex regulatory mechanisms are involved to strike a balance between growth and loss (death), and to respond to trauma by wound healing. These are achieved by a combination of various cell signaling pathways actuated by extracellular and intracellular factors. A few of them will be discussed in this chapter. Summarizing above, the reasons for cell division in multicellular organisms are:

A. Growth;
B. Repair and replacement of damaged parts; and
C. Reproduction of the species.

2.1.1 Cell cycle

In order to divide, each cell passes through a complete cell cycle. Since the instructions for making cellular components are encoded in the DNA, each new cell posses a complete set of DNA molecules. This requires the duplication of DNA (by replication) before cell division. In other words, replication and segregation of chromosomes are the two most important events in cell division. The *cell cycle* is divided into two fundamental parts: the *interphase*, which occupies the majority of cell cycle, and structure the *mitosis*, ending with the division of cell. The genetic blueprints of a cell is structured into chromosomes before cell division. The number of chromosomes in each cell may vary depending upon its type. For example, somatic cells (adult cells) in humans have 23 pairs of chromosomes (total of 46), one of each pair comes from mother, the other from father. Thus, somatic cells in humans are termed as *diploid cells*. On the other hand, reproductive cells in humans (sperms and eggs) have 23 single chromosomes (single copy of each chromosome, total of 23). Thus, the reproductive cells (sperms and eggs) are called *haploid cells*. In transformed cells the number of chromosomes may differ from the diploid cells, and are termed as *heteroploid cells*. During the interphase, DNA in chromosomes is diffusively distributed in the nucleus, therefore the individual chromosomes cannot be distinguished.

In adult cells, the interphase is further divided into various stages depending upon specific work involved. A complete cell cycle of a somatic cell is depicted in Figure 2.1A. Before division, cell must duplicate its chromosomes in a specific part of the interphase called S phase. The duplicated chromosome and the original chromosome remain together by *centromere*, and together called *sister chromatids* (Figure 2.1B). The S phase occurs in the middle of the interphase, preceded by a gap called G_1 phase and followed by another gap called G_2 phase. In G_1 phase, the cell is subjected to stimulation by extracellular mitogens and growth factors. At this stage, cell either progresses into the S phase or exit the cell cycle for cell differentiation. The G_2 phase is the interval between the completion of DNA synthesis and mitosis. The regulation of cell cycle must ensure that the event in each phase is completed before moving to the next. For that the *check-points* are strategically placed in the late G_1 phase and at the G_2/M phase, to monitor the integrity of DNA, in order to prevent progression and propagation of mutated or damaged cells. G_0 phase refers to cells that are quiescent (temporarily or permanently out of cycle). The normal quiescent cell is dependent on external stimuli to exit the G_0 phase and enter the cell cycle through the early part of G_1 phase.

Mitosis is essential for growth and division of cells. It is a dramatic, coordinated change in the architecture of the cell that segregates the replicated chromosomes into two identical sets and initiates the cell division (Figure 2.1C). The entire set of duplicated DNA is segregated prior to the division, giving each cell 23 pairs of chromosomes. Mitosis is divided into four phases: 1) **Prophase:** chromosomes condense into visible structures, spindle fibers begin to extend

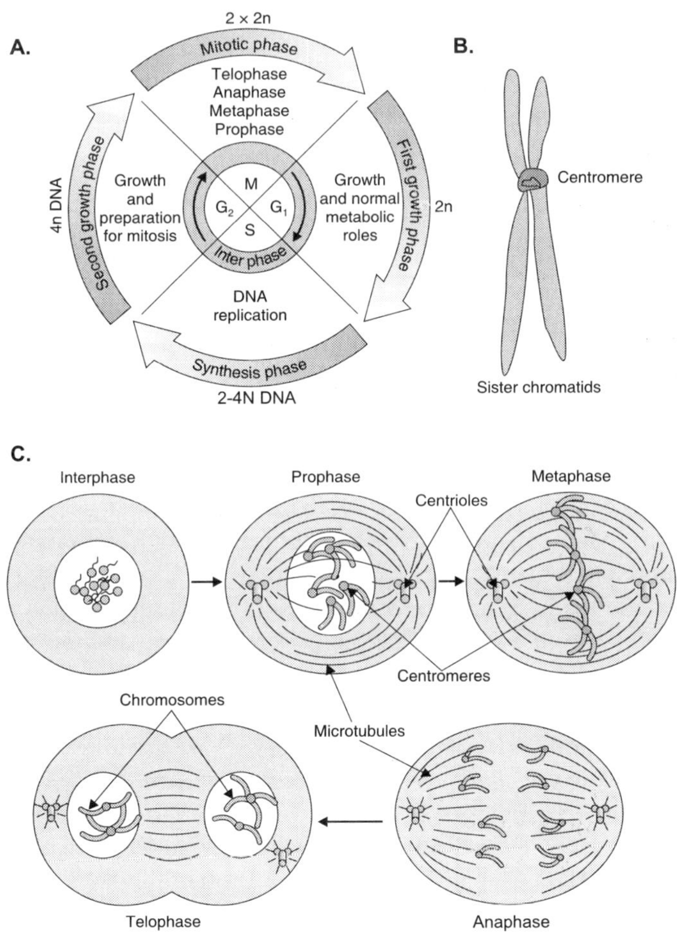

Figure 2.1 Diagrammatic presentation of different cellular events during the division of animal cell. **A.** Cell cycle; **B.** Sister chromatids; **C.** Different phases of mitosis in somatic cells.

from *centrioles*, *centrosomes* move to opposite poles of cell and the process ends with the disappearance of nuclear membrane. 2) **Metaphase:** each pair of sister chromatids (joined by a common centromere) attach to microtubules originating from opposite poles of the spindle and align along the equatorial plane of the cell (midway between the two spindle poles). 3) **Anaphase:** centromeres split, freeing the sister chromatids (now called daughter chromosomes) that move along the microtubules to opposite poles of the spindle. 4) **Telophase:** immediately after anaphase, the physical process of cell division, called *cytokinesis* begins when a contractile ring pinches the cell into two daughter cells, each containing a complete set of chromosomes and a spindle pole, the chromosomes decondense and acquire a nuclear envelope, reforming an interphase nucleus, and the microtubule array returns to its interphase pattern, and finally cell splits into two daughter cells.

In mammals, the length of cell cycle may vary from cell to cell. In general, cells do not differ much in the duration of S, G_2, and M phases, which vary between 2-6 h, 2-4 h, and 1-2 h, respectively. The longest phase in the cell cycle is the G_1 phase, which may vary between 12 to 96 h, depending on the type of cell. The embryonic cell cycle is in general much shorter than somatic cell cycle, because G_1 and G_2 phases are largely suppressed during early development of embryos. It is known that, transformed cells grow more rapidly than normal somatic cells, both in culture (*in vitro*) as well as in the body. It is important to mention here that G_0 phase does not exist in transformed cells, and therefore mitotic signals are not required for them to grow. On the other hand, in absence of mitotic signals, somatic cells remain in G_0 phase. Somatic cells may also undergo *differentiation* or *apoptosis* in the absence of mitotic signals.

2.1.2 Study on cell cycle

Biochemical studies on cell cycle require a population of cells that proceed synchronously through the cell cycle, which means that cells progress in the same phase of the cell cycle. However, somatic (adult cells) and transformed cells are not synchronous. The DNA content of a cell reveals its position in the cell cycle. The study of cell cycle not only explains the position of cells in the cycle, but also assists to determine the *generation time* and the *mitotic index* of the particular cell type, and the presence of *apoptotic* cells and *heterokaryons*, if any. The cells in S phase can be detected by their ability to incorporate labeled DNA precursors whose presence can be detected in cell lysate. To determine the length of cell cycle, the cells are labeled with high specific activity [^{3}H]-thymidine (5 μCi/ml, 2 Ci/mmol) for 30 min. The unincorporated isotope is washed-off, and the appearance of the labeled cells (mitotic) is determined *autoradiographically* at 30 min intervals for a period of 48 h (Maciera-Coelho, 1973). In theory, the first labeled mitotic cell will appear after a time equal to the length of G_2 phase and the percentage of labeled mitotic cells should rise to 100% over M phase (Figure 2.2). After S phase, the percentage of labeled mitotic cells fall, which again rise after the total time required for G_1 and G_2 phases. The time between the midpoint of first ascending and descending curve is the measure of time taken in S phase. On the other hand, the time taken between the midpoint of first and the second ascending curve measure the total cycle time, or the generation time. The duration of autoradiograph varies from cell to cell, but 48 h incubation is sufficient for most of the primary and established cell lines.

In general all cells do not proliferate together. Therefore, the *proliferation index* of growth fraction is calculated by the following equation:

$$N_C/N = [\text{number of proliferating cells/total number of cells}]$$

The time taken to double the number of cells (cell doubling time, T_D) is therefore not equal to the cell cycle or cell generation time. Thus, it is important to carefully distinguish between cell cycle or generation time and the T_D. If all the cells do not have the same generation time, then the T_D is shorter than the generation time. Alternatively, if all cells in the population are not growing, i.e., the growth

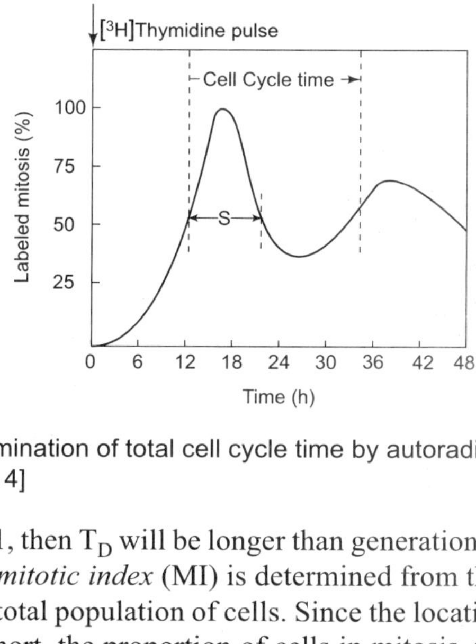

Figure 2.2 Determination of total cell cycle time by autoradiographic technique. [Adapted from Ref. 4]

fraction is less than 1, then T_D will be longer than generation time. The fraction of cells in mitosis, the *mitotic index* (MI) is determined from the autoradiograph by comparing with the total population of cells. Since the location of mitosis is fixed and its duration is short, the proportion of cells in mitosis, the mitotic index can be determined by,

$$MI = \ln 2\ [\text{duration of mitosis/total cycle time}]$$

The most convenient technique to determine the stage of cell division in the cell cycle is carried out by measuring the amount of DNA content in each cell. Progression through S phase and mitosis results in changes of cellular DNA content. The position of cell in the cycle can therefore be determined on the basis of the measurement of DNA content. The cells are treated with a DNA-binding fluorescent dye (e.g., Propidium Iodide, Hoechst 33342, acridine orange) and passed one at a time through a device called *flow cytometer* (see Chapter 9). More the DNA content of a cell, higher is the binding of dye. Thus, by measuring the amount of dye and hence DNA, the stage of each cell in cycle can be determined. Figure 2.3A (left plot) shows pulse width versus area, which distinguishes between single cells and aggregates. Single cells (G_0/G_1 or G_2/M) will have comparable pulse width values. Whereas, aggregates will have larger width values and can be easily seen on the plot to the right of the single cell region. Single cells have been gated and a typical profile of DNA content of asynchronous mammalian cells is shown in Figure 2.3A (right plot). The cells in G_2 phase have twice as much DNA as the cells in G_1 phase, and the cells in S phase have the intermediate amount. There are clear peaks of G_1 and G_2 cells, but as the cells in S phase are spread over the range of DNA contents, the peak between G_1 and G_2 is quite low. The relative areas of the different peaks, calculated with the help of a dedicated software program, reveal the fractions of the cells in different phases of the cell cycle. The position of cells in cycle explain the physiological status: (a) the unique combination of G_1, S, G_2, and M phase cells signify normal status of cells (Figure 2.3A); (b) presence of quiescent cells is denoted by cells in the G_0 phase (Figure 2.3B); (c) apoptotic cells can be seen by examining the presence of cells in sub-G_0/G_1 phase (Figure 2.3B); (d) heterokaryons can be determined from the position of G_2-M phase cells, for example *tetraploid* cells appear in 4N position, *octaploid* cells appear in 8N position, etc. (Figure 2.3C).

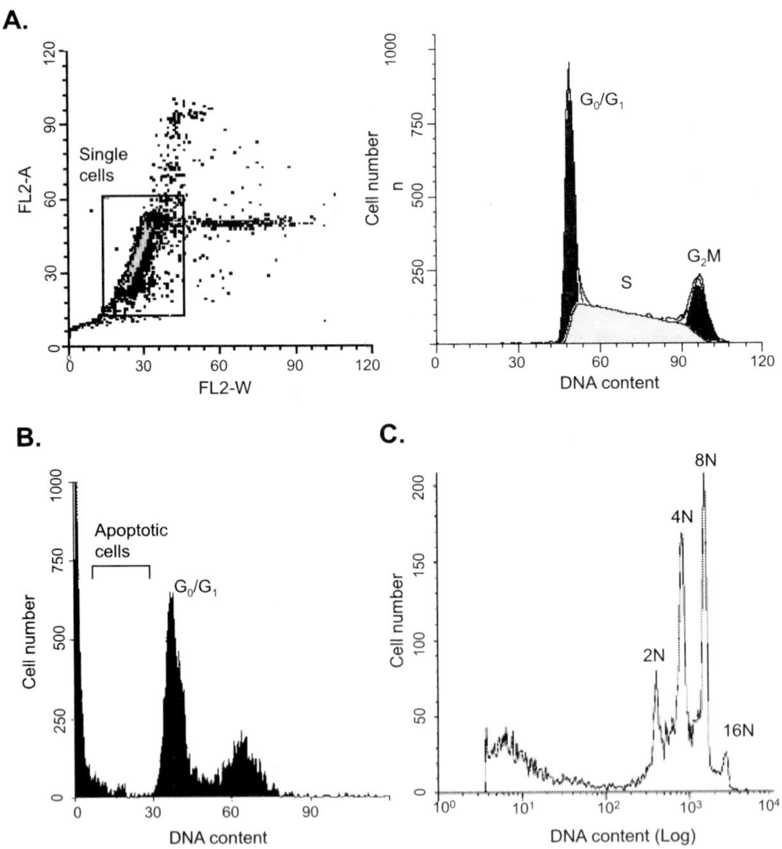

Figure 2.3 Cell cycle analysis of animal cells. **A.** Establishing position of cell cycle histogram; **B.** Resting (G_0) and apoptotic cells (sub-G_0/G_1); **C.** Heterokaryon (2N, 4N, 8N, etc.) in tumor cells.

2.2 GROWTH REGULATION

The duration of cell cycle in somatic cells, is tightly regulated by various factors. In early embryos, the inactivation of maturation promoting factor (mitosis promoting agent) triggers the events that lead to DNA replication and division of microtubule organizing center. In growing somatic cells, the transition in G_1 phase regulates the initiation of DNA replication and division of microtubule organizing center. The growth of mammalian cells is controlled at three different levels [Riley, 1981]:

A. *Intrinsic*: intracellular regulation (cell mass and conformation) to decide whether mitosis will take place or cells will maintain in resting state of G_0 phase.
B. *Intercellular:* space and nutritional factors within a population.
C. *Interpopulation*: hormones, growth factors and inhibitors that act between populations.

Cells divide to control their volume, otherwise their size will become too large. For a long time, it was believed that there is a constant relationship between mass per volume of the cytoplasm and the mass of chromatin that the cells contain. Similarly, in terms of the conformation, it has been observed that in case of normal adhered cells, the growth is related to the degree of flattening on the substratum. For normal *anchorage dependent* cells, there is an optimum height-to-mass ratio for cells to grow in culture. On the other hand, the *anchorage independent* cells (suspension culture) do not change in conformation.

The regulation of growth in mammalian cells can be observed in tissue culture. In establishing primary culture, a fragment of tissue is placed in a plastic dish containing serum-supplemented medium. Within a few days of culture, some cells will migrate away from the tissue and remain attached on the surface of the culture plate for proliferation. The cells growing out of the tissue are called primary cells, and have properties that closely resemble to that of cells *in vivo*. The cells will proliferate only if they are firmly attached to and spread out over the surface of the plate— a property of primary and diploid cell strains termed as anchorage dependence. The cells continue to proliferate until the plate is covered with a single layer (monolayer) of cells (Figure 2.4). Once the dish is fully covered (100% confluent culture), cells stop dividing by a mechanism known as *contact inhibition*. Normal fibroblast (MRC5, WI38) cells are most sensitive to contact inhibition, whereas epithelial and endothelial cells do not stop proliferation even when two growing cells touch each other. As an exception, transformed cell lines (e.g., HepG2, HeLa, MCF-7, etc.) lose contact inhibition and grow in multilayer (Figure 2.4). It is generally believed that when extending cytoplasmic extrusions (*pseudopodia* or *lamellipodia*) of two cells contact each other, they cannot move and the respective cells extend lamellipodia in other directions. For the growth of normal fibroblasts, cellular movement is most important. In confluent culture, since lamellipodia activity and movement are ceased, cells cannot grow. Cessation of growth can also occur due to overcrowding of cells, and this phenomenon is known as *density dependent inhibition* of growth. Each cell type exhibits a typical density-dependent inhibition of growth. The density-dependent inhibition of growth can be avoided by subculturing cells into a larger flask or scraping-off cells from the confluent culture.

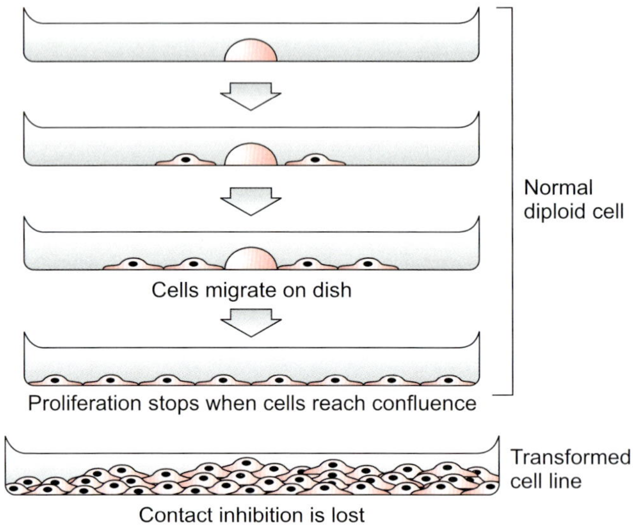

Figure 2.4 The contact inhibition of normal diploid cells. Normal diploid cell strains (e.g. MRC-5) stop proliferation, if they contact each other while growing. In transformed cells (e.g. HepG2), the inhibition is lost and cells keep on growing in multilayers, until the culture medium is exhausted.

Three mechanisms for density-dependent inhibition of growth have been proposed due to:

(a) Competition between neighboring cells for the same limiting growth factors
(b) Accumulation of toxic metabolic end products
(c) The limitation of space for the attachment of cells

In case of sub-critical level of seeding, the cells do not enter into the proliferation stage, despite the fact that no reasons for density dependent inhibition exist. This type of growth inhibition is believed to occur due to the limitation of *autocrine factors* liberated by the cells themselves.

2.2.1 Molecular basis of growth and proliferation

How cells control their growth and proliferation is not clearly understood. When cells are not in a growing stage, it is believed that the rates of protein synthesis and degradation are balanced. We know that removing growth factors reduces the rate of protein synthesis. Whether it also increases the rate of degradation of protein is not known. A cell may leave the cell cycle, temporarily or permanently, and also enter G_0 phase where growth is arrested. The main reason for entering cells into G_0 phase is starving from normal growth factor, or depriving from serum. A G_0 cell is often called *quiescent cell*, but this is probably a mere reflection of the interests of the scientists studying the cell cycle than the cell itself, which has got no fixed meaning. There is other biological significance for a cell entering into G_0 phase. Often cells enter into G_0 prior to terminal differentiation, they never reenter the cell cycle but instead carry out their function until die. For other cells, G_0 proceeds reentry into the cell cycle. Most of the hematopoietic cells in mammalian systems are found in G_0 phase. However, with proper stimulation, such as encountering with appropriate cytokines or antigen, they can be stimulated to reenter in the cell cycle (G_1) and proceed on to new rounds of mitosis. G_0 phase simply does not represent the absence of signals for mitosis, but also explains the active repression of the genes needed for mitosis. Interestingly, cancer cells do not enter G_0 (ES cells have very short G_0 phase) and are thus destined to repeat the cell cycle indefinitely.

Each normal cell has inherent capacity to rectify DNA damage, if any, during growth process. This observation led to the definition of the *restriction point* or *checkpoint* in the cell cycle. The cell has several stages of checkpoint to interrupt the cell cycle. DNA damage checkpoints occur before the cell enters S phase (a G_1 checkpoint), during S phase, and after DNA replication (a G_2 checkpoint). In spindle checkpoints, the cell detects any failure of spindle fibers to attach *kinetochores* and arrest in the metaphase (M checkpoint), and also detects improper alignment of the spindle itself and blocks cytokinesis. The net outcome of checkpoints is reentry of cells into cycle if the damage is repairable; otherwise triggers apoptosis.

In multicellular organisms, individual cells communicate with each other for coordinated growth, differentiation, and cellular functions. Cell communicates in many ways, most importantly by secretion of soluble growth factors by one cell that interact with the cognate receptor present on the other cell. The binding of growth factors induces many receptor-mediated events (signals) inside the cells, such as activation of *protein kinases*. As a result, many intracellular proteins are phosphorylated and this leads to the transcription of *early response genes* and expression of *transcription factors*. The transcription factors in-turn induce the synthesis of the *delayed response genes*. The most common steps for any cell signaling pathways are described in Figure 2.5A, where activation of cellular response refers to the function mediated by delayed response genes. In response to growth signals, the undifferentiated cells move out of G_0 and enter G_1 phase. The cycle begins in G_1 with increased expression of D1, D2, D3 cyclins, products of delayed response genes. In absence of growth factor or serum, the early and delayed genes are not transcribed, thus the cells do not exit from G_0. In most cells, addition of growth factors/serum induces transcription of early genes, which is maximized within 30 to 60 minutes. However, serum-induced early response genes sharply drop with concomitant increase in the levels of delayed response genes and the corresponding proteins. Thus, turning-on early response genes, followed by delayed response genes and synthesis of corresponding D cyclins are close and interdependent events to activate cell cycle. The profiles for synthesis of early and delayed response genes in mammalian cells, in the absence or the presence of growth factor/serum, are shown in Figure 2.5B.

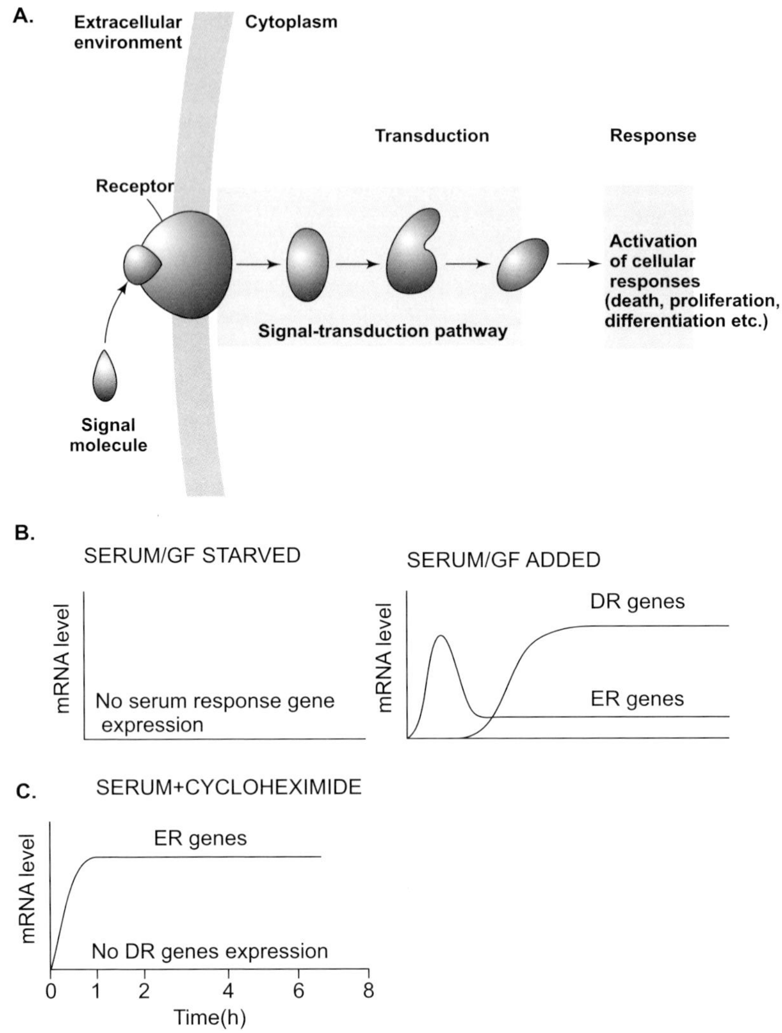

Figure 2.5 Initial events of growth factor/serum mediated proliferation of animal cells. **A.** Major events in cell signaling pathways; **B.** Expression of early response (ER) and delayed response (DR) gene in the presence and the absence of serum. **C.** Drug cycloheximide blocks protein synthesis, thus, DR genes are not turned-on.

To begin cell cycle, D cyclins are associated with cyclin-dependent kinase 4 and 6 (Cdk4 and Cdk6). Cdks are protein kinases that require association with cyclins followed by the phosphorylation for their activity. One of their substrates is believed to be the retinoblastoma gene product (Rb), one of the critical negative regulators of the G_1/S transition. Rb sequesters the E2F transcription factor and inhibits its transactivation (Figure 2.6). Phosphorylation by Cdk4/6 makes Rb dissociate from E2F, allowing it to transcribe number of responder genes that allows the cell to enter S phase. As the cell progresses through late G_1 phase, the expression of cyclin E is increased, which binds with Cdk2 (cyclin E/Cdk2) for transition from G_1 to S phase. Growth factor withdrawal activates GSK-3β (glycogen synthase kinase, a homologue of Zw3 kinase), which phosphorylates cyclin D, leading to its rapid *ubiquitination* and *proteasome* mediated degradation. Ubiquitination, nuclear export, and degradation are three mechanisms commonly seen to modulate concentrations of cell cycle proteins. The expression of cyclin A is increased throughout S phase, which binds with Cdk2 to carry out DNA synthesis. The G_2/M DNA damage checkpoint prevents cells entering into M phase, in case of any genomic abnormity. The cyclin B/cdc2 complex is pivotal in regulating this transition. The cyclin proteins are degraded in mitosis at two different points— cyclin A during metaphase and cyclin B in the anaphase.

Thus, the molecular basis of the regulatory cyclical events that control transitions between phases depends upon three activities: (i) G_1 phase cell committed to division and the key event is phosphorylation of Rb protein, as its nonphosphorylated form represses transcription of genes that are needed for the cycle to progress, (ii) the period of S phase which is marked by the presence of S phase activators, such as protein kinases, and (iii) mitosis, which depends upon the activation of M phase kinase. A striking feature of cell cycle regulation is that similar regulatory activities are found in almost all eukaryotic cells.

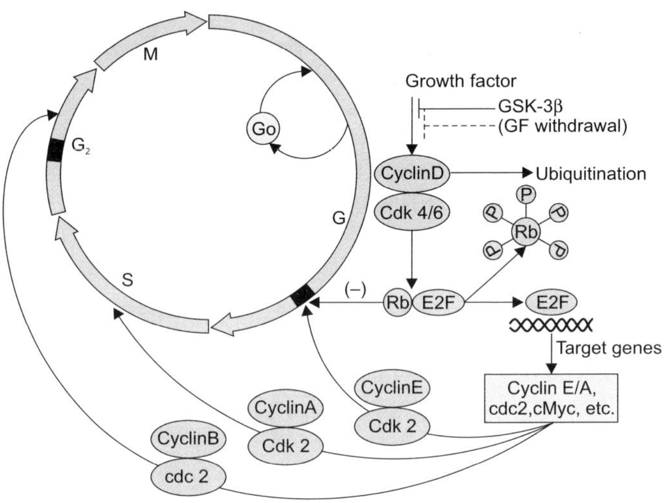

Figure 2.6 Cell cycle and activation of various stage specific kinases and genes. Cdk: Cyclin dependent kinase, Rb: Retinoblastoma protein, GSK-3β: Glycogen synthase kinase 3β.

2.3 HAYFLICK'S PHENOMENON

In culture, mammalian cells can reach to a certain point beyond that they simply stop dividing and will eventually die, even if are provided with adequate nutrients. In normal diploid cells, this usually occurs after around 50 cell divisions or cumulative population doublings, which generally termed as cellular senescence. In primary cells, senescent may ocurs within few replication cycles. This is also known as *replicative senescence* (RS), the *Hayflick limit*, or the *Hayflick phenomenon* after the name of Dr. Leonard Hayflick, who proposed in 1965.

Hayflick observed that after many doubling this limit is correlated with the length of telomere at the end of a DNA strand. Telomere is a repeating DNA sequence (e.g. TTAGGG) at the 3′ end of DNA, the length of which could be 15,000 base pairs. The function of telomeres is preventing chromosomes from losing base pair sequences at their ends. In the same time, they also stop chromosomes from fussing to each other. At each time of DNA duplication in normal cells, a small strand of DNA 100-200 nucleotides are lost. When the telomere becomes too short, the chromosome reaches a critical length and can no longer replicate. It means that RS is attained and cell dies by a process called apoptosis. The activity to telomere is controlled by two mechanisms: erosion and addition. Erosion takes place each time a cell divides. On the other hand, addition is determined by the activity of telomerase, which elongates chromosomes by adding TTAGGG sequences at the end. In somatic cells, telomerase activity is almost undetectable, whereas in tumor cells and during development fetal cells show its activity. The most common pathways that can trigger senescene in mammalian cells are the p16/Rb and p53/p21 mediated.

2.3.1 Does RS equal for all cell types

RS can be seen in all cell types; however time to enter in this phase is varied among the cell types within a species and from one species to other. The initial work on RS was conducted in fibroblasts, later the phenomenon was seen in other somatic cells, such as lymphocytes, endothelial cells, vascular smooth muscle cells, keratinocytes, etc. Those cells is the body is constantly proliferating may enter senescence much early that the cells that doubles once in a while. *Cumalative population doubling* (CPD) is the number of total doublings is experienced by a cell type before entering into RS. It is believed that CPD has correlation with the longevity of the species, for example somatic cells of long-lived tortoise can divide about 110 times, while hepatocytes of human liver can divide about 50 times and in case of mouse it will be roughly 15 times. The embryonic stem/germ cells and most tumor cell lines (e.g. U938, HeLa, A549, etc.) evade RS, so termed as immortal cells.

2.3.2 Identification of cells undergoing RS

Laboratory grown cultures are heterogeneous, some cells arrest their growth. The proportion of growth-arrested cells is progressively increased until cells stop divide due to RS. Senescent cells can be identified by following properties:

(a) Nonresponsive to growth factors- RS is an irreversible phenomenon, so addition of growth factors cannot stimulate cell division.

(b) Cellular morphology- senescent cells are bigger and a senescent population has more diverse morphotypes than growing cells. They are sensitive to cell-cell contact inhibition, so confluent culture show lesser cell density than a confluent growing culture.

(c) Expression of senescence-associated β-galactosidase activity. The most widely used biomarker for senescent and aging cells is RS-associated β-galactosdase (SA-β-gal). The enzyme β-galactosidase, a lysosomal hydrolase, is normally active at pH 4, but in senescent cells its activity is detectable at pH 6.0, which permits the identification of senescent cells in culture and mammalian tissues (Figure 2.7). SA-β-gal activity is expressed by GLB1 gene encoding lysosomal β-D-galactosidase. The expression of SA-β-gal is not required for cells to enter into senescence, however specifically in senescent cells it is over-expressed and accumulated.

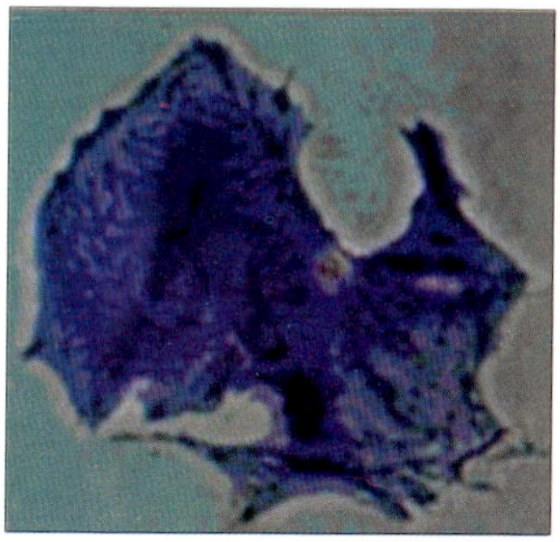

Figure 2.7 Fibroblasts showing senescent morphology by detection of SA-βgal activity with X-gal.

2.4 CELL SIGNALING

In biology, *signal transduction* is any process by which a cell converts one kind of stimulus (signal) into another. Signal transduction often involves a sequence of biochemical reactions inside the cell, which are carried out by enzymes and other molecules that become engaged in the events that proceed from the initial stimulation. The subsequent chain of steps is termed as a signaling cascade or 'second messenger pathway', resulting a small stimulus elicits a large response. The net result of signal transduction in mammalian cells is either proliferation, death, or differentiation.

Signal transduction usually involves the binding of 'extracellular' signaling molecules to receptors on cell surface and trigger various intracellular events. This takes place via a change in the conformation of the receptor that occurs when the signaling molecule (ligand) binds to it. Often, the intracellular events (intracellular signaling) are triggered by the external stimulus, but distinct from the event of transduction mediated by extracellular signals. The intracellular signaling molecules in mammalian cells include heterotrimeric G protein, small GTPases, cyclic AMP (cAMP), cGMP, Ca^{2+}, etc. Intercellular signaling is another way by which one cell communicates with the other, and is primarily accompanied by extracellular signal-

Table 2.1 Growth factors, target cells, and signaling pathways for proliferation of animal cells

Growth factor	*Primary activity*	*Major pathways activated*
Platelet-derived growth factor (PDGF)	Promotes proliferation of connective tissue, glial and smooth muscle cells	Ras/MAPK, JAK-STAT
Epidermal growth factor (EGF)	Promotes proliferation of mesenchymal, glial and epithelial cell types	Ras/MAPK, JAK-STAT
Fibroblast growth factor (FGF)	Promotes proliferation of many cell types	Ras/MAPK
Nerve growth factor (NGF)	Promotes outgrowth and survival of neuronal cells	PI3K/Akt
Erythropoietin (Epo)	Promotes proliferation and differentiation of erythrocytes	JAK-STAT
Insulin like growth factor-I (IGF-I)	Promotes proliferation of many cell types	PI3K/Akt, MAPK
Insulin like growth factor-II (IGF-II)	Promotes proliferation of many cell types, primarily of fetal origin	PI3K
Interleukin-3 (IL-3)	Growth of hematopoietic progenitor cells	JAK-STAT
Colony stimulating factors (CSFs)	Growth of hematopoietic progenitor cells	Ras/MAPK, JAK-STAT
Vascular endothelial growth factor (VEGF)	Endothelial cells	PI3K/MAPK

ing. Intercellular signaling is most common in multicellular organisms, and is divided into four types: *endocrine*, *paracrine*, *autocrine*, and *juxtacrine*.

2.4.1 Extracellular signaling and growth factor receptors

Growth factor receptors are named on the basis of ligands, which are usually small polypeptides also termed as growth factors or cytokines. The growth factors have a variety of biological activities. One such activity is stimulation of cell cycle followed by cell division, this will be elaborated here. Some of the commonly used growth factors, their target cells, and corresponding signaling pathways are shown in Table 2.1. The general pathway of cell signaling through ligand-receptor interaction has been shown in Figure 2.5A. Cell signaling process can be divided into three major phases: a) binding of ligand to its cognate receptor, b) cytoplasmic activities where various *adaptor* molecules and kinases bind to the cytoplasmic domain of the receptor and carry out the signaling events, and c) cellular response in the form of gene expression in the nucleus.

The binding of ligand to the cognate receptor monomers causes them to dimerize by the interaction between the extracellular domains. Receptor dimerization takes place by ligand-induced conformational changes in the N-terminal region of type I receptor. The transmembrane domains diffuse laterally bringing the cytoplasmic domains into juxtaposition. Ligand-receptor binding may trigger internalization of the complex by *endocytosis*. In due course of time, the receptors and the ligands are separated; receptor is recycled back to the cell surface for another cycle of receptor-mediated endocytosis. Dimerization of receptors activates the signaling pathway by triggering autophosphorylation of tyrosines at multiple sites present in the cytoplasmic domains of the receptor tyrosine kinase (RTK). The kinase activity of each monomeric receptor phosphorylates the other monomer. Phosphorylation of intracellular domains provides a means for the 'docking proteins' or 'adaptor proteins' to bind to activated receptors. The target of the activated receptors could be a signaling molecule that binds to the adaptor protein, or the target could be an inactive enzyme, which activates upon attachment with the receptor (phosphoinositide-3-kinase), or it could be a substrate (enzyme) that may be activated by phosphorylation (e.g., c-Src). The initial events are depicted in Figure 2.8.

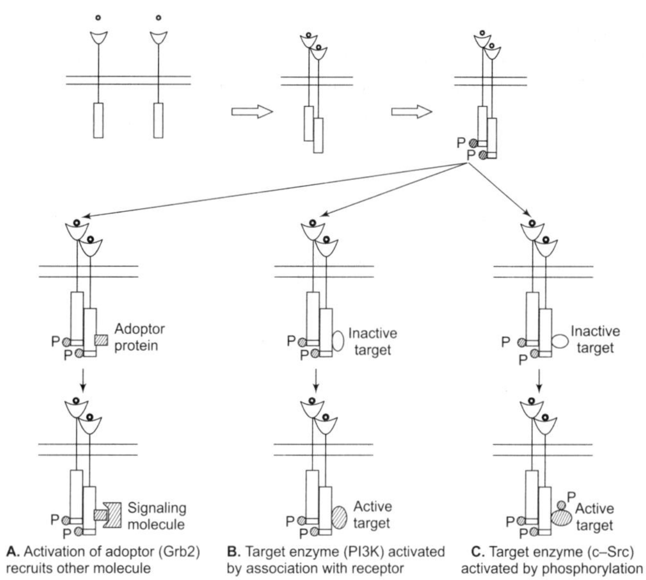

Figure 2.8 Initial events in protein tyrosine kinases induce phosphorylation cascades in animal cells. Cell surface receptors dimerize by interacting with ligand, followed by activation of different cascades. **A.** Activated adaptor protein recruits other signaling molecule(s). **B.** Inactive enzyme gets activated by association with activated receptor. **C.** Activated receptor recruits SH2 domain kinase for its phosphorylation and subsequent activation.

In mammals, EGF and/or PDGF are known to activate different mitogen activated protein kinase (MAPK) pathways through receptor-mediated activation of a cascade of tyrosine phosphorylation and protein-protein association. This activation involves an adapter protein Grb2. SOS is activated by binding with Grb2. Activated SOS in-turn activates small GTP binding protein, Ras. Activation of Ras leads to the activation of the c-Raf (Ser/Thr kinase). The Ras cascade (SOS→Ras→c-Raf) of activation events occur on the cytoplasmic face of the plasma membrane. Raf activates the kinase MEK (formerly known as MAP kinase kinase), which in turn activates MAP kinases (ERK1 & ERK2). The activated MAP kinase (MAPK) enters in the nucleus and phosphorylates specific transcription factors that trigger biological functions, as illustrated in Figure 2.9. RTKs alone are not involved in the activation of Ras/MAPK pathway; integrins and ion channels receptors also take part in this activation process. RTK-mediated activation of MAPK results in either cellular proliferation or differentiation. Cellular proliferation prevails, if the activation is transient, whereas continuous activation of MAPK triggers differentiation. Activation of RTKs in Ras-independent pathways involves second messengers, where phosphoinositide-3 kinase (PI3K) is activated by docking to the cytoplasmic domain of RTK in response to PDGF. The activated PI3K in-turn activates Rac, which stimulates membrane *ruffling*, formation of lamellipodia, and progression into the G_1 phase of the cell cycle (Innocenti *et al.*, 2002). Activation of Ras and MAP kinase pathway by EGF ultimately leads to the phosphorylation of transcription factors, such as c-Fos to create AP-1 and Elk-1 that contribute to cellular proliferation. Transcription of c-Fos appears to be regulated by two proteins, the serum response factor (SRF) and Elk-1, both of which bind to the promoter of the *c-fos* gene. These two proteins are bound to the promoter in quiescent cells, but the promoter does not activate unless cells are stimulated with serum (or growth factor), which phosphorylates Elk-1 by MAPK pathway.

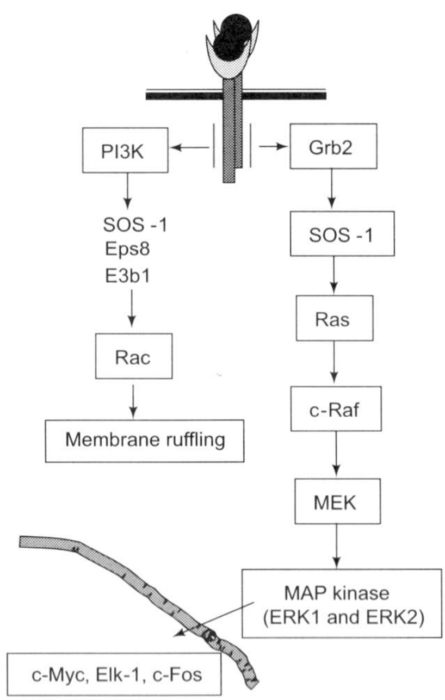

Figure 2.9 Activation of receptor tyrosine kinase (RTK) in Ras-dependent and independent pathways by EGF/PDGF. Ras-dependent pathways involve activation of MAPK pathways, resulting in cellular proliferation or differentiation. Ras-independent pathways involve membrane ruffling, formation of lamellipodia, etc. Grb2/Eps8/E3b1: adaptor proteins; PI3K: phosphoinositide 3 kinase; SOS-1: son of sevenless guanine nucleotide exchange factor-1; MEK: MAPK/ERK kinase; MAPK: mitogen-activated protein kinase; ERK: Extracellular signal-regulated kinase; Ras/Rac: small GTP binding proteins; Raf: MAPK kinase kinase.

2.5 STRESS-INDUCED CELL DEARTH

Protective stress responses to mammalian cells leading to the activation of various pathways that promote survival of cells. For example, *unfolded protein response* increases *chaperon* protein synthesis, which enchances protein folding capacity of the cells thereby promotes cell survival. Beyond certain limit of stress, cells can not cope with the conditions as a result die. The forms of cell death can be explained by programmed (*apoptotic* and *autophagic*) and non-programmed (*necrotic*) mechanisms. Apoptotic and autophagic cell death are induced by activating cellular machinery, whereas necrotic death is induced by injurious agents. Necrosis is an accidental death caused by mechanical damage or exposure to toxic chemicals. Necrotic death is accompanied by distinct characteristic changes including swelling of mitochondria and changes in the ability of the plasma membrane to regulate osmotic pressure. Initially, a progressive hydration of the cytoplasm occurs followed by membrane and organelle disruption, allowing the leakage of the lysosome content into the cytoplasm. Finally nuclear disintegration and complete disruption occur in cell (Figure 2.10). Coagulation necrosis (*oncosis*) may occurs due to denaturation and coagulation of cellular proteins. It can be triggered by factors from outside the cells, such as hypoxia, hyperthermia, physical damage, and toxic injury (Al-Rubeai, 1997). It is characterized by swelling, increase in membrane permeability, and also blebbing.

Apoptosis, by contrast, is a regulated and controlled process operated by preprogrammed manner. This self-demise process occurs in most mammalian cell types and is controlled by a putative 'yin-yang' mechanism involving the activation of killer genes and the deactivation of survival genes (Al-Rubeai, 1997). There are distinct physical changes that occur in apoptotic cells, which include exposure of the phospholipid, phosphatidylserine on membrane surface, shrinkage of cells, bubble-like blebs on their surface, chromatin condensation, release of cytochrome c from leaky mitochondria, and formation of membrane-wrapped fragmented cellular materials (Figure 2.10). Transmission electron micrograph (TEM) of apoptotic cells is shown for better visualization of the changes of organelles in ultrastructure level (Figure 2.11). Apoptosis or programmed cell death is needed for proper development of whole organism or a part of the organism. For example, the formation of the fingers and toes of the fetus requires the removal of the tissue between them by apoptosis. Similarly, the sloughing-off the inner lining of the endometrium layer of uterus at the beginning of menstruation cycle occurs by apoptosis. Programmed death is also needed to destroy cells that represents threat to the integrity of the whole organism. By inducing apoptosis, organism clears cells infected with the viruses: in cell-mediated immune responses the target cells are killed by the process of apoptosis. Apoptosis clears cells containing damaged-DNA, in some in cancer cells are destroyed by apoptosis induced by radiation and chemotherapeutic agents.

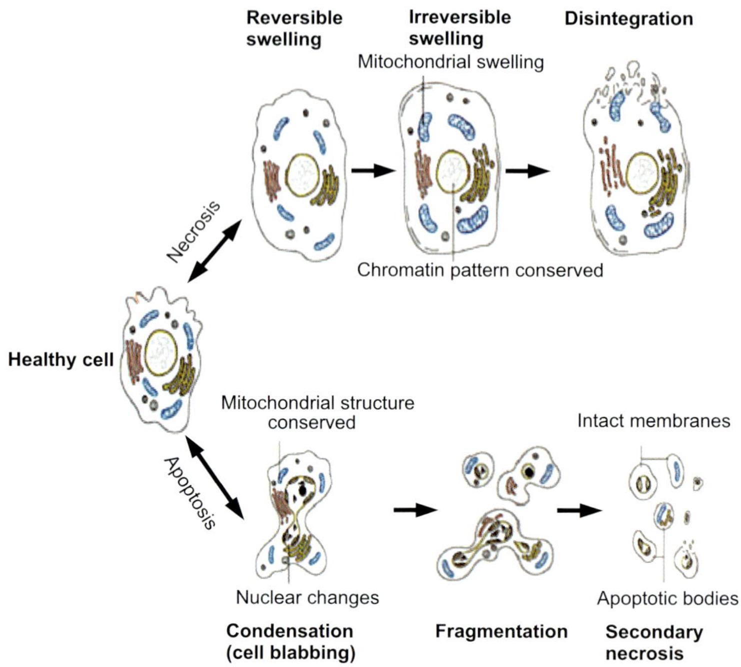

Figure 2.10 Morphological changes in animal cells at different stages of apoptotic and necrotic death [*Source*: Apoptosis, Cell Death, and Cell Proliferation Manual, 3rd edition, Roche Diagnostics, Germany].

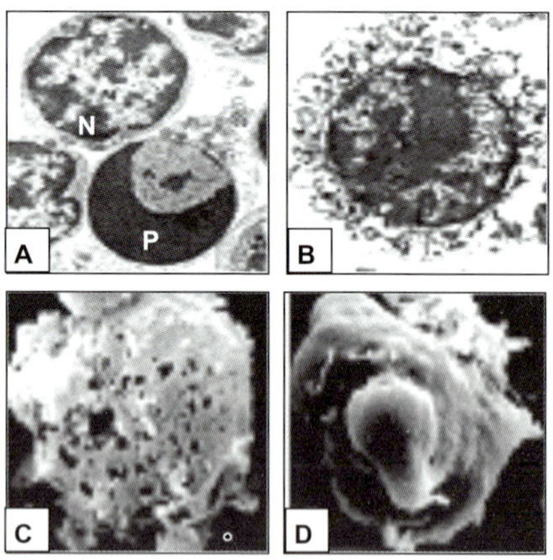

Figure 2.11 Transmission electron micrograph of an apoptotic and a necrotic cell. **A.** Apoptotic (P) and a normal (N) cell. The characteristic chromatin rearrangement appears in P, which is different from its normal organization (N). The membrane and organelles are maintained intact (original magnification: × 8,000). **B.** Necrotic cell, the disruption of plasma membrane and organelles are maintained intact (original magnification: × 10,000). **C.** Numerous lesions appear in the necrotic cell surface (original magnification: × 5,000). **D.** Surface blebbing is evident in apoptotic cell (original magnification: × 5,000). [Falcieri E, et al., (1994) Scanning Microse. 8, 653; Stuppia L, et al. (1996) Cell Death Differ. 3, 397]

Autophagic cell death or autophagy (self-eating) is an evolutionarily conserved process characterized by lysosomal degradation of cytoplasmic proteins and organelles (e.g. mitochondria, etc.). Authophagy is a survival cellular response to large number of internal and external stresses, like nutrient limitation, oxidative stress, hypoxia and ER stress. During the autophagy, double- or multi-membrane vesicles formed, sequestering cytoplasmic components, are know as *autophagosome* or *autophagic vacuoles*. The formation of authophagosomes involves the following multistep processes: initiation, nucleation, elongation and sealing. The sequestered cytoplasmic components are delivered to the lysosomes for degradation.

2.5.1 Induction of apoptosis in vitro cultured

A wide variety of physiological and non-physiological factors are responsible for inducing apoptosis in cell culture system (Al-Rubeai, 1997; Simpson *et al.*, 1997). These are as follows:

(i) Growth factor withdrawal: Withdrawal of positive signals is an indirect way of inducing apoptosis in many primary cells in culture. The survival of most anchorage-dependent cells require continuous stimulation from other cells or from the same cells for remaining them adhered to the surface on which they are growing. The sensitivity of a cell to growth factors depends on the level of expression of a ranges of genes that either enhance resistance or susceptibility to the apoptosis. Serum is essential for the growth and survival of many mammalian cells. The removal of serum has been shown to induce apoptosis in many cells like stem cells, hybridomas and plasmocytomas. Fibroblasts express *c-myc* protooncogenes when cultured in presence of serum supplemented medium, but undergo apoptosis if serum is removed. In absence of growth factors, *c-myc* is down-regulated; initially cells are arrested in G_0 phase but later undergo apoptosis. Apoptosis also occurs when cells are deprived of energy substrates such as glucose and glutamine. In absence of glucose, DNA synthesis is prevented and cells are blocked in G_1 phase, thereby trigger apoptosis.

(ii) Receipt of negative signals: Increased levels of reactive oxygen intermediates (ROI) damage DNA and cell membrane, which causes natural induction of apoptosis. The hydrodynamic regime of bioreactors or during transfer of cells from one place to another plays an important role in determining whether cells will dye or survive. Hybridomas are found to be sensitive to mechanical agitation, where significant apoptotic death has been observed due to imparted shear force on cells. Many protein molecules (ligands) bind to the specific cell surface receptors and induce apoptosis. These molecules are known as death activators, such as tumor necrosis factor-α (TNFα), Fas ligand (FasL), etc.

(iii) Cell cycle and protein load: The molecular mechanisms that regulate cell cycle and apoptosis are interlinked. Hybridomas produce more antibodies if growth rate is maintained slow; however, the cell viability is quickly reduced. The decline in cell viability is due to induction of apoptosis, as cells assume the state 'grow or die', as they lack G_0 state. High expression of recombinant proteins in cell lines also induces toxic effects to the cells and causes apoptosis. Finally, the accumulation of proteins in side cells due to improper folding leads to apoptotic cell death.

(iv) Virus infection: Like induction of apoptosis in virus-infected cells *in vivo*, mammalian cells undergo apoptosis during production of virus particles in culture.

2.5.2 Mechanisms of apoptosis

Apoptosis is mediated by three possible mechanisms, which are: (i) signals arising within the cells, (ii) signals generated by death activators, and (iii) a mechanism different from (i) and (ii). In all the three mechanisms, mitochondria plays pivotal role in triggering apoptosis. The classical pathway for apoptosis in mammalian cells that are activated by wide variety of stimuli is based on Fas/TNFR1 pathways, as shown in Figure 2.12. Both Fas and TNFR1 have cytoplasmic domains that recruit Fas associated death domain (FADD) and TNF receptor associated death domain (TRADD), respectively, upon activation with the respective ligands. FADD also binds on TRADD. In either case, FADD recruits

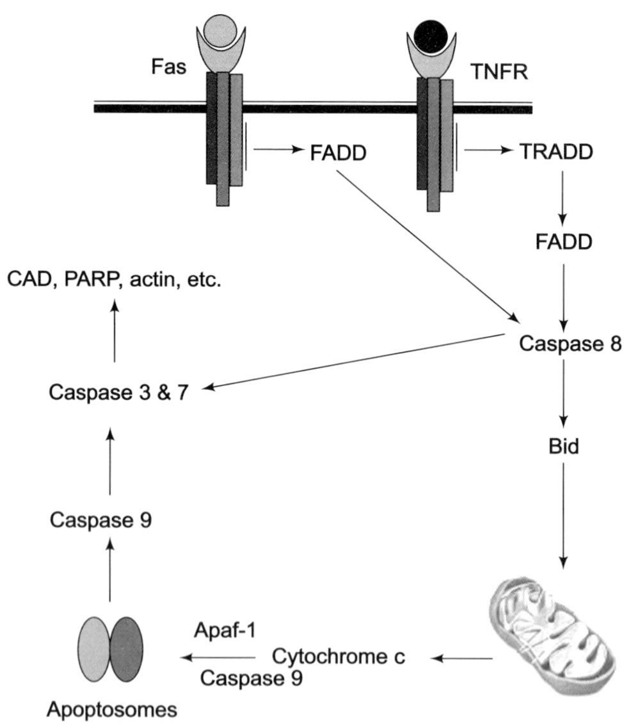

Figure 2.12 Fas/TNFR pathways for apoptotic death of animal cell. CAD: Caspase-activated deoxyribonuclease; Bid: BH3 domain-only death agonist protein; Fas: TNF super family receptor 6; TNFR: TNF receptor.

caspase-8 (previously known as FLICE). Caspase-8 has both death domain and protease catalytic activity, which trigger two pathways of apoptosis. Caspase-8 cleaves a protein called Bid, the C-terminal domain of which (tBid) translocates into mitochondria. Thus, mitochondrial membrane becomes leaky and cytochrome c is released in the cytoplasm. The released cytochrome c binds with apoptotic protease activating factor-1 (Apaf-1) in the cytoplasm and forms a complex known as *apoptosome*, which finally activates Caspase-9. Activated Caspase-9 in-turn activates executioner caspase-3 and caspase-7. Caspase-3 is also directly activated by Caspase-8. One known target of caspase-3 is poly [ADP-ribose] polymerase (PARP). Besides, the executioner caspases digest structural proteins in the cytoplasm and degrade chromosomal DNA. The outcome of stimuli by death activators is apoptosis, and finally death of cells. Neurons and probably few more cell types do not follow the caspase-dependent pathways for self-destruction, instead apoptosis-inducing factor (AIF) initiates cell death. AIF is located in the intermembrane space of mitochondria, which is released and migrated to the nucleus, binds with DNA and triggers the degradation of DNA and subsequent cell death.

The process of apoptosis is genetically regulated. Few genes involved in the regulation of cell proliferation have been implicated in control of apoptosis. For example, *p53* tumor suppressor gene, which plays a central role in the regulation of cell proliferation, has been shown to mediate apoptosis by inducing DNA damage (Al-Rubeai, 1997). It functions by arresting cell in G_1 phase after DNA damage, and allows cell to repair the damage. If damage is not repaired, the cell commits apoptosis. In many cancerous cells, *p53* gene is mutated and thus this pathway is ineffective. The *c-myc* gene has also been found to be involved in apoptosis under a specific condition of cell culture. There are many other genes and their products, such as *bax*, *bad*, etc. are shown to be involved in inducing apoptosis. These genes are known as proapoptotic genes. Similarly, there are few genes familiar as antiapoptotic or survival genes, such as *bcl-2, bcl-xL, etc.*

2.5.3 Detection of apoptosis

There are many techniques that can be followed to determin apoptosis in a cell. These techniques are based on certain characteristic features that are attained by a cell undergoing apoptosis. Following four techniques are most popular in identifying apoptotic cells:

1. Annexin V staining of cell membrane: A number of changes in cell membrane mark apoptosis. One of the significant changes noticed in apoptotic cells is loss of asymmetry in cell membrane phospholipids, altering both the hydrophobicity and charge of the membrane surface, which provides an assay for apoptotic cells. In normal cells, the distribution of phospholipids is asymmetric, that in the inner membrane containing anionic phospholipids (e.g., phosphatidylserine) and the outer membrane projected neutral phospholipids. However, in apoptotic cells, the amount of phosphatidylserine on the outer surface of the membrane increases. Annexin-V, a calcium-dependent phospholipid-binding protein, has a high affinity for phosphatidylserine, and binds to the apoptotic cells only (Figure 2.13A).

2. Study of DNA fragmentation by agarose gel electrophoresis: Apoptosis is marked by the change in morphology of the nucleus, together with a distinctive biochemical event like the endonuclease-mediated cleavage of nuclear DNA. Formation of DNA fragments of (180-200 bp) size is a hallmark of apoptosis in many cell types. In apoptotic cells, specific DNA cleavage becomes evident by electrophoresis in agarose gel yielding a typical ladder due to multiple DNA fragments (Figure 2.13B).

3. Analyzing sub-G_0/G_1 fraction of cells: DNA content of apoptotic cells is reduced due to degradation and subsequent leakage from the cells. The lower DNA content means that cells will be stained less with DNA intercalating dye (e.g., PI, Hoechst 33342). Thus, the cells with lower DNA-stained than the normal G_0/G_1 cells (sub-G_0/G_1 peak) are considered apoptotic. The reduction in DNA content of these cells is measured by flow cytometry (Figure 2.13C).

4. In situ DNA fragmentation study by direct terminal transferase DNA labeling reaction (TUNEL): The low molecular weight DNA fragments as well as single strand break (nicks) in high molecular weight DNA can be identified by labeling free 3′-OH termini with modified nucleotides in an enzymatic reaction. Terminal deoxynucleotidyl transferase (TdT), which catalyzes polymerization of nucleotides at the free 3′-OH DNA and ends in a template-independent manner, is used to label cells that have oligonucleosomal nicks/strand breaks in their DNA. The incorporated nucleotides are detected by peroxidase conjugated secondary antibody. After reacting with substrate, the stained cells are identified under a light microscope. Alternatively, fluorescein-labeled nucleotides (FITC-dUTP) can be detected under a fluorescence microscope.

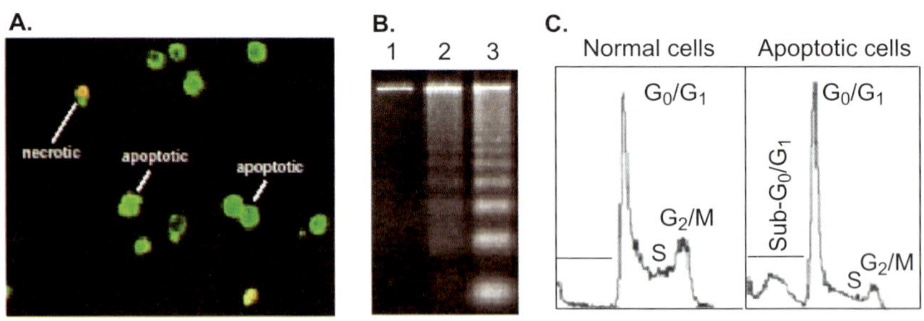

Figure 2.13 Detection of apoptosis in animal cells. **A.** Apoptotic cells appear green due to staining with Annexin-V-FLUOS (green); the necrotic cells take up propidium iodide (PI) and stain orange/green (stained with both Annexin-V-FLUOS and red PI). **B.** DNA fragmentation of apoptotic cells. 1. Control cells, 2. Apoptotic cells, 3. Positive control. **C.** Flow cytometry profiles of DNA content in normal and apoptotic cells stained with propidium iodide. [*Source:* Apoptosis, Cell Death, and Cell Proliferation Manual, 3rd edition, Roche Diagnostics, Germany]

2.6 SUMMARY

1. In multicellular organisms, cells divide to grow, repair or replace the damaged tissues and reproduction.
2. Each somatic cell has four phases in cell cycle: G_0/G_1, S, G_2, and M.
3. DNA intercalating dye, propidium iodide or Hoechst 33342 determines cell cycle position.
4. Growth of animal cells is controlled at three levels: intrinsic, intercellular, and interpopulation.
5. 'Rb' protein is a negative regulator of G_1/S transition; its phosphorylation and subsequent dissociation from E2F transcription factor is necessary for delivering cells into S phase.
6. Ligand mediated signal transduction relates extracellular environment with the activation of intracellular kinases and other proteins for cellular response: proliferation, differentiation and death.
7. Primary cells do not grow beyond certain population doubling due to replicative senescence.
8. Apoptosis is induced by withdrawal of growth factor, negative signals, infection with viruses, and intracellular protein load.
9. Executioner caspase-3 and caspase-7 are involved in the degradation of various structural proteins in apoptotic cells.

ABBREVIATIONS

AIF	Apoptosis-inducing factor
Apaf-1	Apoptotic protease activating factor-1
Cdk4/Cdk6	Cyclin-dependent kinase 4/6
CSF	Colony stimulating factor
cAMP	Cyclic AMP
cGMP	Cyclic GMP
Epo	Erythropoietin
EGF	Epidermal growth factor
FADD	Fas associated death domain
FasL	Fas ligand
FGF	Fibroblast growth factor
GSK-3β	Glycogen synthase kinase, a homologue of Zw3 kinase
GTPases	Guanine triphosphate binding enzymes
MI	Mitotic index
IGF-I/II	Insulin like growth factor-I/II
IL-3	Interleukin-3
JAK	Janus kinase
MAPK	Mitogen activated protein kinase
NGF	Nerve growth factor
PARP	Poly [ADP-ribose] polymerase

PI	Propidium iodide
PKB/Akt	Protein kinase B
PDGF	Platelet-derived growth factor
PI3K	Phosphatidyl inositol-3 kinase
RTK	Receptor tyrosine kinase
Rb	Retinoblastoma
SOS1	Son of sevenless guanine nucleotide exchange factor-1
SRF	Serum response factor
STAT	Signal transducer and activator of transcript
TEM	Transmission electron microscope
TNFα	Tumor necrosis factor-α
TdT	Terminal deoxynucleotidyl transferase
TRADD	TNF receptor associated death domain
VEGF	Vascular endothelial growth factor

REFERENCES

1. Al-Rubeai M (1997) *Biochem. Engg. Biotechnol.* 59, 225.
2. Innocenti M, Pierlugi T, Frittoli E, *et al.* (2002) *J. Cell Biol.* 156, 125.
3. Martin R (1998) *Science* 396, 119.
4. Maciera-Coelho A (1973) In: *Tissue Culture Methods and Applications* (PF Kruse, MK Patterson, eds.), pp. 412. Academic Press, New York.
5. Riley PA (1981) In: *Regulation of growth in neoplasia* (GV Sherbet, ed.), pp. 131. Kargel, Basel.
6. Simpson N, Milner AN, Al-Rubeai M (1997) *Biotechnol. Bioeng.* 54, 1.

SUGGESTION FOR FURTHER READING

1. Bortner CD, Oldenburg NBE, Cidlowski JA (1995) *Trends Cell Biol.* 5, 21.
2. Hayflick L, Moorhead PS (1961) *Exp. Cell Res.* 25, 585.
3. Israels ED, Israels LG (2001) *Stem cells* 19, 88.
4. Johnson DG, Walker CL (1999) *Ann. Rev. Pharmacol. Toxicol.* 39, 295.
5. Lohrum MAE, Vousden KH (1991) *Cell Death Diff.* 6, 1162.
6. Lewin B (2000) *Genes VII*, Oxford University Press, Oxford.
7. Murray A and Hunt T (1993) *The Cell cycle.* WH Freeman and Company, New York.
8. Alberts B, Bray D, Lewis J, *et al.* (1989) *Molecular Biology of The Cell*, 2nd edition, Garland Publishing Co, New York.

Section II
Basic Cell Culture Technology

3

Sources of Animal Cells, Preservation and Control of Cell Lines/Tissues

In adult animals, about 200 different cell types are present; most of them are functionally distinct and have specific growth characteristics. The growth and maintenance of these cells require knowledges on complex cellular process. Inside the body, cells grow and maintain in the form of tissue under the influence of multiple intrinsic (within the cell) and extrinsic (outside the cell) factors. Cells isolated from tissue and cultured in a new environment may lead to death, or change in normal growth pattern or may undergo transformation. Thus, it is most crucial to understand the behavior of an isolated cells *in vitro* for their exploitation in manufacturing of biological.

3.1 SOURCES OF CELLS

Embryonic, fetal and adult normal tissues are three major sources for the derivation of therapeutically and clinically useful cells. In general, embryonic tissues are superior to those of adults in terms of survival and proliferation potential. Embryonic tissues are less specialized and contain more precursor stem cells than the adult tissues. On the contrary, adult tissues contain more differentiated cells of limited lifespan, which are difficult to cultivate *in vitro* under normal conditions. Examples of some widely used fetal originated cell types are mouse embryonic fibroblast (MEF), MRC5, and WI38.

3.1.1 Classification of cells based on growth characteristics

Cells isolated from a normal tissue, whether embryonic or adult source, grown in culture for the first time without subculture are termed as a primary culture. Once the cells are isolated from a tissue, they experience altered cellular microenvironment.

The cell-matrix interactions are reduced *in vitro*, and nutritional requirement and surrounding growth factors are also considerably altered in comparison to *in vivo*. Therefore, establishing a primary culture from a tissue is a complicated process unless specific growth factors and/or hormones are externally added in the culture medium. In many cases, the primary cultures lack sufficient number of cells to attach to the culture dish and to survive. This could be mainly due to inadequate amount of autocrine growth factors produced by the cells. The first stage of natural selection of cells occurs in the primary culture; cells that survive the tissue dissociation process and adhere on the culture plate will form the basis of the primary culture. The primary culture may be of heterogeneous population of many cell types present in the tissue of origin. When a primary culture is *subcultured* or *passaged* for the first time, it is known as *cell strain*. A cell strain can be propagated and subcultured several times. In successive subcultures, a more homogeneous cell type emerges with the ability to proliferate rapidly, and simultaneously gradual dilution of contaminating slow growing cells occur. Cell strains have a finite life span in culture; most of them can be propagated for a limited number of *generations* (*population doublings*) in an unaltered form. Many cell strains, as for example human fibroblasts, die after finite life span; some of them however transformed to become a *continuous cell line*. Usually, cell strains grow between 20 and 80 population doublings before entering into *senescence* or death phase. The actual

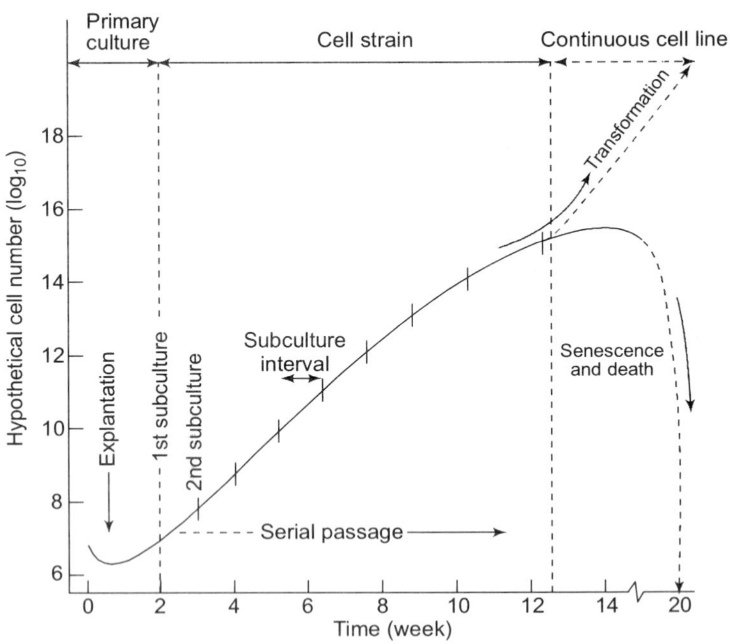

Figure 3.1 Evolution of a cell strain and a continuous cell line from primary cells isolated from a tissue. [Adapted from Ref. 1]

number of population doublings may vary from cell strain to cell strain, and the culture conditions they are exposed to. For example, human fibroblasts (MRC5, WI38) can be cultured for around 50 population doublings before they stop dividing (see 2.3). The non-dividing fibroblasts remain alive for about one-and-a-half year before they die. The evolution of a continuous cell line from the stage of *explant* culture is shown in Figure 3.1. Transformation may occur spontaneously, or induced with chemicals or viral infection. It occurs at various stages of selection processes such as during isolation of tissue in primary culture, during subculture for propagation as cell strain, and in senescence (Figure 3.1). In normal diploid fibroblasts, transformation may occur after 12 weeks, and senescence within 10-20 weeks of culture.

Transformation results in acquisition of an infinite life span and alteration of growth control in a cell. It has been shown that many mutational events are taken place before cells transform with the activation of *oncogenes* and simultaneous inactivation of tumor suppression genes. The characteristic features of primary cells, cell strains, and continuous cell lines are elaborated in Table 3.1. All three different classes of cells are used for manufacturing of biological. The most commonly used cell types, their origin, characteristics, and possible applications are shown in Table 3.2. The photomicrographs of primary cells, cell strain, and continuous cell line are shown in Figure 3.2.

Table 3.1 Characteristic features of primary cells, cell strains, and continuous cell lines

Special characteristics	*Primary cells*	*Cell strains*	*Continuous cell lines*
Life-span	Finite	Finite	Infinite
Ploidy	Diploid, Euploid	Diploid, Euploid	Heteroploid, Aneuploid
Growth rate	Slow	Slow	Rapid
Anchorage dependency	Present	Present	Considerably reduced
Serum dependency	High	High	Reduced
Mode of growth	Monolayer	Monolayer	Monolayer, Multilayer, or Suspension
Contact inhibition	Absent	Present	Absent
Cloning efficiency	Low	Low	High
Tumorigenicity	Non-tumorigenic	Non-tumorigenic	Non-tumorigenic/ Tumorigenic
Genetic alteration	Absent	Present	Present

Table 3.2 Different cells used in manufacturing biological

Name of the cells	*Origin*	*Nature*	*Characteristics and Applications*
Vero	Monkey kidney cell line	Adult, Continuous	Non-tumorigenic; used for making human viral vaccines
BHK-21 kidney	Syrian hamster uous cell line	Newborn, Contin-	Virus transformed; used for producing vaccines against foot and mouth diseases (FMD) in cattles
MRC5	Human lung	Embryonic, Finite cell line	Cell strain; used for production of vaccines against viral diseases in humans
WI38	Human lung	Embryonic, Finite cell line	Cell strain, used for making viral vaccines for humans
Primary Kidney cell	Monkey	Adult, Primary cell	Primary cell, used for production of vaccines against viral diseases in human
CHO-K1 ovary	Chinese hamster cell line	Adult, Continuous	Non-tumorigenic; used for producing recombinant proteins for humans
BHK-21 kidney	Syrian hamster uous cell line	Newborn, Contin-	Non-tumorigenic; used for making recombinant proteins for human
NSO	Mouse	Myeloma	Recombinant proteins for humans
HEK-293	Human kidney	Embryonic, Continuous cell line	Recombinant proteins for humans

3.2 SELECTION OF CELLS FOR MANUFACTURING BIOLOGICAL

On the basis of end usage, biological derived from animal cells are classified into three broad categories: (a) products intended to be used for humans, (b) products intended to be used for animals, and (c) products intended to be used for *in vitro* diagnostics. There is no specific criterion by which mammalian cells are selected for the production of reagents for *in vitro* applications. The most stringent selection criteria to be followed are if the derived products intended to use for therapeutic applications in human. Further, the selection criteria become less stringent or more on the basis of the nature of products; in case of vaccines more stringent criteria are followed than therapeutic proteins. If the product is vaccine, the question arises whether it will be used for children or adult.

3.2.1 Mammalian cells used for manufacturing of vaccines

The suitability of animal cells as a substrate for human viral vaccines depends on the following properties:

(a) They should be easily acquired and cultivable
(b) They should be homogeneous and able to grow for many population doubling
(c) They should allow easy propagation of target virus
(d) They should be free from microbial contaminants and oncogenie materials
(e) They should not be transformed spontaneously
(f) The *karyotype* of the cell should remain stable.

Not many cell types, available today, qualify above properties. Certain donor animal species such as monkey, pig, chicken have inherent problems of (i) animal viruses that are normally absent in humans, (ii) undergo limited propagation *in vitro*, and (iii) restricted susceptibility to human viruses. In contrast to non-human cells, studies on human diploid cells (HDCs) derived from pre-natal tissue showed fewer problems. Therefore, HDCs are generally accepted as the substrate of choice for the production of human viral vaccines. So far, 10 different HDCs of embryonic origin are reported; only two of them are used for the production of vaccines (Jacobs, 1978) in the various parts of the world (Table 3.3).

Table 3.3 HDCs used for the production of human viral vaccines

Cell	*Source*	*Vaccine*
MRC-5	Fetal lung	Polio (live), Rubella (live), Rabies (inactive), Measles (live), Mumps (live), Hepatitis A, Chickenpox, Smallpox.
WI-38	Fetal lung	Polio (live), Rubella (live), Rabies (inactive), Measles (live), Mumps (live), Chickenpox.

Source: http://www.cogforlife.org/fetalvaccines.htm

The alternate sources of cell substrates for making human vaccines are primary cells derived from chick-embryo or monkey kidney. Despite animal source, these primary cells allow easy propagation of few human viruses. Polio, mumps, rabies, and measles vaccines are being manufactured in these cells. The third category of cells used for making human vaccines in some parts of the world is generally known as continuous cell lines (CCLs). There are specific advantages of using CCLs over both HDCs and primary cells for making vaccines. However, the safety of the vaccines produced in CCLs has been the focal attention while performing quality control of the same. The most concerned cellular contaminants are:

(i) *Transforming proteins*- A number of proteins encoded by cellular transforming genes can induce proliferation of many cell types. Fortunately, these proteins are not oncogenic, their growth promoting activities are transient, and are secreted in extremely low quantities to produce observable effects on human. One such protein is transforming growth factor- α (TGF-α).

(ii) *Viruses*- Viral contaminants from cells have been a problem for quite some time. It is important to note that simian virus 40 (SV40), the most pathogenic human virus, was isolated from a primary cell. Besides, many endogenous viruses are concern to human, but along with the inactivation of vaccines, these viruses are also inactivated. For example, inactivated influenza vaccine is produced in fertilized hens' eggs, which is likely to contain avian leucosis virus.

(iii) *DNA*- Concern about cellular DNA in CCLs are those portions of the genome, which differs from that of normal cells.

In 1981, World Health Organization (WHO) had made a major revision of the requirements for inactivated polio vaccine by incorporating provisions for the use of non-tumorigenic and virus-free CCLs as substrate for vaccine production. Vero cells (African green monkey kidney) has been approved for the production of inactivated polio vaccine, provided that purified product contains cellular DNA within 10 pg/dose, and formalin is used for inactivation of the virus. Since then, two more inactivated vaccines have been produced in Vero cells, these are rabies and smallpox vaccines.

3.2.2 Mammalian cells used for manufacturing recombinant proteins

The selection criteria for mammalian host cells for expression of recombinant therapeutic proteins are not stringent like that followed for the production of viral vaccines. The reason being quite simple that the therapeutic proteins are purified in multi-step processes which are not possible in case of vaccines. Even host cell transforming proteins, viruses, and DNA contaminat are present in crude products that can be separated out in the purification steps. Because of infinite life and easy cultivation, continuous cell lines are used for manufacturing therapeutic proteins. The choice of cells for the production of therapeutic proteins is governed by the following factors:

(a) Structural complexity of protein (e.g., number of disulfide bonds, oligomerization, extent and type of glycosylation, etc.

(b) Stability and the product yield

(c) Proposed usage (therapeutic or diagnostic).

Idially, recombinant proteins should be comparable to their counterparts purified from the natural sources in terms of biological, chemical, and pharmacological properties (Bhatia and Mukhopadhyay, 1999). Recombinant glycoproteins often alter biological properties and functions and increase immunogenicity and protease sensitivity when compared with their native counterparts. The mammalian expression systems are preferred over yeasts and baculovirus expression systems in case glycosylation imparts special properties to the protein. Glycosylation of protein may modulate four basic properties: biological activity, antigenicity, immunogenecity, and circulatory half-life, which are critical for the efficacy of a therapeutic protein. Thus, proper glycosylation is warranted for biological activity of a protein. Students are suggested to consult reference books, as details of protein glycosylation are beyond the scope of this book. Since the structure of glycan in a recombinant glycoprotein may differ with the change in host cells, it is essential to consider the end use of the product before deciding about the host cells. Mutant of Chinese hamster ovary cells (CHO-K1) has proven useful in expressing glycoproteins with minimal heterogeneity. Maximum expression (136.4 mg/l) was obtained in a high cell density perfusion culture of CHO-K1 cell line for tissue plasminogen activator (Kopp *et al*, 1994). Interferon γ (IFNγ) and erythropoietin (EPO), two more therapeutic proteins, are also produced in CHO-K1 cell line. The rodent cell lines such as mouse NSO or rat YO myeloma are used to produce humanized antibodies. The highest yield (1g/l) of antibody was reported in NSO cells (Werner, 1994).

3.3 GENERATION OF CELL BANKS

The purpose of generating cell banking is to ensure that: (a) an adequate supply of the cells will be available over the entire life span of the product, (b) properly characterized cells can be used in the production process, and (c) better process control and quality assurance of the final product(s) are possible. Thus, in spite of the nature of the products to be produced, once a cell strain/cell line is established, a good cell banking system is indispensable. According to the Food and Drug Administration (FDA), the cell banking is maintained in two tiers system: a *master cell bank* (MCB) and a *manufacturer's working cell bank* (MWCB). The MCB is defined as the collection of cell stocks of uniform composition derived from a single tissue or cell lot. On initial arrival into the laboratory, the new culture stock may be associated with potential source of contamination from bacteria, fungi and mycoplasma, and thus are handled under quarantine conditions until proven negative for microbial contamination. Following initial expansion, 3-5 ampoules are frozen as a token stock before a MCB is established. One of the token stock ampoules is thawed and expanded to produce a MCB of 20-100 ampoules, depending upon the projected level of use. The ampoules are frozen and stored in the liquid nitrogen. The MWCB stocks are derived from one or more ampoules of the MCB. The MCB should be thawed and cultured to produce a MWCB. The size of this bank will again depend on the envisaged level of demand. The MCB source cells are expanded by serial subculture up to a passage number selected by the manufacturer and approved by the regulatory authority. At that point the cells are combined into one pool, dispensed into individual ampoules and cryopreserved to form the MWCB. One of such ampoules from the MWCB is used for the production of biological. The general scheme for quality checking of cells in MCB and MWCB stages are shown in Figure 3.2.

3.3.1 Master cell bank

The tests for the quality of MCB should be done either on aliquots (at random) of the cell bank or on cultured cells from the aliquot. The cell stocks are tested in terms of viable counts, authenticity and freedom from adventitious organisms. The authenticity tests of the MCB should be done once during the preparation of master stock. It is presumed that significant properties of the cells and the stability of these properties will not change during manufacturing process. Different tests to be carried out are as follows:

***A. Morphology*:** It is determined by light and electron microscopy. In case of fibroblastic cell stain, the morphological feature will be spindle shaped (bipolar), usually arranged in parallel arrays at confluence, if contact inhibited (Figure 3.3). On the other hand, epithelial cells are polygonal shaped with clear sharp boundaries between cells (Figure 3.3).

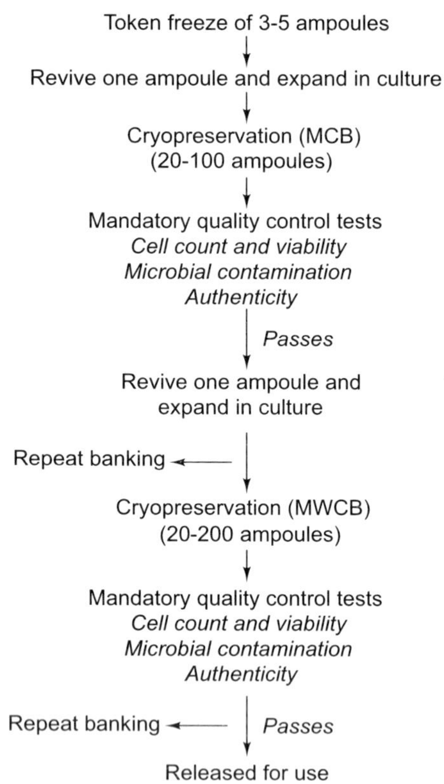

Figure 3.2 General scheme for quality control of master cell bank (MCB) and manufacturer's working cell bank (MWCB).

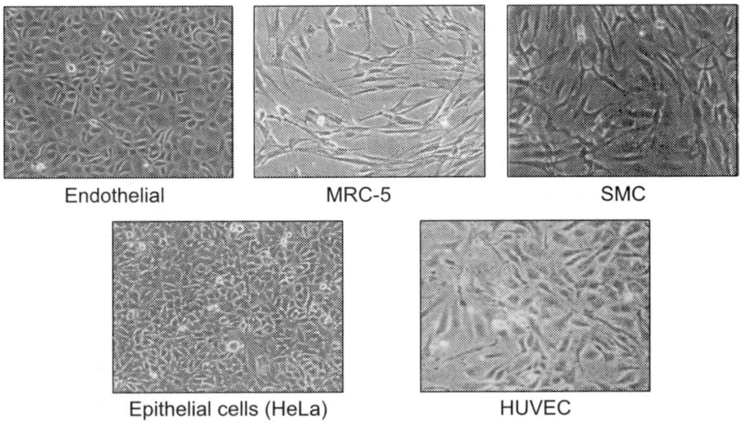

Figure 3.3 Morphological distinction between fibroblasts, epithelial, endothelial, and smooth muscle cells.

***B. Tissue of origin*:** The first criterion used to authenticate a cell is to identify the species of origin. Various methods, depending upon genetic, immunological, and biochemical techniques are followed to identify the origin of a cell.

1. DNA fingerprinting: The application of *DNA fingerprinting* provides a straightforward means of cell line identification. Details of DNA fingerprinting are beyond the scope of this book. Students are advised to consult references given at the end of the Chapter (Jeffreys *et al.* 1985a,b; Hay, 1992). In brief, the specific DNA probes for tandem repetitive *mini-satellite* regions of the genome are hybridized to the fragments of genomic DNA of the test cells, and visualized by autoradiography after *southern blotting*.
2. Immunological test: The indirect fluorescent antibody technique is followed to verify the species from which cell lines were obtained. First, injecting species-specific cells in rabbit ear vein generates antiserum. Antiserum is collected, heat inactivated, and stored at –70°C. The test cells are labeled with appropriate dilution of antisera, washed to remove excess antisera, and counter stained with FITC/PE-conjugated secondary antibodies (goat anti-rabbit). Cells are further washed and examined under a fluore-scence microscope. Appropriate positive and negative control antisera are used to authenticate the test procedure.
3. Biochemical tests: Biochemical tests are the powerful methods to identify a cell line. Electrophoresis of Isoenzyme is the most popular test by which same tissue of different species, different tissues of same species, different cell lines and different clones, or monolayer and suspension cells of the same cell line can be discriminated. Gel electrophoretic mobility of three isoenzyme systems, glucose-6-phosphate dehydrogenase (G6PD), lactic dehydrogenase (LDH), and nucleoside phosphorylase (NP) in vertical starch gel are different, thus species of origin for the cell lines can be

identified without ambiguity (Hay, 1992). The functional properties of cells, such as cytochrome P450 enzyme system (hepatocytes), and cytoplasmic oestrogen receptors (mammary epithelial cells), etc., are also utilized for identification.

***C. Plasmid copy number*:** Plasmid copy number and the physical state of the expression system should be available for cells containing an expression system for producing recombinant DNA derived protein.

***D. Productivity of the cells*:** The productivity of a cell is defined by the amount of product formed by unit cell number in unit time. When cells are used as substrate to produce viral products (vaccines), productivity refers to the number of plug forming unit produced by one million cells per hour (pfu/10^6cells.h). In case of a cell expressing recombinant protein, or a hybridoma producing monoclonal antibody, the productivity is expressed as microgram product produced by one million cells per hour (μg/10^6 cells.h). Productivity may be significantly changed if the tissue of origin in different.

***E. Quality of the product*:** Quality means the functional properties of the product. In case of viral vaccines antibody response, *bioneutralizing capacity* of the antisera is most important to be known. For a protein product, *biological activity* is the most concerned factor. Quality of the product may change if the host cells are mistakingly contaminated with other cells.

***F. Growth characteristics*:** This is the most crucial information to the technologists for manufacturing cell-derived products. Some of the important characteristics are: growth pattern— whether the cell type is anchorage dependent or not; life-span— immortal (transformed cells) or finite (cell strains); plating efficiency— split-ratio to be used in subculture (low plating efficiency means low split-ratio to be used in subculture); freeze-thaw data— condition of freezing and retrieval of cells from frozen stock to maintain maximum viability of the cells; medium and serum requirements— type of medium and concentration of serum; fragility— shear sensitivity of the cell; stability of gene expression— any deletion of genes during culture and drug selection pressure to avoid reversion to the original host.

***G. Test for microbial contamination*:** The adventitious agents tested include bacteria, fungi, mycoplasma, and viruses. Test for adventitious viruses should include routine *in vivo* and cell culture inoculation tests based on the assay history of the cell line to detect possible contaminating viruses.

3.3.2 Manufacturers' Working Cell Bank

The MWCB is derived from the MCB and propagated for an approved number of passages in tissue culture. When needed, only spot-checking of contaminants, which may have been introduced from the culture medium, is carried out. Recommended tests are similar to that of MCB.

3.4 CRYOPRESERVATION OF ANIMAL CELLS AND TISSUES

The aim of cryopreservation is to maintain stocks for future use. Cells have a tendency to lose, or change their properties over a period of time because of continuous division in culture. Therefore, cells are used for research or manufacturing biological must be properly preserved. The reasons for preservation of animal cells are: (a) to avoid genotypic drift due to genetic instability, (b) to avoid senescence and transformation in case of cells having finite life-span, (c) to avoid phenotypic instability (selection and dedifferentiation), (d) to avoid microbial contamination, (e) to avoid cross-contamination with other cell lines, (f) to save cells from incubator failure, (g) to minimize time and materials (when cells are not in immediate use), and (h) to distribute among other users.

3.4.1 Physico-chemical aspect to improve recovery of cells

During freezing and thawing, cells are damaged primarily due to the formation of intracellular ice crystals and osmotic imbalance. The addition of cryoprotectant (e.g., DMSO, glycerol) and following optimum freezing and thawing rates ensure minimum cellular injury. Some cells are more sensitive than others during freezing and thawing cycle, so cell recovery may not be same for all cell types. The most popular storage medium for animal cells is liquid nitrogen (LN_2), where they can be preserved for more than 15-20 years.

During freezing of a cell, three physical events may occur. These are formation of ice crystals, removal of water from cells, and increase in solute concentration. Due to the presence of salts in the freezing medium, freezing temperature is depressed and the whole cell is exposed to super-cooled surrounding liquid. Ice crystals are beginning to form between –3° and –11°C (Griffiths, 1985), and the crystals continue to grow until all the free water is frozen. During this freezing process, the salt concentration in the remaining water of the freezing medium is progressively increased, resulting in osmotically induced water loss (dehydration) from the cells and their subsequent shrinkage. If appropriate freezing medium is used and optimum cooling rate is maintained, a cell is dehydrated with minimum injury to the membrane, resulting in less damage due to freezing (Figure 3.4, pathway 1-2-3-4). If the cooling rate is too fast, there will not be enough time for complete dehydration of cell, resulting in the formation of intracellular ice crystals. This situation is more dangerous than the intracellular increase of salt concentration, and consequently the recovery of viable cells is reduced upon thawing (Figure 3.4, pathway 1-6). If the cooling rate is lower than optimum, excessive salt concentration may cause damage to the cells.

Any type of cell, tissue, and embryo can be stored frozen; however, the optimum-cooling rate differs from cell type to cell type. The recommended cooling rates for lymphoid may vary between 0.5° to 2°C/min, for epithelial cells it is between 2° to 10°C/min, and for fibroblastic cells it is between 1° to 3°C/min (Griffiths, 1985).

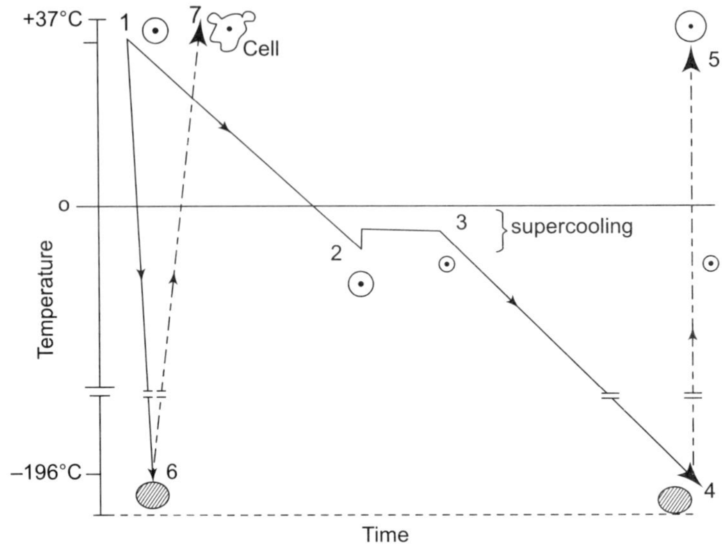

Figure 3.4 Freezing and warming profiles of animal cells. During slow cooling rate (path 1-2-3-4), the formation of ice crystals starts at point 2, water is removed from the cells and cells shrink. After a brief phase transition, cells are dehydrated and reach at –196°C without intracellular ice formation. Upon warming (path 4-5), cells are rehydrated and regain their original volume. During fast cooling (path 1-6), cells do not get sufficient time to dehydrate, as a result intra-cellular water form ice crystals. Upon warming (path 6-7), leads to damage of cells. [Adapted from Ref. 7]

During thawing, the cell is rehydrated to regain its normal size by drawing water from its surrounding. During rehydration stage, cell is further subjected to hypertonic stress that may also causes cellular damage. The adverse effect on cell can be reduced by short exposure of hypertonic stress. This can be achieved if cells are thawed at a faster rate. The extent of damage (dead) to the cells during freezing-thawing cycle is examined by trypan blue dye exclusion method or by propidium iodide or Sytoxgreen staining.

3.4.2 Preparation for cell-freezing

Besides cooling and thawing rates, the optimum recovery of viable cells from a frozen a stock relies on few other critical factors as follows:

Formulation of freezing medium

The freezing medium is composed of basal growth medium, serum or serum albumin, and a *cryoprotectant*. Most commonly used cell membrane penetrating cryoprotectants are DMSO and glycerol, which are highly soluble in water. DMSO or glycerol freely penetrates the membrane, strongly binds with intra-cellular water and lowers the freezing point. Depression of freezing point reduces ice crystal formation and thus provides enough time to leave intracellular water favoring dehydration of cells, and prevents membrane damage. Non-penetrating cryoprotectants like hydroxyethyl starch and polyvinylpyrrolidone (PVP) are also used in the freezing mixture. Despite the fact that DMSO is highly toxic to the cells, it is most popular among all cryoprotectants. DMSO in the range from 7.5 to 10% (v/v) is normally used in the freezing medium. Since DMSO is highly toxic to the cells and toxicity increases with the temperature, pre-cooled freezing medium (4°-10°C) is used to suspend cell pellet before freezing. Again, to minimize toxic effect of DMSO, the thawed cell suspension is immediately diluted to 10-15 times its original volume with the cold medium.

Design of storage container and its volume

For optimum recovery of the viable cells, lesser temperature gradient is desirable in the storage vial during cooling and thawing the content. Glass ampoules are preferable as they improve heat transfer rate, but plastic screw cap vials are more popular for storing cells in ultra low temperature. In general, the capacity of the plastic vial is 2 ml, in which 1.5 ml cell suspension can be stored. For storing larger volume of cells, bigger plastic containers are not recommended; in place, plastic bags are used. To facilitate heat transfer, the surface area to volume ratio is maintained high by uniform spreading of cell suspension in the bag.

3.4.3 Equipment for freezing and storing of vials

Constant rate or programmable freezers are more popular for freezing mammalian cells. LN_2 programmable freezers are available in the brand names of Cryo-Med and Planer (Model: Kryo 320-1.7, 520-16, 750-30). Since cooling rate significantly influence the viability of the cells, the controlled rate freezers ensure maximum viability for a wide range of cells, including stem cells, embryos, heart valves, cord blood cells. The precise cooling rate may vary from cell to cell. As a general practice, cooling rate may vary from 1° to 3°C/min. Mammalian cells are freezed either by mechanical freezer or by liquid or vapor phase nitrogen. The advantages and disadvantages of these storing methods are summarized in Table 3.4.

The single most important safety consideration in LN_2 storage methods is the potential risk of *asphyxiation* due to leakage of nitrogen from the storage tank that can lead to oxygen depletion in the room. This is critical as oxygen depletion may cause loss of consciousness of workers without warning. Consequently, LN_2 freezers and storage vessels should be placed in a well-ventilated area to minimize this risk. The LN_2 storage vessels should be equipped with an alarming device to warn low LN_2 levels. This is particularly important in case of vapor phase storage systems. All ultra-low temperature storage vessels include a racking/inventory system designed to organize the contents for easy location and retrieval. The photographs for LN_2 storage systems are shown in Figure 3.5.

Table 3.4 Advantages and disadvantages of ultra-low temperature cell storage methods

Method	*Advantages*	*Disadvantages*
Electric freezing (–140°C)	• Low running costs in small-scale operation, steady temperature • Easy in maintenance	• Chances of breakdown are more • Storing temperature higher than LN_2 • LN_2 backup needed in case of electricity failure
Vapor Phase N_2 (–196°C)	• Simple and reliable • Low temperature can be achieved • No cross contamination due to LN_2	• High running costs • Vertical temperature gradient in vapor phase • Require regular supply of LN_2
Liquid Phase N_2 (–196°C)	• Simple and no mechanical components • Steady ultra-low temperature	• High running costs • Require regular supply of LN_2 • Chances of cross-contamination due to LN_2

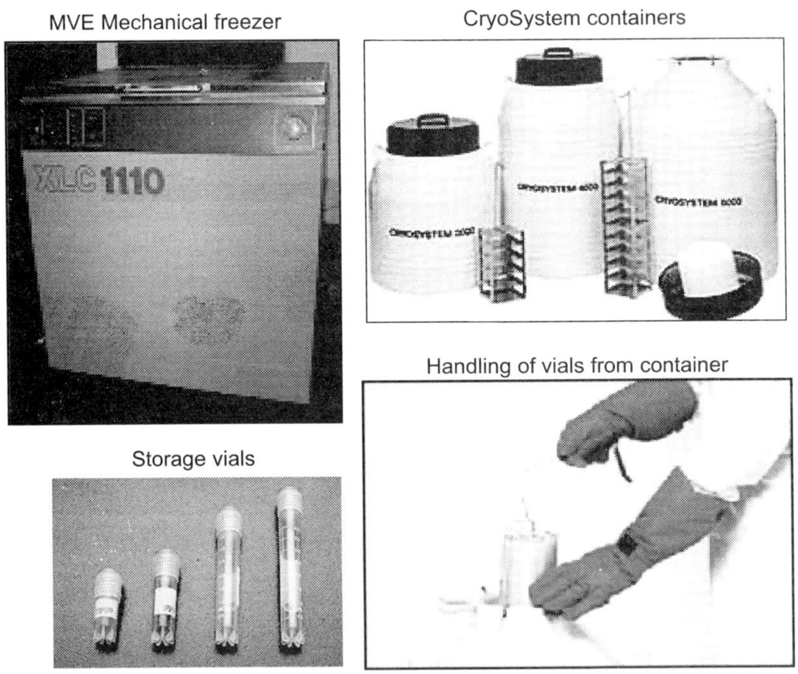

Figure 3.5 Storing and handling of cryo-vials in ultra-low temperature storage systems. The mechanical (–140°C) and LN_2 storage (–196°C) systems are shown. [*Source*: LABEQUIP, Ontario, Canada; Planer PLC, United Kingdom]

3.5 SUMMARY

1. Cells isolated from a tissue and cultured for the first time (without subculturing) is known as primary culture.
2. Transformed cells are heterokaryons; they could be either tumorigenic or non-tumorigenic.
3. Selection of a cell line for manufacturing biological depends on the specific properties that it can add to the products. These are, improve specific activity and productivity, provide right kind of glycosylation, improve infectivity (in case of viral vaccine), and does not contribute harmful contaminant in the final product.
4. Animal cells are stored in two stages: master cell bank and manufacturer's working cell bank.
5. Cells are properly characterized before the frozen cell stocks are declared safe for use in manufacturing process.
6. Mammalian cells are sensitive to freezing and thawing steps: freezing should be slow, whereas frozen cells are normally thawed rapidly in water bath set at 37°C.
7. Animal cells are successfully stored in LN_2 for 15-20 years, provided that cell-freezing steps are optimally followed.

ABBREVIATIONS

DMEM	Dulbecco's modified Eagle medium
ES	Embryonic stem
EC	Endothelial cell
FBS	Fetal bovine serum
HBSS	Hank's balanced salt solution
CCL	Continuous cell line
CHO	Chinese hamster ovary
DMSO	Dimethylsulphoxide
EPO	Erythropoietin
HDC	Human diploid cells
IFNγ	Interferon γ
LN_2	Liquid nitrogen
MCB	Master cell bank
MWCB	Manufacturer's working cell bank
SV40	Simian virus 40
TGFα	Transforming growth factor α

REFERENCES

1. Griffith JB (1985) *Animal Cell Biotechnology* (RE Spire, JG Griffith, eds.) vol. 1, p. 494. Academic Press, Yew York.
2. Bhatia PK and Mukhopadhyay A (1999). In: *Adv. Biochem. Engg. Biotechnol.* (T Scheper ed.), vol. 64, p. 155. Springer-Verlag, Berlin.

3. Griffiths JB (1985). In: *Animal Cell Technology* (RE Spier and JB Griffiths, eds.) vol. 1, p. 50. Academic Press, London.
4. Hay RJ (1992). In: *Animal cell culture. A practical approach* (Freshney RI, ed.), 2nd edition, p. 95-148. IRL Press, Oxford.

SUGGESTION FOR FURTHER READING

1. Center for Biologics Evaluation and Research (HFM-20), (1993). *Draft of points to consider in the characterization of cell lines used to produce biological.* FDA, Rockville, MD, USA.
2. Warner RG. (1994). In: *Animal cell Technology-products of today, prospects for tomorrow* (RE Spier, JB Griffith and W Berthold, eds.). p. 573. Butterworth-Heinemann UK.
3. Freshrey RI (1987) Culture of Animal Cells. A manual of basic techniques; 2nd edition, pp. 207. Alan R. Liss Inc. New York.

4

Design of Cell Culture Media

Optimum composition of media has a pivotal role in any successful animal cell culture process. The primary functions of medium are to provide physical environment to the cells by maintaining proper osmotic pressure, temperature, pH, viscosity, O_2 and CO_2 levels, and finally nutrition. Nevertheless, the culture medium is also involved in the removal of toxic metabolic end products. Over the part few decades, various defined basal media have been developed, which are now commercially available. When a cell is isolated from its original tissue and cultured *in vitro*, the medium must provide all the essential chemical constituents in which the cell was exposed *in vivo*. Therefore, it is expected that the culture media should meet all the requirements for survival and growth of the cells. It has been observed that the productivity of many animal cell-derived products depends on the composition of the medium in which they are cultured. Since, culture medium constitutes a major part of the manufacturing cost of the product, a precise formulation of medium is imperative in cell culture process.

4.1 SELECTION OF CELL CULTURE MEDIUM

The selection and/or formulation of cell culture medium requires precise attention in terms of cell specific requirement, technology to be followed to manufacture the desired product, and finally its availability. Following are the few important factors that need to be clarified for selection of a medium:

1. *Application*: The objective of cell culture process governs selection and formulation of medium. In case of the production of recombinant proteins, a medium that promotes high cell growth is initially desirable. Later, a different medium that can maintain cells in slow growing rate, yet in the product-generating state can be used. If differentiated cells are the final product, the medium should be formulated with specific differentiating agents.

2. ***Cell type***: Primary and specialized cells require specific growth factors or high serum in the culture medium. On the other hand, transformed and established cell lines require simple and low serum containing medium. The composition of medium and pH buffering chemicals (HEPES and $NaHCO_3$) sometimes dictate its selection, as certain cell types are sensitive to these chemicals. In general, the choice of the medium should be such that it is widely applicable in different cell types and products.

3. ***Commercial availability***: Low cost, reliability in formulation and use, made from readily available simple components are few critical points to be considered in selection of medium. It is important to note that the formulated media should be acceptable by the regulatory agencies.

4. ***Upstream processing***: Media constituents should be readily soluble in water and easily sterilizable by membrane filtration.

5. ***Culture system***: Nutrient requirements for a specific cell type are influenced by the culture system. High cell density culture process requires enriched medium. Diverse strategies to control nutrient feeding rate and the pH of the medium are also deciding factors for the selection of the medium.

6. ***Productivity***: Optimum medium for cell growth should stimulate cell division by providing growth factors and removing growth inhibitors. High productivity can be achieved if, cells are grown to very high densities ($5 - 10 \times 10^6$/ml), able to retain cellular characteristics in the production phase, and able to yield high product in unit volume of medium.

7. ***Minimize load in protein purification step***: Low serum containing or serum free medium require low capacity and less stringent protein purification steps for manufacturing therapeutic proteins.

4.2 BASIC CONSTITUENTS OF MEDIA

Basic media refers to energy sources, amino acids, vitamins, inorganic salts and trace elements, fatty acids and lipids. From time to time, different cell culture media have been developed to meet with the specific requirements by the cells and the technology adopted for culturing them. Basic media can be divided into three broad groups: basal media (MEM, DMEM, GMEM, etc.), complex media (αMEM, RPMI-1640, IMDM, Leibovitz L-15, Ham F12, etc.), and serum-free media (special composition). The basic nutrients that fulfill cellular requirements are included in the basal media. The wide ranges of media being used today are based on a medium designed more than 50 years ago (Eagle, 1955). From a cell culture point of view, a nutrient is defined as a chemical that enters the cell and is used as a structural component, as a substrate for biosynthesis, as a substrate for energy metabolism, or has a catalytic role in such metabolism (Bettger and Ham, 1982). Various components of the cell culture media are described below.

4.2.1 Carbohydrates

The main source of energy in cell culture media is carbohydrates, generally in the form of sugars. The most popular carbohydrates sugar used in the media are glucose and galactose; however, in some media maltose or fructose has been used. The concentration of glucose varies from media to media, and is generally maintained between 1 g/l to 4.5 g/l in some complex media. Media containing higher concentrations of sugars are able to support the growth of wider range of animal cells. The other major source of energy in the cell culture media is glutamine (0.3 g/l to 0.75 g/l). Despite the fact that glutamine is an amino acid, there are considerable evidence to prove that its utilization as a carbon source is also important for rapidly dividing cells. Human diploid fibroblasts utilize both glucose and glutamine as energy source. It has been found that the utilization of glucose and glutamine is reciprocally controlled in diploid fibroblast cells (Zielke *et al.*, 1978). Besides glutamine, several amino acids and α-keto acid (pyruvate) are also utilized as energy source by many cells.

4.2.2 Amino acids

Amino acids are the nitrogen sources in cell culture media, but their numbers and concentrations may vary from media to media, depending upon the specific requirements by the cells. According to the basic requirements of the whole organism, amino acids are classified into 'essential' and 'nonessential' forms. All thirteen essential amino acids are mandatory in the basal media. The requirements for the essential amino acids are reduced when nonessential amino acids are included in the medium. The nonessential amino acids commonly used are alanine, asparagine, aspartic acid, proline, glycine, and serine. Glutamine plays a vital role in metabolism; it acts as a precursor of nucleic acid besides being an important source of energy. Amino acids are normally used in the purer form derived from synthetic or natural sources. However, enzymatically digested plant or animal proteins are often used as complex sources of amino acids.

4.2.3 Vitamins

Media contain mainly water-soluble vitamins, most of which act as cofactors in enzymatic reactions. Many vitamins, especially belonging to B group are necessary for cell growth and proliferation (e.g., biotin, folic acid, nicotinamide, pantothenic acid, pyridoxine, riboflavin, and thiamine), and for some cell lines the presence of B12 is essential. Use of fat-soluble vitamins (e.g., A, D, E, K) in cell culture medium is not common. Vitamin A and α-tocopherol (vitamin E) are the two constituents of M199 medium.

4.2.4 Minerals and trace elements

The inclusion of inorganic salts in media has several benefits. Primarily, salts maintain osmotic balance in the cells, buffering of media, and help to modulate membrane potential by the preserve of sodium, potassium and calcium ions. Divalent Ca and Mg-ions are involved in the extracellular matrix (ECM) components-mediated cell attachment, and are also required as enzyme cofactors. In most media, one of the two principle inorganic salt bases are used. These are Earle's base salts and Hank's base salts.

Trace elements (e.g., Co, Cu, I, Fe, Mn, Mo, Zn, Se, Cr, Ni, V, As, F, Sn) are essential for the growth of animal cells. The biological roles of a few trace elements are clearly understood. The requirement of iron is essential as a component of cytochrome. The deficiency of iron leads to low growth rate and yield of cells. Selenite is a detoxifier, which helps to remove oxygen free radicals.

4.2.5 Fatty acids and lipids

Fatty acids (e.g., linoleic and oleic acid) are essential components of the media for culturing certain cell types. Cholesterol and steroids are also important components of media for culture of several specialized animal cells. The cholesterol and phospholipids are used in the basal medium of serum-free media formulations.

4.2.6 Nucleic acid precursors

Nucleic acid precursors (purines and pyrimidines) are not essential elements in animal cell growth media, as cells are capable of synthesizing them (Kelley, 1972). However, the growth of some cells can be improved by supplementing media with these precursors. The most commonly used nucleic acid precursors are adenine, cytidine, thymidine, and hypoxanthine. The nucleic acid precursors are used in α-MEM.

4.3 OTHER IMPORTANT DEFINED MEDIUM SUPPLEMENTS

4.3.1 Antibiotics

Although antibiotics are routinely used in cell culture media, their addition may lead to the development of resistant microorganisms, lower growth rate of cells and low yield of the products. The toxic effects of antibiotics on the cells may increase in the low protein-containing medium. Thus, it is recommended to avoid the use of antibiotics in cell culture media, if possible. In large-scale cell culture, the use of antibiotics has been proved beneficial to manage losses due to contamination. The most commonly used antibiotics in cell culture media are penicillin (50 IU/ml), streptomycin (50 μg/ml) or gentamycin (50 μg/ml) as antibacterial agent, and amphotericin B (2.5 μg/ml) as an antifungal agent. It should be kept in mind that antibiotics are unstable in the normal culture condition.

4.3.2 Oxygen

Oxygen is an essential nutrient for the growth of organisms. There is no exception for it in case of animal cells. The demand of oxygen by animal cells is not very high, despite cultures often fail due to the shortage of dissolved oxygen in the media. This is due to low solubility of oxygen in the culture medium and other reasons, the details of which will be discussed in the Chapter 6.

4.3.3 Hormones, growth and attachment factors

The presence of hormones, growth and attachment factors in culture medium is crucial for certain cell types. The common source of these constituents in culture media is serum. However, for many primary cells (e.g., stem cells, hepatocytes, epidermal cells, etc.), in addition to serum, media is supplemented with hormones and growth factors. The proliferation of cells is controlled by external growth factors (mitogens), which help them to exit from G_0/G_1 phase and enter into S phase. In tumor cells the growth factors are encoded by different oncogenes (*c-onc*), whereas in normal cells these oncogenes remain in inactive state. In certain situation where the growth of a tissue is essential to repair its own damage, growth factors are produced just for few divisions of cells. Once cellular *homeostasis* attains, expression of growth factors is stopped. The requirements for hormones and growth factors may vary from cell to cell. The most commonly used hormones in culture media are estradiol, hydrocortisone, dexamethasone, progesterone, glucagone, and insulin. Similarly, PDGF, FGF, and EGF activate the cell cycle. The attachment factors (e.g., fibronectin, laminin, collagen type I, IV, ėtc.) are required by the anchorage dependent cells for proper attachment and spreading. Many cells synthesize attachment factors *in vitro* once they attach on a compatible substratum. The attachment factors are involved for growth and differentiation of many specialized cell types.

4.4 SERUM AND ITS ROLE IN ANIMAL CELL CULTURE

Serum is one of the most important ingredients of cell culture media. It is composed of albumin and other serum protein, growth factors, and micronutrients. The most commonly types are fetal bovine serum (FBS) and fetal calf serum (FCS). Other types of sera used in cell culture media are newborn calf serum (NBCS) and horse serum (HS). The concentration of sera used in culture media ranges between 0.5 to 30% (v/v), depending upon the type of the cells. The biological roles of major components of serum are described in Table 4.1. In addition to that, serum may also contain cytokines like TNFα and TGFβ, the presence of which at higher concentrations are found to be harmful for the growth of many primary cells.

4.4.1 Selection of serum lot

The type and the concentration of serum can affect the growth of cells. Therefore, it is important to screen different batches of sera for their ability to support cell growth and/or differentiation. There are a few standard tests available that may be adopted to select sera lots, which include *cloning efficiency*, *plating efficiency*, *growth curve*, and the *preservation of cell characteristics*.

Table 4.1 The major constituents of serum and their biological roles (Maurer, 1992)

Nature	*Constituents and functions*
Proteins	*Albumin:* carriers for lipids, maintain viscosity, remove toxic metabolic end products; *Fibronectin, Fetuin, Laminin and Collagen, Serum spreading factor:* promote attachment and spreading; *Transferrin:* facilitates iron transport; α_2*-Macroglobin*, α*1-antitrypsin:* protease inhibitors; *Hemoglobin:* transports oxygen
Polypeptide	*IGF-I & II, EGF, FGF, PDGF, Somatomedin A & C, EPO:* stimulate cell division, elicit wide range of cellular responses
Lipids	*Linoleic acid, Phospholipids:* membrane synthesis
Amines	*Amino acids:* protein synthesis
Ions	*Trace elements, Minerals:* maintain osmotic balance, act as cofactors, inactivate free radicals
Hormones & Steroid	*Insulin, Hydrocortisone:* stimulate protein-carbohydrate-lipid synthesis, potentiate mitogenic effects of growth factors
Sugar	*Glucose:* energy source

Cloning efficiency is determined by plating 100 cells/ml in culture plate (35 or 60 mm diameter) in the presence of test serum (2-20% supplement) for 1-2 week culture. Colonies are stained with crystal violet and the number of colonies are counted. The cloning efficiency (number of colonies formed from 100 cells) and the colony size (cell proliferation) are determined and compared between the test with the standard serum. The quality of test sera is considered satisfactory if the number of colonies and their sizes at lowest concentration of test serum are comparable with the standard values. *Plating efficiency* is determined by the percentage of cells seeded (at very low seeding density) which give rise to colonies. It is assumed that each colony is originated from individual cell. Higher plating efficiency obtained in lowest test sera concentration is considered better for the culture. In *growth curve* studies, long lag phase implies that the culture needs adaptation in the test sera. In general, a shorter doubling time is preferable if rapid cell proliferation is desirable. A high saturation density will provide more cells from a given amount of test sera. The selection of sera will be economical if either/or both criteria are followed. *Preservation of cell characteristics* means maintenance of desired functions of the culture, whether it is virus yield in unit cell number, or differentiation into specific cell type. After testing new sera lots, the qualified sera lot are normally reserved for about 1 year in the supplier's premises to make them available through out the year.

There is a risk of contamination associated with serum. The risk of contamination can be minimized if serum is procured from a reputed supplier who can provide a test certificate. Serum is screened for the presence of bovine viral diarrhea virus (BVDV) and mycoplasma. Heat inactivation of serum at 56°C for 30 min reduces the risk of contamination, since it inactivates some viruses. However, the routine use of heat-inactivated serum is not followed for cell culture. FBS is used for manufacturing a number of therapeutic products. Thus, an excellent test record are to be maintained for the serum. The spread of Bovine Spongiform Encephalopathy (BSE) in Europe, and the probable link between BSE and a new variant of Creutzfeldt Jakob disease in Humans, has renewed the concern about the safe sourcing of all bovine materials. In 1993, FDA prohibited the use of bovine derived materials from the animals that have resided in, or originated from countries where BSE has been diagnosed. The use of FBS in production processes is acceptable, provided proper documentation on source, age of the animals, and other test reports for the absence of adventitious agents are available. It is mandatory for the suppliers to make the documentation available to the users of serum.

4.4.2 Problems associated with serum-supplemented culture media

Despite serum has several benefits for mammalian cell culture, its use in media may cause many disadvantages in manufacturing biological. Some of them are described below:

1. Serum is the most expensive component in cell culture medium (about 70% cost in case of FBS), and it requires storage at –20°C.
2. Periodic crises in the supply of serum may lead to compromise with the productivity.
3. Batch to batch variation of serum needs expensive screening procedure to select.
4. Being highly viscous, serum slows down the filter sterilization of medium.
5. Risk of contamination by virus, fungi, and mycoplasma may increase.
6. Serum is the physiological fluid of blood cells, but not for many tissues in the body. It may suppress the growth of many cells, for example epidermal keratinocyte.
7. Presence of serum in cell culture media increases the complexity of the downstream processing of therapeutic protein.
8. Serum may be potential cytotoxic to some cells due to the presence of complements and other bacterial toxins and lipids.

4.5 PHYSICOCHEMICAL PROPERTIES OF CELL CULTURE MEDIA

The physicochemical properties of culture media are discussed below:

4.5.1 Buffering systems

Most animal cells optimally grow in the pH ranging from 7.0 to 7.4, therefore close control of pH is required in any cell culture process. Normal fibroblasts prefer higher pH range (7.4-7.7), whereas, continuous or tumor cell lines prefer more acidic pH (i.e., <7.0). On the other hand, the epidermal cells optimally grow at pH 5.5. The control of pH of the medium is most important, immediately after seeding of cells. Most of the media contain phenol red as a pH indicator, the color of which changes depending upon the pH of the media (yellow at pH 6.5, orange at pH 7.0, red at pH 7.4, slightly bluish red at pH 7.6, and purple at pH 7.8). The pH of the media needs to be buffered for two different reasons:

(1) Release of CO_2 from medium causes pH to increase
(2) Excessive formation of CO_2 and lactic acid in the culture medium cause pH to drop.

Buffering of media is usually achieved either by gaseous CO_2 balances with the CO_2/bicarbonate content of the culture media, and/or by zwitterionic buffer using chemical, N-2-hydroxyethyl-piperazine-n′-2-ethane sulfonic acid (HEPES).

In CO_2/bicarbonate based buffer, the culture needs to be maintained in the atmosphere containing 5-10% CO_2 in air. CO_2/bicarbonate buffer system is low cost, non-toxic and also provides other chemical benefits to the cells. Usually, 26 mM $NaHCO_3$ is used in the culture medium and 5% CO_2 is maintained in the CO_2 incubator to get comparable pH effect to that of circulating blood. In CO_2/bicarbonate buffering system, dissolved CO_2, atmospheric CO_2 and $NaHCO_3$ level determine final pH of the medium. The dissolved CO_2 produces H_2CO_3, which dissociates into HCO_3^- (Eq. 5.1). As HCO_3^- has low dissociation constant (pKa 6.1) in medium, it tends to reassociate to form acidic H_2CO_3. Thus, an increase in atmospheric CO_2 reduces pH, which can be counteracted by elevating bicarbonate concentration in the medium (Eq. 4.2). The increased HCO_3^- concentration shifted equilibrium of equation 4.1 to the left corresponding to pH 7.4. CO_2/

$$H_2O + CO_2 \longrightarrow H_2CO_3 \longrightarrow H^+\,HCO_3^- \qquad 4.1$$

$$NaHCO_3 \longrightarrow Na^+\,HCO_3^- \qquad 4.2$$

bicarbonate buffer is suboptimal throughout the physiological pH range. HEPES is a much stronger buffer in the physiological pH range as its pKa is 7.3. HEPES buffer may be stronger than bicarbonate buffer, but it is costlier and toxic to some cells. For accurate control of pH over physiological range, 20-30 mM HEPES, 12 to 36 mM $NaHCO_3$, and 5-10% atmospheric CO_2 are recommended. In some media, pyruvate is used to control pH making them independent of $NaHCO_3$ and atmospheric CO_2, as it is metabolized by the cells to produce more CO_2. Leibovitz L15 medium contains 550 mg/l sodium pyruvate for optimum control of pH in the absence of atmospheric CO_2.

4.5.2 Osmolality

Osmolality is a measure of the total osmotic activity contributed by ions and non-ionized molecules in the medium. Most cultured cells have a fairly wide tolerance for osmotic pressure. The osmotic pressure of blood plasma is about 290 mOsm/kg, thus it is considered optimum for culture of animal cells. In general, osmolality ranging from 260 to 320 mOsm/kg is tolerated by most of the cultured cells. The osmolality of the media is maintained by changing NaCl concentration.

4.5.3 Viscosity

The viscosity has little effect on growth, when cells are cultured in static condition. However, it is much concerned in case of agitated culture system (e.g., bioreactor). Animal cells lack cell membrane, thus they are extremely sensitive to shear forces (see Chapter 6). By increasing viscosity of the medium it is possible to minimize shear-induced damage to the cell membrane. Usually, serum (10% v/v) maintains the optimum viscosity of the culture medium. Carboxymethyl cellulose or polyvinyl pyrollidone can be used in serum free medium as viscosity modulator.

4.5.4 Surface tension

Protein solution lends to froth when air is spurged through it. Cell culture medium supplemented with serum often froths due to aeration. The surface tension at air-medium interface reduces in the presence of protein, resulting in the formation of stable foam. Excessive foaming may have following adverse effects on cell culture:

(1) It reduces oxygen transfer through the headspace.
(2) It may causes damage in the cell membrane
(3) It may leads to contamination.

Thus, foaming is considered detrimental to animal cell culture process. The addition of a silicone antifoam helps to prevent foaming by increasing gas-liquid interfacial surface tension.

4.5.5 Temperature

Temperature is most crucial for the optimum growth of animal cells. Besides, temperature has an indirect role in controlling medium pH. The pH of the medium is influenced by the solubility of CO_2, which increases with the lowering of temperature. Thus medium should be equilibrated in CO_2 at lower temperature than that recommended for cultivation. The optimum temperature for growth of most of the animal cells is 37°C.

4.6 DEVELOPMENT OF SERUM-FREE MEDIA

An emerging trend in designing cell culture media for production of recombinant proteins is use of serum-free and elimination of constituents of animal origin. Substitution of animal proteins in culture media will facilitate downstream processing and ensure better biosafety of the products.

4.6.1 Potential merits and demerits in low serum or serum free media

There are many pros and cons for using low serum or serum free media in culturing animal cells, which are mentioned in Table 4.2. It may be noted that serum free media have both distinct advantages as well as disadvantages in animal cell culture process.

Table 4.2 Merits and demerits of low serum or serum free media

Potential advantages	*Potential disadvantages*
1. Avoid batch to batch variation of sera, thus improves reproducibility	1. Less tolerance of physiological variations
2. Improve economy due to low serum or no serum in the culture media	2. Supplementation of growth factors may be as expensive as using serum
3. Downstream process become easier due to low protein load	3. Reliable serum free medium is not available commercially
4. Avoid serum mediated cytotoxicity and interference in bioassay	4. Cells growth often slow down, so saturation cell density will be low
5. Selective growth of particular cell type is possible by using specific growth factor(s)	5. Due to specific nature, same medium cannot be used for different cell types
6. Absence of serum proteases degrade prevents degradation of protein products	6. Protease release by the cells may affect the products
7. Contamination due to virus, fungi, and mycoplasma can be minimized	7. Shear sensitive cells may be damaged in the absence of serum

4.6.2 Basic supplements of serum free media

Serum free media (SFM) are divided into three classes:

(a) *Class-I*: Complete synthetic media are prepared from the components of known molecular structures, do not contain any protein.

(b) *Class-II*: Partial substitution of serum in media with low cost nitrogen sources, e.g., enzymatically digested peptones, bovine pituitary extract, lactalbumin hydrolyzate, yolk emulsion, vegetable proteins, etc.

(c) *Class-III*: Media composed of purified proteins, hormones, and growth factors (e.g., albumin, insulin, transferring, PDGF, etc.).

Each of this class of media have own limitations. Class-I media are cheapest, but support growth of a few transformed cells with limited cell densities. Class-II media are low cost, autoclavable, but undefined by nature and suitable for large-scale industrial applications. But all cells cannot be grown in these media. Class-III media are costlier, but widely used for culturing many primary and specialized cells.

Supplements of SFM are selected on the basis of various factors present in the serum and their respective roles in cellular growth and metabolism. The components that are not present in the serum but are essential for growth are provided through the basal medium. Serum supplementing factors are grouped on the basis of their functions as follows:

Antioxidants

Serum has major roles to protect cells from the cytotoxic effects of oxygen radicals. Thus, SFM is supplemented with (a) selenite, which is known to activate peroxidase, a key enzyme involved in detoxifying oxygen radicals; (b) ascorbic acid (vitamin C) and α-tocopherol (vitamin E) as antioxidants; and (c) albumin as carrier of toxic metabolic end products.

Lipids

Serum provides intermediate constituents for the biosynthesis of membrane components. Few cells require exogenous source of lipids for optimum growth, so occasionally SFM are supplemented with cholesterol, oleic acid, and/or with pyruvate as precursor for the biosynthesis of acetyl-CoA. Lipids are insoluble in water; they need special delivery vehicle to make them available to the cells. Various alternate methods for delivering lipids in the cell culture media have been discussed in the literatures (Maurer, 1992; Blasey and Winzer, 1989; Iscove, 1984).

Hormones and growth factors

Insulin is the most common hormone used in SFM (class III media) in the concentrations ranging from 1-10 μg/ml, depending upon the cell type. It is required to promote the uptake of glucose and amino acids by the cells. Many class III media require cytokines for growth or differentiation. For example, EGF (1-10 ng/ml) is used in media for growing human epidermal keratinocyte and normal human endothelial cells; bFGF (0.5-10 ng/ml) is used for culture of normal human dermal fibroblast and epidermal melanocyte, etc. The growth factors normally used to supplement SFM are EGF, PDGF, IGF-I, NGF, IL-3 and 6, GM-CSF, SCF, etc., depending upon the specific requirement by the cells.

Proteins

Albumin (10-20 mg/ml) is the major serum protein used in class III SFM due to its many functions. It acts as carrier protein for lipids, hormones, minerals, vitamins, etc., as protective agent against mechanical damage to the cell membrane, and as a buffering agent. Transferrine (5-10 μg/ml) is another indispensable protein used in class III SFM, as it is involved in transporting Fe^{3+} to the cells.

Attachment and cell spreading factors

Most primary cells (expect lymphoid cells) and solid tumor cells are anchorage dependent, require proper attachment on a surface prior to proliferation. On the other hand, typical transformed cells of blood lineage and all hybridomas are anchorage-indepedent cells, which can grow without attachment to the solid surface. For anchorage-dependent cells produce their own attachment factors when cultured, so are not separately added in the culture media. Typical attachment factors used in culture of anchorage-dependent cells are fibronectin (cells of mesenchymal origin), laminin (for ectodermal and endodermal cells), collagen IV (for neuronal, endothelial cells etc.).

4.6.3 Selection of optimum base media

Media that include serum free, protein free, chemically defined, and animal component free formulations are commercially available in the platform of few commercially important animal cells, such as CHO, NSO, HEK293, hybridomas, embryonic stem cells, and other stem cells of hematopoietic, neuronal, mesenchymal nature. The compositions of these media are available in the commercial catalogues. The most popular media used among the serum free or low serum formulations are DMEM, IMDM, Ham F12, α-MEM, McCoy's 5A, MCDB202, RPMI-1640.

4.6.4 Adaptation of cells to SFM

Adaptation of animal cells from serum containing to serum free or low serum medium is a laborious task. Cells often change morphology, surface markers (charge of phenotype), karyotype, growth characteristics, and capability of metabolizing drugs during adaptation period. All such cellular alterations may lead to the change of productivity and functional properties of the cells. Therefore, before adaptation in serum free/low serum media, it is essential to gather maximum information on the cell type to be adopted in SFM. Adaptation process may require weeks to months, in fact the process may results selection of a subclone or mutant clone. Thus, it is customary to monitor such changes in the cell strain, if any, during adaptation, and to determine the productivity of the clone. Two methods are generally followed for adopting cells from serum-supplemented to serum-free media: sequential adaptation and adaptation with conditioned medium. In sequential adaptation method, cells are cultivated in passage '1' using 75% serum-supplemented with 25% SFM. In the following passages, serum-supplemented media are reduced from 50, 25, and 0% with corresponding increase in percentage of SFM. Most of the cell lines are believed to be adapted in 100% SFM within passage '4'. For adaptation with conditioned medium, cells are cultured in 100% serum-supplemented medium in passage '1'. In passage '2', cells are cultured in 50% conditioned medium of passage '1' and 50% SFM. In passage '3', cells are cultured in 50% conditioned medium of passage '2' and 50% SFM. Similarly, cells are cultured in passage '4' using 50% conditioned medium of passage '3' and 50% SFM. Finally, in passage '5', 100% SFM is used to culture the cells.

It should be remembered that serum-free cultures are more sensitive to temperature, pH, shear force, osmolality, and proteolytic enzymes. It is also important to be noted that in SFM, antibiotics are used at very low concentrations (5 to 10-fold lower than serum-supplemented medium).

4.7 DESIGN APPROACHES TO OPTIMIZE SFM

Four different strategies are followed to replace serum in cell culture medium. The measures of the success in each strategy are based on cell growth rate, cell attachment efficiency, lack of cell aggregation, maximum cell density, resistance to shear damage, yield of the product, stability of the product, extent of glycosylation, maintenance of differentiated or undifferentiated state (in case of products related to cell therapy/tissue engineering) of the cells.

(i) ***Analytical approach*****:** This approach to design SFM is most laborious and time consuming. Each serum factor involved for growth, survival, and productivity is identified, and tested individually for performance characteristics as above. This method is not quite successful, as in many cases the growth factors and hormones act synergistically.

(ii) ***Molecular approach*****:** The microarray analyses help to identify medium components. The basal medium is supplemented with different combination of growth factors performing equivalent functions of the serum. This technique allows testing of growth factors and receptors, hormones, attachment factors, etc. that can affect cell growth or other cellular processes in culture system leading to optimum performance.

(iii) ***Limiting factor approach*****:** Limiting factor method can be applied, where in the existing medium, the serum concentration is reduced step by step, unless the cell growth and the productivity is affected. The concentration of each component of nutrient is adjusted until growth and productivity is restored (Ham, 1981). Serum concentration is reduced from 10% to 0% in successive steps. In each step, cells are cultured for minimum of 5 passages, and for each serum concentration frozen stocks of cells are prepared. For example, in one such stage (say at 1% serum level), where the growth and the productivity of the culture is significantly reduced than the previous stage (say at 3% serum level), it becomes necessary to workout on the individual components of serum. Initially, a suitable basal medium is selected for the cell line under investigation. The medium is supplemented with a mixture of insulin-transferrin-selenite (similar to 10% FCS containing medium) and 1% serum. This formulated medium is used to culture the cells. If growth is still poor, add growth factors depending upon the cell type (0.5-10 ng/ml). At this stage the growth and the productivity many be restored; if so, the serum level is further reduced to one-third (0.3%). In case where no improvement of growth and productivity of the culture is observed, the above medium is further supplemented with lipid mixtures (1-10 μg/ml). It may be possible that the cell growth is restored at this stage. Otherwise, it is considered that the particular cell type cannot be grown below 3% serum level. In addition to the supplementation of medium, other experimental variables need to be considered in designing low serum or serum free medium. These are seeding density, trypsinization procedure, and handling of cells.

(iv) *Statistical Approach*: It is a multidimensional statistical method for designing of medium, popularly known as design-of-experiment (DOE). It is most popular in optimizing microbial culture media. By this technique, multiple variables can be tested simultaneously. Thus, the number of test condition reduces without a significant loss of information and speed of the development process. A statistical method for designing medium for recombinant CHO cell line producing antibody has been described by Gifford and colleagues (Sigma-Aldrich Biotechnology, Saint Louis). In brief, three best performing media are initially selected from the six animal protein-free media. The next step is to perform media mixing experiments at various proportions (17, 33, 50, 67, and 100%) between above three media. Design expert software determines ideal combination of media at which growth rate and antibody productivity is maximized, or antibody productivity is highest. Final step is to test the performance in bioreactor with feeding of various supplements, like vitamins, amino acids, protein hydrolysates, etc.

4.8 METABOLISM AND CATABOLIC CONTROL

Animal cells require both glutamine and glucose for generation of ATP and NADPH to drive biosynthesis of cellular materials and to obtain the cell derived products. If the product of interest is from the expression of recombinant gene, the cellular energy requirement for such synthesis process could be a substantial fraction of the total energy capacity of the cell. So, if ATP synthesis is rate limiting, high growth rate may reduce product formation by restricting availability of ATP for it.

Glucose and glutamine are simultaneously catabolized to produce ATP (Figure 1.9). If glucose is converted to lactate by anaerobic glycolysis, 2 mol of ATP are produced per mol of glucose catabolized. On the other hand, if glucose is converted into CO_2 by TCA cycle, typically 36 mol of ATP are synthesized per mol of glucose catabolized. Glutamine is also catabolized to produce energy, the yield of which depends on the relative flux of glutamine through different possible pathways (Glacken *et al.* 1985; Glacken, 1988). The relative amount of cellular energy provided by glucose or glutamine varies in the cell type and environmental conditions. Again, aerobic glycolysis rate is high in transformed cells as compared to the normal cells (McKeehan, 1982). Together, carbohydrate levels in the culture medium can affect energy metabolism, cellular productivity, and finally, production of lactic acid. Again, ammonia is the end product of glutamine catabolism by animal cells and also its thermal degradation. Both lactic acid and ammonia are inhibitors of animal cells. The formation of these two inhibitors can be minimized, and hence volumetric ATP production rate can be improved if the concentrations of glucose and glutamine are maintained at low levels during culture of the cells. This can be achieved by feeding glucose and glutamine at controlled rates. Alternatively, lactate level can be maintained at low concentration by providing slowly metabolizing carbohydrate (e.g., fructose, galactose) in the culture medium instead of glucose.

4.9 MEDIA PREPARATION AND TESTING

Animal cells are sensitive to the toxic effects of impurities, normally present in the purified grade of chemicals and distilled water. Therefore, the viability of cells, their growth rate, the expression of specialized products, and functions may be affected, if cell culture tested chemicals and water are not used in the preparation of media. In-house media preparation is not advisable for any unestablished laboratory. Ready-made liquid media for tissue culture, which are open and use type, are available from the suppliers. Best alternate is to procure ready-mix powdered media and reconstitute in the laboratory. These media are made from cell culture tested chemicals, cheaper, and have better shelf-life than liquid media. Before use in culture, the powdered media are reconstituted in cell culture grade water, filter sterilized, and tested for sterility. Various precautionary measures to be taken during media preparation are discussed below:

Quality of water**:** Endotoxins are heat-stable lipopolysaccharide complex that present the cell wall of grame-negative bacteria. They are released when bacteria die and fragmented. Endotoxins may result unwanted responses to the cullture cells. Endotoxin interferes with growth and proliferation of animal cells. Its effects are not completely predictable, and in some cases endotoxin enhances cell growth. For therapeutic grade products, the presence of endotoxin beyond specific limit can elicit a febrile response in the patient. Thus, it is important to keep cell culture free from contaminating endotoxin. The assurance that the key ingredients, water, are free of endotoxins not only restrain concern about the reagent's physical properties but also assures that the reagent has been prepared under cGMP manufacturing process. The water purification procedures is adopted for cell culture grade water is rather complicated. It is recommended that the cell culture grade water should be free from dissolved inorganic, organic substances and endotoxin. In general, the quality of water used in media preparation must meet a nondetectable limit of endotoxin (<0.005 EU/mL) and salts in terms of resistivity (18 mΩ.cm).

***Dissolution of powdered media constituents*:** The quality of the ingredients used in formulating powdered media is normally in pure form and readily soluble in working strength (1 ×). It is difficult to formulate concentrated powdered media (5 × strength), due to low solubility of some of the constituents in the water. Commercial powdered media are supplied as complete media without $NaHCO_3$, and media without glutamine and $NaHCO_3$. Liquid media are normally available in working strength, complete with glutamine and $NaHCO_3$. Powdered media are available from 1 L to 50 L pack. Depending upon the consumption rate, storage facility, and delivery schedule, powdered media can be ordered. During dissolution of media, care should be taken so that $NaHCO_3$ is added just after complete solubilization of other media constituents in water. $NaHCO_3$ is alkaline in nature; certain amino acids lend to precipitate at transiently high local pH of the solution during solubilization process. Normally, pH of the media is adjusted by equilibrating media with CO_2 gas at 36.5°C, prior to filter sterilization. In alternate, sterile media is equilibrated in 5% CO_2 incubator. Majority of the media contain $NaHCO_3$, so the color of the complete media in general is red to bluish red (7.4–7.6 pH). In rare cases, the pH of the culture media is adjusted with NaOH or HCl.

***Sterilization*:** Cell culture media are generally filter sterilized using 0.10/0.22 μm membrane filter (see Chapter 5). However, few autoclavable media are commercially available, which can be heat sterilized prior to the addition of thermolabile components and $NaHCO_3$. Class II SFM is autoclaved at 121°C for 15 min. Filter sterilization of medium could be pressure or vacuum driven, though pressure filtration is commonly used in the laboratory and in large-scale applications. Sterile media are supplemented with appropriate concentrations of serum before use. Filter sterilized serum are sold in supplementing cell culture media.

***Sterility tests*:** Medium is subjected to sterility tests for bacteria, fungi, mycoplasma, and viruses. Sera are the most common source of contaminants, especially with mycoplasma and viruses, though routinely tested by the suppliers. The test report of sera can be obtained from the suppliers upon request. In case of in-house prepared media, bacterial and fungal contamination is most common. Contamination occurs due to defects in filtration apparatus or techniques. In large-scale sterile filtration, about 1-3% cases media is contaminated due to failure of filters. As a thumb rule, unless medium is tested sterile, it is not supplemented with antibiotics or sera.

***Storage*:** Media components are thermolabile, therefore reconstituted media are better stored at 4-10°C. Glutamine is the most temperature sensitive component used in the basal media, which is degraded even in refrigeration. The shelf-life of liquid medium is not more than 6 months. Powdered media can be stored at 4–10°C for 2 years.

4.10 SUMMARY

1. Selection of culture media is decided by cell type, culture system, nature of the product, down-stream processing cost, etc.
2. The pH of cell culture media is primarily maintained by the combination of $NaHCO_3/CO_2$ buffering system.
3. Serum is the most costliest component of cell culture media. The serum lot is selected on the basis of cloning efficiency, plating efficiency, growth curve of cells, and the preservation of cellular characteristics.
4. Design of low serum or serum free media is a tedious exercise, which requires systematic experimentations. Four different strategies are followed to design SFM. These are based on analytical, molecular, limiting factor, and statistical approaches.

ABBREVIATIONS

ATP	Adenosine triphosphate
BSE	Bovine Spongiform Encephalopathy
BVDV	Bovine viral diarrhea virus
bFGF	Basic fibroblast growth factor
DMEM	Dulbecco's modified Eagle medium
ECM	Extracellular matrix
EGF	Epidermal growth factor
EU	Endotoxin unit
FBS	Fetal bovine serum
FCS	Fetal calf serum
GMEM	Glasgow's minimum essential medium
GM-CSF	Granulocyte macrophage colony stimulating factor
GMP	Good manufacturing practice
HEPES	N-2-hydroxyethyl-piperazine-n□′-2-ethane sulfonic acid
HS	Horse serum
IMDM	Iscove's modified Dulbecco's medium
IGF-I	Insulin like growth factor-1
IL-3/6	Interleukin-3/6
MEM	Minimum essential medium
NADPH	Nicotinamide adenine dinucleotide phosphate, reduced form
NGF	Nerve growth factor
NBCS	Newborn calf serum
PDGF	Platelet-derived growth factor
SFM	Serum free medium
SCF	Stem cell factor
TNFα	Tumor necrosis factor α
TGFβ	Transforming growth factor β

REFERENCES

1. Bettger WJ and Ham RG (1982) In: *Advances in nutritional research* (Draper HH, ed.). p. 249. Plenum Press, NY.
2. Blasey HD and Winzer U (1989) *Biotechnol. Letters* 11, 455.
3. Eagle H (1955) *Science* 122, 501.
4. Gifford J, Albee A, Deeds ZW, Delong B, Kao K, Ross JS and Caple MV. An Efficient Approach to Cell Culture Medium Optimization- a statistical method to medium mixing. *Product literature.* Sigma-Aldrich Biotechnology, 3050 Spruce Street, Saint Louis, MO 63103 USA.
5. Glacken MW, Fleischakar RJ and Sinskey AJ (1986) *Biotechnol. Bioeng.* 28, 1376.
6. Glacken MW (1988) *Bio/Technol.* 6, 1041.
7. Ham RG (1981). In: *Tissue growth factors* (R Baserga, ed.) p. 13. Handbook of Experimental Pharmacology 57.
8. Iscove NN (1984). In: *Cell culture methods for molecular and cell biology* (DW Barnes, DA Sirbasku, and GH Sato, eds.) vol. 4, p. 169. Alan R. Liss. New York.
9. Kelley WN (1972) In: *Growth nutrition and metabolism of cells in culture* (Rothblat GH and Cristofalo VJ, eds.). vol. 1, p. 211. Academic Press, New York.
10. Miller WM, Wilke CR, and Blanch HW (1988) *Bioprocess. Engr.* 3, 103.
11. McKeehan WL (1982) *Cell Biology International Reports.* 6, 635.
12. Maurer HR (1992) In: Animal cell culture- A Practical Approach, 2nd Edition, p. 15. (RI Freshney, ed.), IRL Press, Oxford.
13. Zielke HR, Ozand PT, Tildon JT, Sevdalial DA and Cornblath M (1978) *J Cell Physiol.* 95, 41.

SUGGESTION FOR FURTHER READING

1. Glassy MC, Tharakan JP and Chau PC (1988) *Biotechnol. Bioengg.* 32, 1015.
2. Lambert KJ and Birch JR (1985) In: *Animal cell biotechnology*, vol. 1, p. 86, (RE Spier and JB Griffiths, ed.), Academic Press, London.
3. Mizrahi A and Lazar A (1988) *Cytotechnol.* 1, 199.
4. Ozturk SS and Palsson BO (1990) *Biotechnol. Prog.* 6, 121.

5

Good Manufacturing Practices, Aseptic Techniques, and Cell Culture Equipment

To start a productive cell culture facility, effective laboratory design and careful selection of equipment/instruments are the second most important factor, after the training of manpower. The main reasons for this are: cell culture based processes are prone to microbial contamination, manufacturing processes are cost intensive, and the final products require certain criteria to fulfill the requirement of clinical trials. Therefore, a proper understanding of the process requirements and the knowledge of manufacturing methods to achieve the target are ingenuously important. This Chapter describes basic laboratory design for animal cell culture facility, biosafety criteria, maintenance of aseptic processes and some essential equipment used in animal cell culture process.

5.1 LABORATORY DESIGN

Perhaps, one of the most underrated aspects of animal cell culture is designing the facility to ensure that quality products are produced in a safe and efficient manner. The major requirement that distinguishes cell culture facility from most of the other laboratories engaged in biological research is the need for maintaining asepsis. There are several aspects for the design of a good cell culture facility. The present discussion has been limited to routine cell culture laboratory for research and pilot-scale facility for manufacturing of cell based therapeutic products (excluding viral vaccines) to be used for human clinical trials. In this case the specification of *containment area* has been limited to the *biosafety cabinet* (BSC). The facility design has been made compliance to the *current good manufacturing practice* (cGMP) recommended by the Food and Drug

Administration (FDA), United States of America. GMP regulations are implemented to ensure controlled auditable processes to manufacture products, which are safe and effective in humans. cGMP is composed of various elements like standard operating procedure (SOP), facility and equipment, documentation and labeling, personnel and training, process control and validation, auditing and error management. Due to the limitation of scope in this Chapter, only facility design and equipment are covered.

Table 5.1 Design considerations for regulatory approval of the facility

1. **Design for consistency and reproducibility:** ***Simpler process:*** Minimum number of operations, familiar and proven operations. ***Sensitivity of the cell line:*** Minimize raw materials variation, process delay. ***Contamination control and elimination of potential product cross-contaminations:*** Through on-going training of operators, proper cleaning of equipment, steam sealing, aseptic designing of equipment, heating-ventilation-air conditioning (HVAC) system design of the facility.
2. **Design for biosafety (therapeutic proteins):** ***Risk assessment for biological and chemical hazards*** ***Primary containment of the equipment*** ***Secondary containment of the facility*** ***Bio-waste treatment*** ***HVAC system design of the facility*** ***People and material flows***

GMP requires well-controlled and reproducible processes. Table 5.1 presents a few process design considerations, with particular reference to the reliability and reproducibility of a cell culture based process that may directly affect the quality of the products and their licensing by the drug controlling authority. Ideally, cell culture workspace can be divided into several suites, depending upon the specific activities, inter-relationship between the activities, and the cleanliness demanded by the nature of the activities. Matching with the cGMP requirements, animal cell culture facility can be divided under many headings. The major suites and the corresponding activities are described in Table 5.2. Before planning for a facility design, such activities, their interrelationship, and other mandatory requirements for research and development on cell culture derived products should be considered. Adequate space, defined flow-path to avoid mixing of clean and dirty areas, manufacturing operation in defined area location, easy cleaning, operation, and maintenance are the main aspects of the design.

5.1.1 Facility design

There is no master facility design that could be used in all kinds of biological research and related manufacturing activities. There may be many common steps in SOP; each product is evaluated in terms of how, where, and under what conditions it should be manufactured. Most critical considerations for facility design are as follows:

Table 5.2 Animal cell culture facility: basic research laboratory and pilot scale facility for manufacturing therapeutic proteins.

1. **Service and warehouse:**
   Utility: Typical utilities: Deionized water, reverse osmosis (RO) water, ultra pure water, clean steam, process steam, sterile air, clean air, sterile oxygen/carbon dioxide/ nitrogen, chilled water, process water, hot water, power supply with emergency back-up, air conditioning with class 10,000 clean air supply.
   Maintenance: Mechanical, electrical, and civil maintenance, computation and process control.
   Consumables and raw materials: Perishable, non-perishable materials, sterile/ non-sterile materials.
2. **Support services:**
   Glassware washing, sterilization, and storage; Freezing and cell banking; Cold room; Quality control and quality assurance; Animal breeding and testing facility
3. **Basic research:**
   Cell culture; Molecular biology; Protein purification process development
4. **Upstream processing:**
   Media preparation; Tissue culture process control; Inoculum/seed culture preparation; Bioreactor operation; Biowaste and killing
5. **Downstream processing:**
   Harvesting; Buffer preparation; Purification; Polishing and storage of products; Quality control and quality assurance
6. **Office and administration:**
   Material control; Accounts and personnel; Break room and coffee area; Dispensary

1. Containment areas and its level:
   a. Construction materials in these areas
   b. Airflow and pressure differences
   c. Position of biosafety cabinets;
2. Non-containment areas: proper description, including justification and identification of jobs to be performed there;
3. Adequate separation of clean from non-clean operations;
4. Location for storage of seed lots and running cultures;
5. Media preparation and inoculation in appropriate contained location to avoid contamination;
6. Bioreactor operation, culture harvest, and product purification activities located in restricted, controlled areas regardless of scale; and
7. Movement of man and materials.

Figure 5.1 shows a typical plan for research and process development facilities for manufacturing therapeutic products in human applications. The plan shows earmarked areas on the basis of major activities mentioned above (Table 5.2), the containment level, cleanliness are demanded by the nature of the activities, and the movement of man and materials. It is important to be remembered that maintaining clean conditions is most crucial in planning facility and designing of equipment. Clean environment is maintained by judicial facility design, which includes smooth and hard surfaces of walls, floors, and ceilings that can will be easily cleaned; proper temperature and humidity control; positive HEPA filtered air pressure (Class 10,000); and the implementation system to monitor environmental conditions, cleanliness, routine room sterilization schedule, etc.

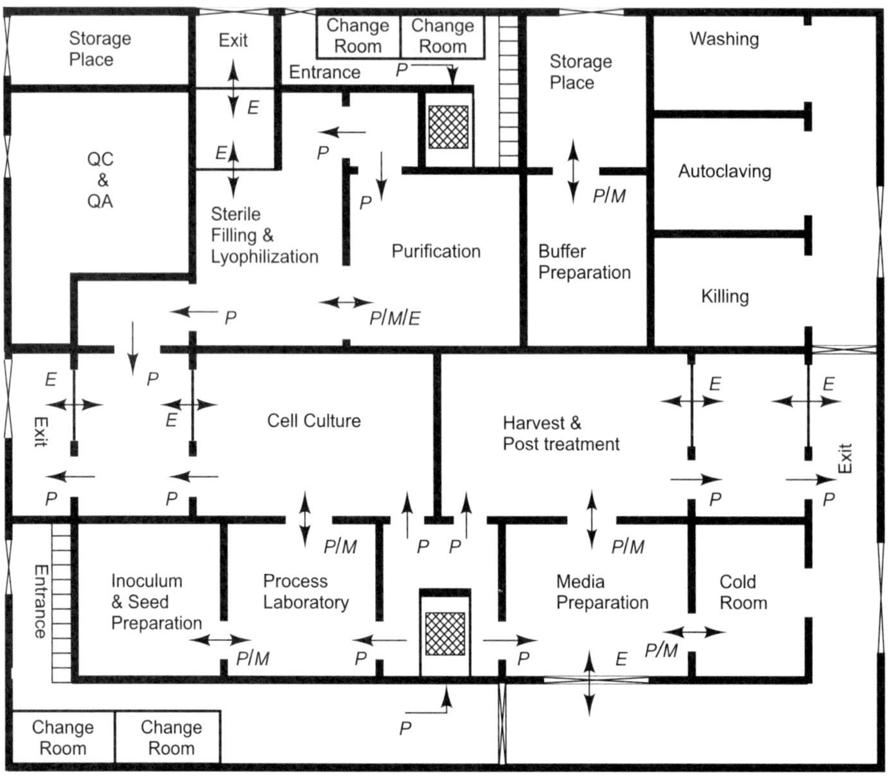

Figure 5.1 Layout of animal cell culture development facility (core units). The sketch showing different suits for basic cell culture and protein purification activities. The movement of personnel and materials are shown in the sketch. P: Personnel, M: Material, E: Equipment. Single-headed arrows indicating unidirectional flow of M/P/E, and double-headed arrows indicating bi-directional flow of M/P/E.

5.1.2 Equipment design

The most important considerations for design and selection of equipment are perfect match and optimum utilization. GMP requirements are considered in designing entire process requirements. Transfer of sterile liquid or culture from one vessel to another should be performed under appropriate conditions to prevent ingress of contamination into the process, as well as egress of the potential hazardous organism in the working environment. The most critical equipment design considerations are as follows:

1. Ease of cleaning, sterilization, and validation;
2. Bioreactors should be able to maintain culture sterility for at least 3-4 weeks;
3. Bioreactors/media preparation tanks should permit adequate mixing and/or aeration;
4. Bioreactors or other vessels should be capable of providing aseptic sampling and addition of nutrients;
5. Equipment and media feeding lines should be made of corrosion-resistant stainless steel 316 and free from cracks and crevices (at least 200 grain polished);
6. Minimum aerosol generation from centrifuge.

The common equipment that require for basic and applied research in animal cell culture process, and also GMP compliance process developments for manufacturing therapeutic proteins are listed in Table 5.3.

Table 5.3 Cell culture and related equipment

Basic research	*Process development*
• Laminar flow hoods • Biosafety cabinets • CO_2 incubators • 37°C incubators • Phase-contrast and fluorescence microscopes • Roller apparatus • Spinner apparatus • Ultra pure water generator • Freezers (–20 and –70°C) • Refrigerators • Control rate freezer and liquid nitrogen storage container • Bench top centrifuges • Sterilizer • Hot air ovens (50 and 160°C) • ELISA reader • Spectrophotometer • Osmometer • Balances • PCR machine • Electrophoresis and western blotting apparatus • Fluorescence-activated cell sorter • Cell counter • Pipette aids • pH meter • Micro-pipettes • Water bath • Pressure type media filtration apparatus • Protein sequence analyzer • Standard chromatography setup • HPLC	• Roller apparatus • Spinner apparatus (1 L) • pH meter • Balances • Bioreactors (5, 20, 100 L) • Fluidized bed bioreactor (20 L) • Media feeding tanks (100, 250 L) • Steam sterilizable vessels for media preparation and holding (250 L, 2 Nos.) • Peristaltic pumps • Media filtration device • Steam sterilizable spent media holding tank (500 L) • Microfilter and ultrafiltration units • Preparative chromatography setup • Killing tank (250 L, 2 Nos.) • Diafiltration unit • Buffer preparation tanks (50 L, 4 Nos.) • Buffer filtration units (4 sets) • Lyophilizer • Sterile filling device • ELISA reader • Refrigerators • Freezer (–20°C) • Spectrophotometer • 2D electrophoresis apparatus

5.2 ASEPTIC PROCESS

5.2.1 Nature of common contaminants in cell culture

Any microbial contamination of cell culture facility could be catastrophic, unless the source and the nature of contaminants are identified. The common microbial contaminants in cell culture laboratory are bacteria, yeasts, molds, and mycoplasma. Cross-contamination of slow growing animal cells with fast growing cells like HeLa is not uncommon. The general characteristic appearances of microbial contamination in cell culture media are as follows:

A B

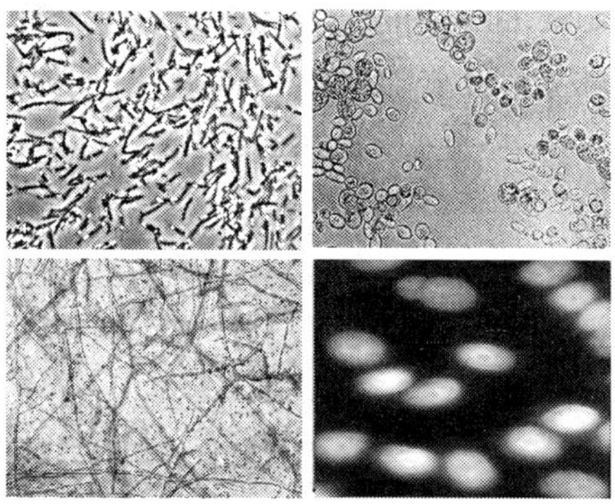

C D

Figure 5.2 Photomicrographs of common contaminating organisms in animal cell culture process. **A.** Bacteria (1000 *x*); **B.** Budding yeast (400 *x*); **C.** Mould (400 *x*); **D.** Mycoplasma (1000 *x*).

(A) *Bacteria* In case of bacterial contamination, sudden decrease in medium pH, cloudiness of medium, and sometime formation of film or scum in culture medium are commonly observed, which can be detected by naked eye. In general, reconstituted cell culture media is stored at 4-8°C, so minor bacterial contamination, if any, cannot be visualized without microscope. Again, media is often supplemented with antibiotics (e.g., penicillin, streptomycin, or gentamycin). So, low-grade contamination, if any, cannot be grown. Early bacterial contamination is identified only under bright-field high-power microscope (1000 ×), with or without staining (Figure 5.2A). Morphologically, two distinct bacterial (bacilli and cocci) contaminants are normally found in cell culture.

(B) Yeast In case of yeast, the color of medium does not change unless heavily contaminated. Since yeasts are larger than bacteria, they can be easily detected in moderate magnification (200 × to 400 ×). Yeast is round or ovoid in shape, they bud-off as smaller daughter cells. Sometimes, yeasts grow in staggered chain like pattern as (Figure 5.2B)

(C) *Mold* Mold contamination does not change the color of the medium. Molds grow slowly, thus in visual observation they appear quite late in the culture.

Molds form thin filamentous mycelia, and sometime denser clumps of spores become visible through naked eye (Figure 5.2C). In general, the mold contamination is rare in case the cell lines are obtained from an established source. The primary cultures derived from rodents or primates are often contaminated with molds.

(D) *Mycoplasma* Mycoplasma is a group of bacteria, which survive inside the cell as parasite. Animal cells contaminated with mycoplasma are considered more problematic than other type of contaminants. The main concern of its contamination is difficulty to detect, as mycoplasma do not always reveal their presence with macroscopic changes in culture medium. In continuous cell lines, mycoplasma may grow slowly, and thus the host cells neither manifest any morphological changes nor do they die. Sometimes, mycoplasma contamination can be evident by their degenerative effects on the host cells. The two most common sources of mycoplasma in cell culture are cell stocks and the serum. Mycoplasma contamination is identified by staining blue of DNA with a fluorescent dye DAPI, which can be detected easily by high-power microscope (1000 ×), as shown in Figure 5.2D (black and white picture).

Often precipitates present in culture media comprising the serum proteins (cold insoluble globulins) and cell debris may mislead as microbial contamination. These materials are easily distinguishable by their irregular shape under the microscope.

There are several ways by which animal cell culture gets contaminated. The possible sources are media, cell lines or tissue samples, working surface, CO_2 incubator, biological safety cabinet, culture flask, pipettes, etc. Therefore, it is always advisable to take extra precaution while cleaning equipment and handling cells inside the safety cabinet.

5.2.2 Aseptic operation

It is essential to identify the possible sources of contamination prior to planning of an aseptic operation. At the same time it is equally important to validate and revalidate aseptic processes in regular basis. Following are the few important sources (other than previously discuss) from which contamination is taken place:

1. Particulate matters are the carriers of bacteria. Proper attention should be paid to remove particulate matters. Sterilization and depyrogenation of each part of the equipment, likely to get into direct contact with the sterile materials is necessary. Equipment and accessories that need to be made particle-free are glass vessels or bottles, storage vessels, bioreactors, tubing, working gowns, masks, and gloves.
2. Environment factors

   (a) Air quality In a typical cell culture process, less stringent conditions can be adopted as the culture/medium is not directly exposed to the air. The Class 10,000 clean room should be adequate for such purpose. The design and planning for maintenance the integrity and efficiency testing of HEPA filters should be included in the routine validation protocol. Installation of air locks in proper locations of the facility, and maintenance of positive air pressure are two integral parts of the aseptic design.

(b) *Personnel:* The presence of working personnel in the sterile or clean room influences the the quality of air. The design and validation study should include the maximum number of people expected to be present at any time during the process. In order to minimize bioburden (load of organisms in air), it is mandatory to follow certain precautions recommended in cGMP. For example, air shower at entry point of the change room, use of sterile functional dress including head cap and gloves, restricted movement of personnel and others. Since the major source of contamination is people, new barrier technology has been introduced in aseptic design. This technology maintains air quality, around the product, equivalent to Class 100, or better while allowing personnel to access through glove ports.

3. Sterile filtration of media and air is the most important factor for maintaining aseptic cell culture process. Sterilization of media and air is carried out by filtration through 0.22/0.1 μm sterile filters.
4. Care and maintenance of laboratory area.

In order to maintain a clean and safe working environment, neatness and cleanliness are the two fundamental requirements. Any spillage should be dealt with immediate attention. Routine cleaning of laboratory space work surfaces and the floors should be undertaken. All major pieces of equipment should be regularly maintained and serviced by qualified technical personnel. For example, bio-safety cabinets should be checked six monthly to ensure that they are safe to use (leakage proof). The test results should confirm desired air velocity through the HEPA filter and the particle count. Humidified CO_2 incubator is an another equipment of concern, as fungus and bacteria can easily grow on the water tray. Regular cleaning of the incubator will minimize the risk of contamination.

Various disinfectants used for equipment decontamination and the methods of their applications have been described. The disinfectants are grouped into four categories, the scope of application are summarized below:

(a) *Hypochlorites*: These are effective general-purpose disinfectant, active against viruses. It is corrosive against metals, so it should not be applied on metal surface. Hypochlorites are easily inactivated by organic matter and therefore should be prepared freshly. The recommended dosage of hypochlorite are 1000 ppm for general surface disinfection, 2500 ppm in discard waste pots for washing pipettes, and 10,000 ppm for tissue culture waste and spillage.

(b) *Phenolics*: These are not active against viruses, and are not inactivated by organic matter, as in the case of hypochlorite.

(c) *Alcohols*: 70% (v/v) ethanol and 60-70% isopropanol is used. The mode of action of alcohals is dehydration and fixation of organisms. Isopropanol is not effective against viruses, whereas ethanol is effective against most viruses except nonenveloped viruses.

(d) *Aldehydes*: These may cause irritation in skin and their application is limited to the problems associated with sensitization. In case of corrosion sensitive equipment (e.g., centrifuge bowls), hypochlorites can be replaced with glutaraldehyde for sanitization purpose. It is highly recommended to change the disinfectants in cyclic order, in every six months interval, to avoid the development of resistant strains of the contaminants.

5.2.3 Clean room/clean zone

Clean room is one of the basic requirements for cell culture facility. It is a part of heating-ventilation-air conditioning (HVAC) system. The specification of clean rooms should be matched with the nature of the work. International Organization for Standardization (ISO) has introduced a new standard, ISO 14644-1 in place of the Federal standard 209 E (US). The basic difference between ISO specifications and Federal standard 209 E is that the former is based on metric system (Table 5.4). In cell culture applications, the most favored Class is 10,000 (ISO 7), where the culture/medium is not directly exposed to the environment. According to this standard, the extent of room cleanliness is determined by the number of air changes carried out in unit time (hour). For example, in Class 10,000 the number of air change may vary from 60 to 90 volumes (equivalent to the volume of the room) per hour, which increases to 150-240 in Class 1000. In properly designed clean room, air filtration is carried out in three successive stages, e.g., pre-filter, bag filter, followed by high efficiency particulate air (HEPA) filter. HEPA filters are protected from early clogging by initial two-stage prefiltrations. In normal situation, HEPA filters are replaced once in every 1 to 2 years, depending upon the load of particulate matters in the air. The pre-filters and bag filters are replaced many times in a year. HEPA filters made up of glass fiber roll into a paper-like material and are 99.99% efficient in removing particles of size 0.3 μm or larger. The whole filter material is pleated to increase the fiber surface area, and is bonded into a frame.

Table 5.4 Specification of clean room and clean zone

Federal Standard 209 E Class			*ISO 14644-1*			*Number change*
Class	Particle counts/ft^3 (≥ 0.5 μm)	Average air flow velocity (ft/min)	ISO Class	Particle counts/m^3 (≥ 0.5 μm)	Average air flow velocity (m/sec)	Air change/ hour
Class 100,000	75000	1–8	ISO Class 8	2,640,000	0.005–0.041	5–48
Class 10,000	1500	10–15	ISO Class 7	52,800	0.051–0.076	60–90
Class 1000	675	25–40	ISO Class 6	23,800	0.127–0.203	150–240
Class 100	25	40–80	ISO Class 5	880	0.203–0.406	240–480
Class 10	7	50–90	ISO Class 4	246	0.254–0.457	300–540
Class 1	1	60–90	ISO Class 3	35	0.305–0.457	360–540

Source: Clean rooms and associated controlled environments - Part 1: Classification of air cleanliness. Complementing ISO 14644-1 (Nelson Laboratories, Inc., USA and Connect 2 Clean rooms. Ltd. Lancashire, UK.).

5.2.4 Containment and biosafety levels

Containment is used to describe safe procedure to manage infectious agents (e.g., viruses, bacteria) or recombinant DNA in the laboratory environment in which they are being handled or maintained. Two most common ways by which pensonnel are exposed to infection are aerosols and skin contact. Therefore, biosafety relies on the combination of physical barriers and special handling techniques to avoid release of infectious agent in the laboratory and in the surrounding environment. Accordingly, *primary containment barriers* and *secondary containment barriers* have been referred in handling infectious agents. Primary containment barriers refer to the protection of workers and immediate working environment from exposure to the infectious agents. Biosafety cabinets, protective clothing, self-contained biosafety suits, and safety respirators constitute primary containment barriers. The secondary containment barriers refer to the protection of wider environment and the community. It is primarily related with the facility and equipment design and has three levels:

(a) Liquid wastes from animal cell culture equipment (e.g., washing of bioreactors, holding tanks, centrifuges, spent media, etc.) should be sterilized at 121°C for 30 min, prior to discharge in the treatment facility.

(b) Dedicated ventilation and exhaust air treatment equipped with HEPA filters.

(c) Shower-out routine, airlocks, and control access. The animal cell culture facility is provided with some of these containments, depending upon the level of hazard.

Different biosafety levels followed for biological applications are described below. There are no specific biosafety regulations for handling animal cells; however, NIH guidelines for handling potentially infectious organisms has been adopted in classifying biosafety levels for animal cell culture.

Biosafety level 1 (BL1)

Similar practice is adopted for work with organisms not known to cause disease in healthy adults. For that, standard open-bench type working area (clean bench in front of gas flame) and routine cleaning of working bench should be adequate. However, open-bench work is not recommended for animal cell cultures due to microbial contamination.

Biosafety level 2 (BL2)

The methods followed and the equipment used for moderate-risk organisms can be adopted here. Use of biosafety cabinets (Class IIA, IIB1, IIB2, IIB3), protective clothing, limited access, decontamination of wastes are some of the recommendations for BL2.

Biosafety level 3 (BL3)

Equivalent for handling high-risk, life threatening organisms. Use of Class III biosafety cabinet, change-in, shower-out procedures, decontamination of wastes, negative internal air pressure gradient towards areas of maximum potential exposure of risk are few recommendations for BL3. Limited access and exhaust air filter/decontamination are two primary requisites for BL3 level.

For handling recombinant DNA molecules in cultures greater than 10 L, additional biosafety guidelines are recommended by NIH. In this case the containment levels are referred to as BL1-LS, BL2-LS, and BL3-LS, depending upon increasing precautions against uncontrolled escape of organism or formation of aerosols. The entire operations require routine validation for containment and decontamination procedures.

For most small-scale animal cell culture works, which do not involve human or animal pathogens, BL2 is adequate. In addition to that restricted access, internal air pressure gradients, and air locks are recommended. In large-scale bioreactor, the cultures should be confined in pressure-rated tanks with sterile filters on gas inlets and outlets. The shaft of the rotating impeller should be housed in double mechanical seals, and the space between the seals is regularly flashed with steam and condensate. During feeding of liquid or removal of samples from the bioreactor, adequate precautions should be taken to avoid the formation of aerosols. Transfer of liquid between two tanks is carried out through sealed pipelines. As per BL-LS requirements, bioreactors and other tanks should be steam sterilizable, and all liquid effluents should be heat-sterilized prior to discharge to the waste.

5.3 BIOSAFETY CABINET

Biosafety cabinet is the most commonly used clean bench in the laboratory that maintains horizontal (Figure 5.3) or vertical (Figure 5.3) laminar flow of air. In laminar airflow, the entire volume of air within a designated space moves at a uniform velocity (60-90 ft/min) in a single direction along parallelflow lines. These clean benches discharge HEPA-filtered air across the work surface in front of users, and therefore are not called biosafety cabinets (BSCs).

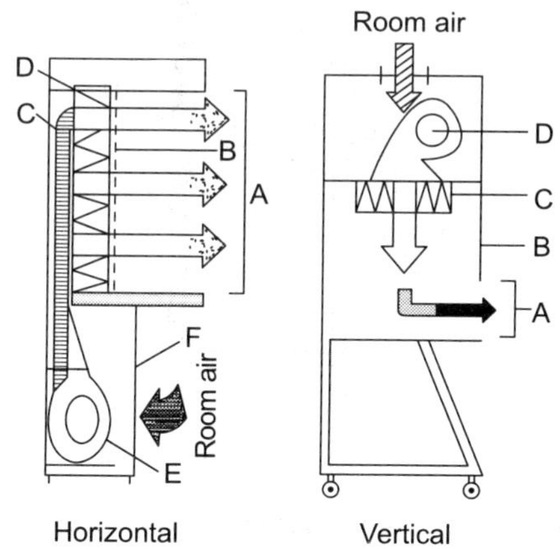

Figure 5.3 Horizontal and Vertical laminar flow cabinet: (A) front opening (contaminated air); (B) supply grille; (C) supply HEPA filter (sterile air); (D) supply plenum; (E) blower; (F) grille.

These types of clean benches only provide product protection. In animal cell culture, they can be used for certain activities like media filtration, dust-free assembly of sterile equipment, etc. These benches are not recommended for cell culture, drug formulations, and manipulating potentially infectious materials.

The biological safety cabinet is a primary barrier that protects the personnel by controlling the aerosols. Selection of BSCs depends on the following:

(a) The hazard classification of the agent
(b) The need for protection of research material or personnel
(c) The extent to which hazardous aerosols are involved. Depending upon the safety levels, BSCs are grouped into Class I, Class II, and Class III (CDC-NIH, 2000).

Class I

The Class I BSC provides protection to personnel and environmental and not to the product. Functionally, it is identical with a chemical fume hood installed with a HEPA filter to protect the environment (Figure 5.4). Here, unfiltered room air is drawn across the work surface at a minimum velocity of 75 ft/min. The potential applications of Class I BSC are housing small equipment (e.g., centrifuge, cell harvester, bioreactor), and conducting certain procedures (e.g., aerating culture/ medium or tissue homogenation), in which chances of aerosol formation is high. The Class I BSC can be connected with building exhaust system, where it exhaust fan provides the static pressure head to draw room air into the cabinet.

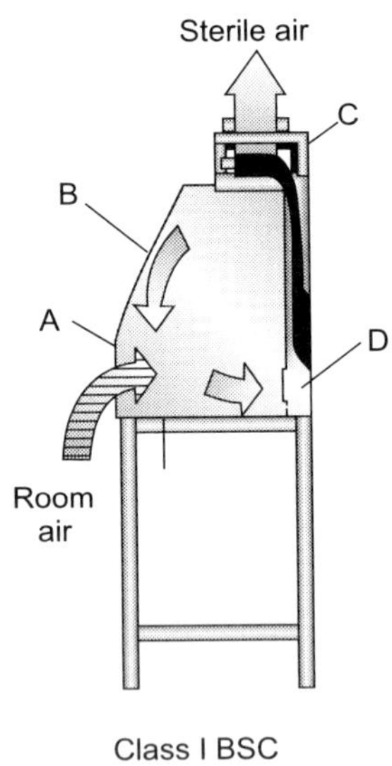

Figure 5.4 Class I biosafety cabinet: (A) front opening; (B) sash; (C) exhaust HEPA filter; (D) supply plenum. Black color arrow indicates the flow path of contaminated air; gray color arrow indicates the flow path of sterile air; hatched arrow indicates flow path of fresh room air inlet. (*Source*: U.S. Department of Health and Human Services, CDC-NIH, 2000)

Class II

The Class II BSCs provide personnel, environmental and product protection. It is further classified into four types (A, B1, B2, and B3) on the basis of strinency of the applications. In Class II cabinet, air is drawn around the personnel into the front grille to protect the users. The downward laminar flow of HEPA-filtered air provides protection to the products. The cabinet air is passed through an exhaust HEPA filter to decontaminate air, which may be recirculated back into the laboratory (Type IIA BSC) or exhausted out of the building (Type IIB BSC). All Class II cabinets are made for the work involving organisms recommended for biosafety levels P1, P2, and P3. The Class II cabinets provide the contaminants-free working environment for cell culture and also may be used for the formulation of drugs. All Class II BSCs have internal blower to handle airflow. The downward laminar flow of sterile air maintains particulate-free environment. The downward air is drwan into two parts, by the blower, one part through the front grille other through the rear grille. In addition, sufficient room air is drawn by the blower through the front grille to maintain an inflow velocity of about 75 ft/min in the cabinet. Finally, blower discharges air to the environment through the exhaust filter. In class II cabinet, approximately 30% of the air passes through the exhaust HEPA filter and 70% is recirculated back into the working zone through the supply HEPA filter (Figure 5.5). Air losses at the exhaust (30%) are compensated by drawing fresh room air though the front panel. The Class IIA BSC is normally directly connected to the building exhaust system through a dedicated exhaust fan.

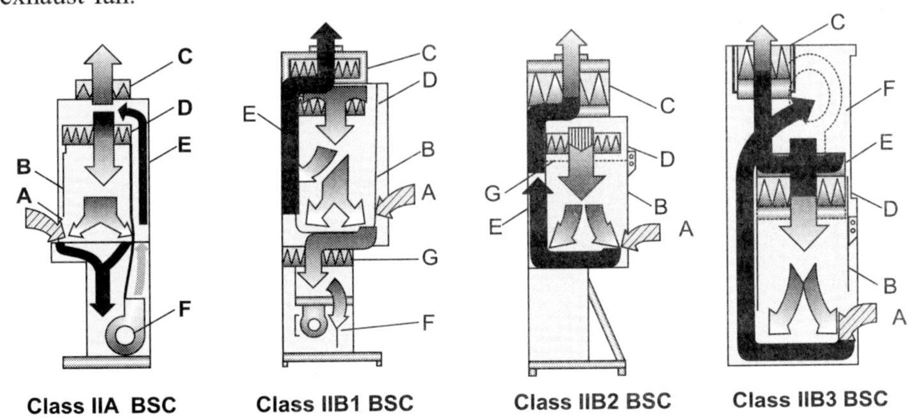

Figure 5.5 Class II biosafety cabinet: (A) front opening; (B) sash; (C) exhaust HEPA filter; (D) supply HEPA filter; (E) negative pressure exhaust plenum; (F) supply blower; (G) filter screen. The carbon filter in the building exhaust is not shown. In Class II B1, B2, and B3 cabinets exhaust should be connected with the building exhaust system. Black color arrow indicates the flow path of contaminated air; gray color arrow indicates the flow path of sterile air; hatched arrow indicates the flow path of fresh room air inlet. [*Source*: U.S. Department of Health and Human Services, CDC-NIH, 2000]

When carcinogens are used in cell or microbial cultures both biological and chemical containment become important. The Class IIB1 BSCs have been designed for manipulation of minute quantities of these hazardous chemicals with *in vitro* biological systems. Like Class IIA, the blower draws room air through the front grille, which gets mixed with a portion of the cabinet's recirculating air and then filtered through the supply HEPA filters. Filtered air flows upward through a plenum at each side of the cabinet and then move downward to the working area (Figure 5.5). In some cabinets an additional HEPA filter is provided to remove particulate matter, generated by the blower/motor system. In this case the room air is drawn through the front of the cabinet at a minimum inflow velocity of 100 ft/min. Like Class IIA cabinet, the down-flow air stream splits into two parts just above the working surface. In Class IIB cabinet, approximately 70% of the down-flow air is exhausted through HEPA filter and discharged out of the building. The remaining 30% of the down-flow air is drawn through the front grille. In this cabinet, customarily all activities involving hazardous chemicals are carried out at the rear of the cabinet. The Class IIB1 cabinets should be connected preferably to a dedicated exhaust system.

The Class IIB2 BSC is a complete exhaust cabinet, where no air is recirculated within it (Figure 5.5). Thus, Class IIB2 cabinet provides simultaneous primary biological and chemical containment. The supply blower draws room or outside air at the top of the cabinet, and blows it through a HEPA filter down into the working area of the cabinet. The building or cabinet exhaust system draws air through both the rear and front grills, capturing the supply air plus the additional amount of room air needed to produce a minimum inflow face velocity of 100 ft/min. The cabinet air is exhausted through the HEPA filter. The Class IIB2 BSC is expensive to operate, as this cabinet is based on 100% exhaust of air.

The Class IIB3 BSC (Figure 5.5) is an exhausted Class IIA cabinet having minimum inward air flow of 100 ft/min. All positive pressure contaminated plenums within the cabinet are surrounded by a negative air pressure plenum, which prevents envirommental contamination due to accidental leakage.

Class III

The Class III BSC (Figure 5.6) has been designed to work with agents that warrant highest order of biosafety level (P4). Class III cabinet provides maximum protection to the working personnel and the surrounding environment. It is a gas-tight enclosure with a non-opening view window on which arm-length rubber gloves are attached in a gas-tight manner. The products inside the cabinet are manipulated through the gloves, thus, direct contact can be avoided. In this cabinet, the materials are transferred through a dunk tank or double-door pass-through box (such as an autoclave), which can be decontaminated. Airflow inside cabinet is maintained by an independent system, installed outside the cabinet, which maintains the cabinet under negative pressure (0.5 inches of water gauge). Depending on the design of the cabinet, the HEPA filtered air become somewhat turbulent in the working area. The exhaust air passes through double HEPA filter, or a HEPA filter and an air incinerator, before being discharged into the environment.

BSCs should be preferably installed with a dedicated duct to exhaust air outside the working space. In cases of Class IIB1, B2, B3 and Class III BSCs, some times in-line charcoal filters are used in the exhaust line. Explosion proof motor and other protective electrical components are recommended for BSCs to be used for handling organic solvents. Table 5.5 describes biosafety levels and applications of different clean air systems.

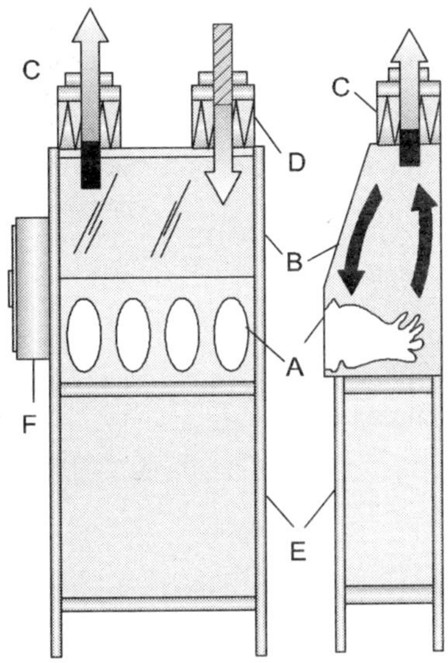

Figure 5.6 Class III biosafety cabinet: (A) glove port with O-ring for attaching arm length gloves to cabinet; (B) sash; (C) exhaust HEPA filter; (D) supply HEPA filter; (E) space to keep dunk tank; (F) double-ended autoclave or pass-through box. The cabinet exhausts should be connected with a dedicatel duct different from building exhaust system. Black color arrow indicates the flow path of contaminated air; gray color arrow indicates the flow path of sterile air exhaust; hatched arrow indicates the flow path of fresh room air inlet. [*Source*: U.S. Department of Health and Human Services, CDC-NIH, 2000]

Table 5.5 Applications of clean air systems

Type	*Containment/ safety*	*Biosafety level*	*Specific applications*
Horizontal/vertical laminar flow hood	Product, but not personnel	Nil	Media filtration, dust-free assembly of sterile equipment
Class I BSC	Environment and personnel, but not product	P1	Handling volatile and nonvolatile toxic chemicals and radiochemical, homogenizing tissues. Running small bioreactor, centrifuge, and harvester.
Class IIA BSC	Personnel, product, and environment	P2/BL2	It can be used for nonvolatile toxic chemicals and radiochemicals, but not for volatile toxic chemicals.
Class IIB1 BSC	Personnel, product, and environment	P2/BL2	It can be used for volatile and nonvolatile toxic chemicals and radiochemicals.
Class IIB2 BSC	Personnel, product, and environment	P2/BL2	It can be used for volatile and nonvolatile toxic chemicals and radiochemicals.
Class IIB3 BSC	Personnel, product, and environment	P3/BL2	It can be used for volatile and nonvolatile toxic chemicals and radiochemicals.
Class III BSC	Personnel, product, and environment; Maximum safety	P4/BL3	It can be used for volatile and nonvolatile toxic chemicals and radiochemicals. Can be used for all types of infectious agents.

5.4 RISK ASSESSMENT

The aim of risk assessment is to anticipating any unwanted incidence to prevent injury to the individual, protect property and avoid harmfule effect to the environment. Risk assessments are to be undertaken prior to start any new activity. It has got two elements:

(1) identifying and evaluating the level of risks; and
(2) defining ways of minimizing or avoiding such risk.

In case of animal cell culture, the level of risk is dependent upon the cell line to be used and whether the cell line is likely to cause any harm (virus infected) to the human. The different classifications of risk are given below:

Low risk

Non human/non primate continuous cell lines (e.g., BHK-21) and some well characterized human diploid lines of finite life span (e.g., MRC-5). Primary continuous cell lines (eg, Vero) also belong to this category.

Medium risk

Poorly characterized animal cell lines.

High risk

(1) Cell lines derived from human/primate tissue or blood, (2) Cell lines with endogenous pathogens, refer to Advisory Committee on Dangerous Pathogens (ACDP) guidelines (1985), and (3) Cell lines used for experimental infection where the categorization is dependent on the infecting agent; refer to ACDP guidelines (1985).

Animal cell lines repository organization (e.g., ATCC, ECACC) recommend minimum containment level for a given cell line based risk assessment. For most cell lines, the appropriate level of containment is Category 2. However, this may need to be increased to Category 3 depending upon the type of manipulations to be carried out and whether large culture volumes are envisaged. For cell lines derived from patients with HIV or HTLV, Category 3 containment is required. Otherwise for most cell lines the laboratory should be designated at least for Category 2 (ECACC recommendation, 1985).

5.5 STERILE FILTRATION

Sterile grade filtration has many applications including sterilization of tissue culture media, buffers, air/gases and decontaminating cell culture supernatants. Animal cell culture is often spoiled by contamination with bacteria, yeast, or mold. The main reasons of contamination are as follows:

(a) Nutritionally fortified cell culture media promote faster growth of the contaminants.
(b) Doubling time of the contaminants are much shorter (bacteria: 20 min, yeast: 30-40 min, mold: 2-3 h) than that of the animal cells (16-36 h). In animal cell culture, contamination normally occurs through media, gases (air, O_2, and CO_2), cell culture equipment, seed culture (cell stock), and breaking of sterile barriers. In this Chapter, we shall discuss on sterile filtration of both liquid (media) and gases.

5.5.1 Air, gas, and vent filtration

Non-liquid sterile filtration is most popular in biotechnology and pharmaceutical industries. Typical applications are bioreactor inlet and exit air filtration, tank venting, sterile filling, pressure fluid transfer, vacuum break in autoclave and freeze drier, containment of airborne dust and powder in spray drier. The primary function of gas filtration is to remove dust particles, bacteria and other microorganisms (e.g., viruses and bacteriophages) from the inlet stream of air/gases to the sterile environment, and also contained cultures (cells and viruses) inside the process equipment. In earlier days, glass wool depth filters were most common in the fermentation industries for sterilization of air. Later, in 80's, hydrophobic membrane filters were introduced for the same purpose.

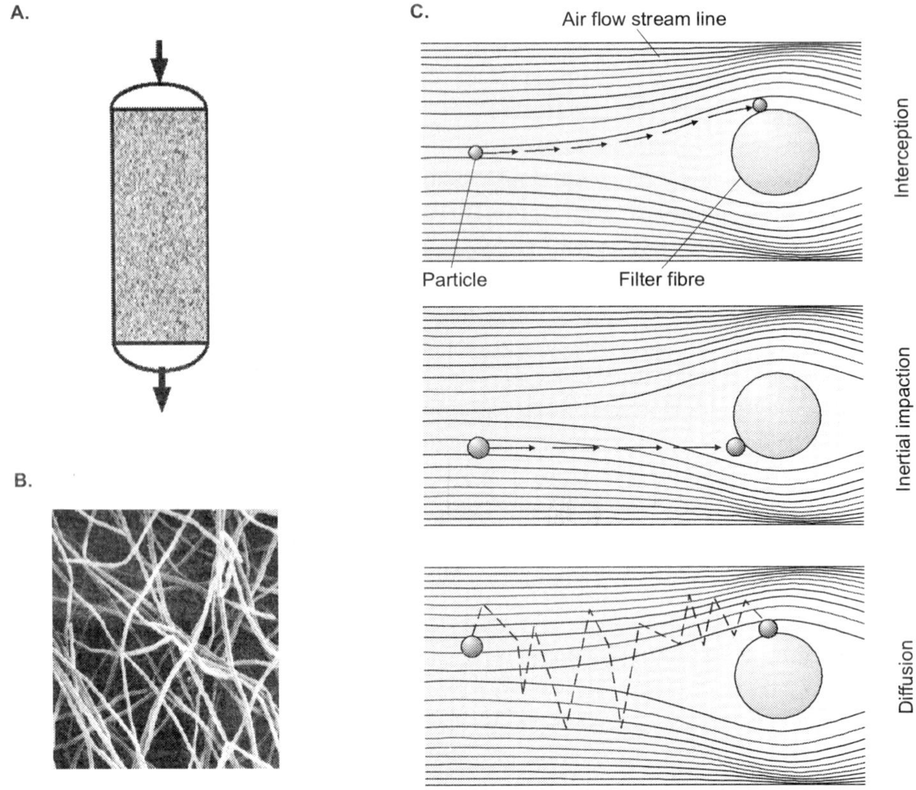

Figure 5.7 Depth filter for sterilization of air. A. depth filter; B. glass wool filter medium; C. particle filtration mechanism: interception, inertial impaction, and diffusion. [*Source:* Ref. 1 & 3]

5.5.1.1 Depth filter

Depth filters are large packed bed (Figure 5.7A) of fibrous filter media made up of glass wool/fibers (Figure 5.7B) of diameter ~18 μm. Dorman (1964) described the mechanisms by which particles are removed by the depth filters. In brief, there are three major mechanisms by which particles are removed. These are as follows:

(1) *Interception*—smaller particles (0.5-5.0 μm) move along the flow path (Figure 5.7C) of air and are finally trapped by direct contact with the fiber.

(2) *Inertial impaction*—one of the efficient methods for collection of larger particles (> 5 μm) from gases. Due to inertia, the particles do not follow the path of the airflow (Figure 5.7C), and instead move directly and ultimately captured by collisions with the fiber.

(3) *Diffusion*—particles smaller than 0.3 μm are subjected to Brownian movement.

Diffusion causes effective diameter of the particle 10 times larger than the actual diameter. The net effect is that the particles diffuse across the flow path of air, increase their chances of collisions with the fiber and are finally collected by interception. The average pore size of the packed bed filter is more than 1 μm; by diffusion, particles smaller than this size can be intercepted. Two other mechanisms, gravitational settling and electrostatic attraction between particles and fibers have some contribution to particle removal, but are less efficient. The collection efficiency of particles by inertial impaction and interception increases with the increase in the size of the particles, whereas reverse is in the case of diffusion. Thus, overall collection efficiency depends on the combined effects of these three mechanisms (Harper, 1985). Besides particle size, the collection efficiency of depth filter also depends on the diameter of the fibrous material, pressure drop across the bed, flow rate of the air, and the dryness in air. Depth filters are sterilized by live steam at 130°C; a good quality glass fiber can be sterilized 10-15 times. After steam sterilization, the depth filter is completely dried by blowing compressed air prior to use in a sterile application.

Air flow rate and hence superficial velocity of air based on empty cross-section of the filter is another factor directly responsible for collection efficiency. The optimum velocity of air in 18-μm glass fiber packed filter is between 20 to 30 cm/s (Aiba and Humphrey, 1973). So, it is important to operate the filter within these range of the air velocity. The presence of condensable moisture in air unexpectedly reduces the collection efficiency of the particles, as diffusion interception does not work under moist conditions. For successful operation of a depth filter, the inlet air to filter must be reasonable dry and preheated so that moisture does not condense in the filter bed. In order to avoid moisture condensation, the depth filter is maintained warm by jacket steam heating.

5.5.1.2 Membrane filter

Membrane filters are recommended for sterilization of gases used in cell culture applications. Hydrophobic filters are most popular for sterile filtration, because water droplets are strongly repelled by the membrane before being carried out with air stream. The primary mechanism for bacteria removal by hydrophobic membrane is based on direct interception. Unlike depth filter, the efficiency of bacteria removal by hydrophobic membrane is consistent over wide operating ranges. Hydrophobic air filters are made of either polytetrafluroethylene (PTFE) or polyvinylidene fluoride (PVDF). Depending upon the nature and the scale of operation, membrane filters of different shapes and sizes are available. Some of the commercially available membrane filters for sterile filtration of air/gas

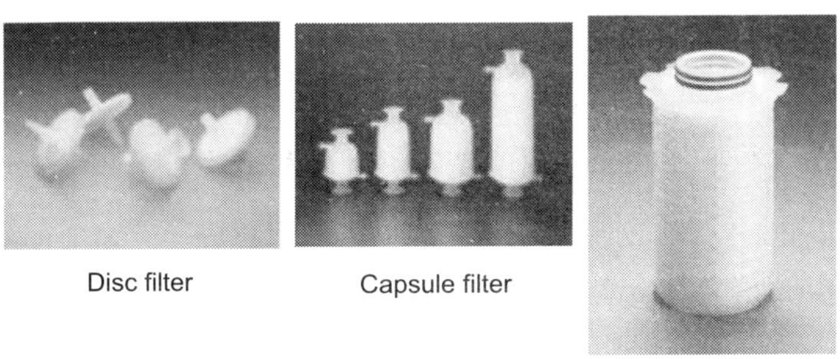

Disc filter

Capsule filter

Cartridge filter

Figure 5.8 PTFE-based hydrophobic membrane filters (pore size 0.2-μm) for sterilization of air. [*Source*: Pall Filtration]

are shown in Figure 5.8. Often, oxygen is directly fed into animal cell culture bioreactor to improve dissolved oxygen level. The membrane filters used to sterilize pure oxygen is different from ordinary air filters in terms of filter design and materials of construction. The membrane filters commonly used for oxygen are exposed to rapid oxidative environment and build-up of static charge and subsequent discharge. As a result, there is a threat of ignition that may damage the filters. Pall Emflow[R] CPRF is recommended for the filtration of pure oxygen, as the filter is made of non-combustible materials, PTFE membrane and polyaramid support layers. In order to control static discharge related failure, it is also recommended to maintain low flow velocities (< 25 m/s) of oxygen for using relatively large surface area of the filter and a prefilter to remove combustible dust particles.

Direct steam sterilization is the most popular choice for sterilization of filters, whether they are used in the laboratory or in the industry. In alternate, γ-radiation (> 25kGy) is used for sterilization of filters, in that case PVDF membranes with special grade of polypropylene is used for support layers. PTFE is unstable in γ-radiation and therefore not recommended for sterilization of the same membrane. In case of sterilization with steam, the life of the filter depends on the quality of the steam, its temperature and pressure, and differential pressure across the membrane. Maximum temperature of 140°C and differential pressure between 300 to 1000 mbar are recommended for Pall PTFE filters. The cumulative steam life of PTFE filters is typically 100-200 h at 140°C. Thus, for a 30 min sterilization cycle, each filter can be used at least for 200 cycles. The scheme for steam-in-place sterilization of a membrane filter is shown in Figure 5.9.

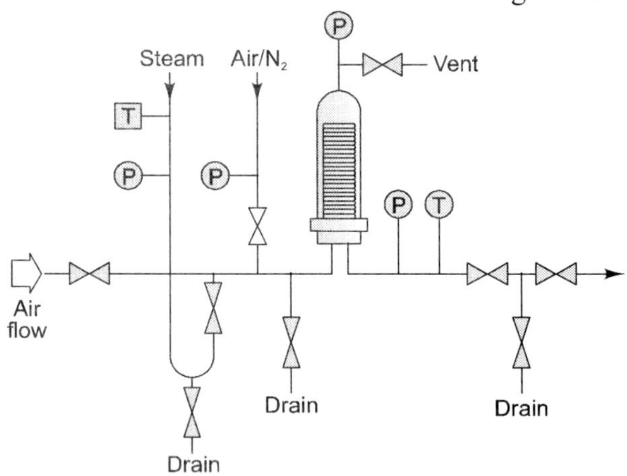

Figure 5.9 Steam-in-pace sterilization of membrane filter. Maximum differential steam pressure across the membrane filter should be within 0.3 to 1.0 Kg/cm^2 (gauge). [*Source*: Dr. Philip Blosse, Pall Life Sciences]

5.5.2 Media filtration

In laboratory, both vacuum and pressure driven media filtration devices are used. However, in large-scale applications only pressure driven media filtration devices are used. Durapore hydrophilic PVDF (Millipore Corporation) based filters can be used in both pressure and vacuum driven filtration. Different symmetric filters used in pressure and vacuum modes are shown in Figure 5.10A. Liquid membrane filters with asymmetric pore structure is also available, where greater throughput can be achieved than ordinary symmetric pore membranes for the same filtration area. The cross-section views of symmetric and asymmetric pore membranes are shown in Figure 5.10B. The asymmetric membranes are more open at the top, where filtration starts, than at the bottom. The mean pore size decreases from the open side to the tight side. This unique architecture of asymmetric membrane extends the useful life of the filters and promotes greater throughput, while 0.1-0.22 μm pores maintain bacterial retention. The asymmetric filters are especially useful for filtering serum or other proteinaceous solutions. The open side acts as

A. B.

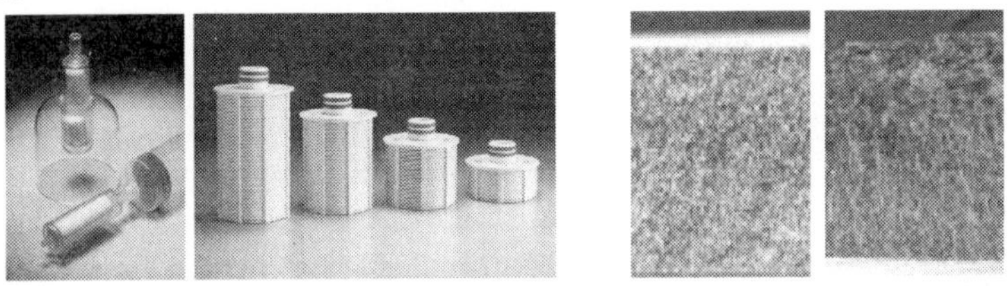

Figure 5.10 Membrane filters for sterilization of medium. **A.** Symmetric Durapore Polyvinylidene fluoride (PVDF^{TM}) membrane (*Source*: Millipore Corporation, Bedford, Massachusetts, USA); **B.** Scanning electron micrographs of cross-section of a traditional microporous (left) symmetric filter (Durapore PVDF) with a uniform pore structure from top to bottom, and a new fast-flow microporous (right) asymmetric membrane (Polyethersulfone, PES). The new PES membrane is more open at the top, where filtration starts.

built-in pre-filter catching large particles before they have a chance to clog the bacteria retentive tight side. The technical specifications of some of the commercially available asymmetric liquid filtration membranes are given in Table 5.6. The suitability of a membrane filter for sterilization of media is based on following criteria:

Table 5.6 Specification of asymmetric membranes

Trade name	*Materials of construction*	*Pore Size (μm)*	*Sterilization*	*In-line sterilization cycle*	*Filtration area (sq.ft)*
Sartopore 2, Sartorius	Polyethersulfone (PES)	0.1	In-line at 134°C for 20 min	25	6-18
Express PES, Millipore	Polyethersulfone (PES)	0.22	NA	NA	NA
SterASSURE, Cuno Incorporated	FlexN Nylon 6,6	0.1-0.2	In-line at 135°C for 20 min	15	1.5-30

NA: Not available

Compatibility assessment The filter cartridge must be both chemically and physically resistant to the process stream and operating conditions. Laboratory tests should be conducted to assess the chemical and pH compatibility.

Product viability study Many pharmaceutical products are bactericidal. This test verifies the survival capacity of the organisms selected for microbial challenge test, in the whole product, for the process time.

Microbial retention study This is to qualify the ability of the filter media to produce sterile filtrate. A solution of the product or equivalent is inoculated with the challenge organism at a concentration of 1×10^7 viable organisms or above per square cm of filter area. The challenge is carried out separately in three filter membranes from different production batches, with at least one being at minimum specification. Products are tested for bacterial endotoxin using the Limulus Amoebocyte Lysate (LAL) test to ensure safety.

Extractables analysis Both quantitative and qualitative information on extracted materials must be generated. Commonly, extractable materials cannot be evaluated directly in the actual product, but a "Model Solvent" approach can be developed. Products are optimized to reduce extractable, ensuring that unwanted materials are not introduced into filtered media.

Evaluation of Adsorptive Effects It is critical that filters are selected on the basis of minimum adsorption and loss of product components. Laboratory scale tests are required to generate adsorption profiles.

Generation of product wet integrity test parameters Integrity tests are carried out on critical filters immediately before and after batch filtration for the confirmation of performance as per specification. In order to evaluate the integrity of the membrane filters, it is recommended to conduct the test in cartridge soaked with the product (www.novatec.co.cr).

5.5.3 Process-specific filter validation

Microporous filters used for various applications warrant required levels of purity for both gases and liquids. The verification of filter performance is identified as an important process monitoring in those applications where product sterilization or microbial bioburden reduction is required. This has led to the development of a number of integrity test procedures to validate the filters. Integrity test offers non-destructive methods for proving the capability of a filter product to meet its stated performance when it is installed for an application. This gives confidence to the filter users of that the installed filter will achieve the required performance. In particular, integrity tests enable users to verify that no filter damage has occurred during storage, installation into the filter housing or following procedures such as chemical cleaning or *in situ* steam sterilization prior to use or during the subsequent manufacturing process. Integrity test methods used for gas filters differ from that in case of liquid filters. Currently, following three main integrity tests are adopted for liquid filters: (www.domnickh.co.uk)

(a) **Bubble Point:** this is the minimum applied differential pressure required to vent the largest pore in a wetted filter membrane.
(b) **Diffusional Flow:** this is the gas flow rate resulting from gas diffusion across a wetted filter media at an applied differential pressure of approximately 80% of the bubble point for that media.

(c) **Pressure Decay**: this is the drop in gas pressure (due to diffusion) measured over time from a sealed volume connected to the upstream side of a wetted filter membrane. This is simply another way of measuring flow due to diffusion. In hydrophobic air/gas filters, in addition to above three, following two more integrity tests are performed:

(d) **Water Intrusion:** this is the volume of water that penetrates (intrudes) into the structure of a hydrophobic media at a given applied pressure (typically held for 10 minutes). In reality, this is a combination of water intruding into the membrane and the evaporation losses caused by the difference in pressures at the liquid/air interface.

(e) **Aerosol Penetration:** this is the percentage aerosol that penetrates to the downstream side of a test filter during a gas-borne aerosol challenge using concentrated sub-micron aerosol.

5.6 ULTRA-PURE WATER FOR CELL CULTURE

In cell culture laboratory, pure water is used for rinsing vessels and glassware, diluting reagents, preparing media, and generating steam. The quality of water required for all four applications is not the same. Distilled or *reversed osmosis* (RO) water is adequate for rinsing and steam purposes. However, for media preparation, ultra-pure grade of water is used. Ultra-pure grade water is produced by combination of both physical and chemical methods of purification. Therefore, the system design is most important in production of ultra-pure water. The final quality of the product water obviously depends on the quality of the purification media used in the system, quality of the materials used to make the hardware of the system through which the water flows or is stored, and last but not least the sanitization schedule followed to clean the water purification system. Water purification media, containers, and other hardwares which are in direct contact with water, should not be made of any component that are extractable in water. The quality of ultra-pure water should be regularly monitored on the basis of the following few parameters:

***A. Suspended matter and microorganisms*:** The water should be free from suspended particles and any form of living matter.

***B. Resistivity*:** This parameter measures the ionic contamination level present in water, and offers no information regarding other contaminants such as microorganisms, particles, and perhaps more importantly, organic substances. The resistivity of ultra-pure water above 18 MΩ.cm at 25°C is considered suitable for making cell culture media.

***C. Total organic carbon (TOC)*:** This parameter measures combine organic impurities in water. The total permissible loading of TOC is ultra-pure water is 10 ppb (μg/l).

***D. Pyrogen*:** The presence of pyrogen in ultra-pure water can affect in many animal cells. The level of pyrogen is measured in terms of endotoxin by LAL assay.

Since no single purification method generates ultra-pure grade of water, the commercially available systems for generating tissue culture grade water are based on combination of different methods normally used to purify water.

Millipore ultra-pure grade of water is generated in successive stages, in which impurities are depleted step by step. The basic scheme for obtaining ultra-pure water is shown in Figure 5.11A. In first stage, feed water is subjected to micro and active carbon filtrations to remove particulate matters and organic contaminants, respectively. Particulate matters are removed by the principle of sieving when water is passed through multilayer synthetic polymeric filter cloths made of either decron or nylon. The active carbon filter removes organic contaminants by the principle of adsorption. Naturally occurring activated carbon from wood or coconut shell is used in this stage. These materials allow slow adsorption of low molecular weight organics; however, some ionic contaminants are released in the product water. The main purpose of pretreatment is to remove the major load of suspended particles and organic impurities from the feed water.

The second stage of treatment removes soluble salts present in the pretreated water. The most popular method used to remove salts is *reverse osmosis* (RO). Here, water at a high pressure (> 5 kg/cm^2) is pumped through the membrane filter designed to reject 95-99% NaCl. At the end of second stage of purification, water become highly pure, but not enough for use in cell culture media preparation. Purified RO-water is suitable for rinsing culture vessels/equipment and to generate ultra-pure steam for sterilizing bioreactors and other cell culture equipment. Synthetic activated carbon is used in the third stage of water treatment, in which remaining low molecular weight organics are rapidly adsorbed. Millipore uses synthetic activated carbon made from controlled pyrolysis of styrene beads. It has the same high porosity and surface area as natural activated carbon. The residual ionic substances that escape the third stage of treatment are removed with the help of *ion exchange resins*. Ion exchange resin retains both cations and anions by electrostatic forces. The final stage of treatment in generating ultra-pure water is *ultra-filtration*. Asymmetrical pore (lowest pore size 0.22 μm) membrane filters made of PVDF material are used in this stage. The membrane retains particles and bacteria larger than the minimum pore size on their surface by sieving mechanism. A laboratory scale ultra-pure water generating system is shown in Figure 5.11B.

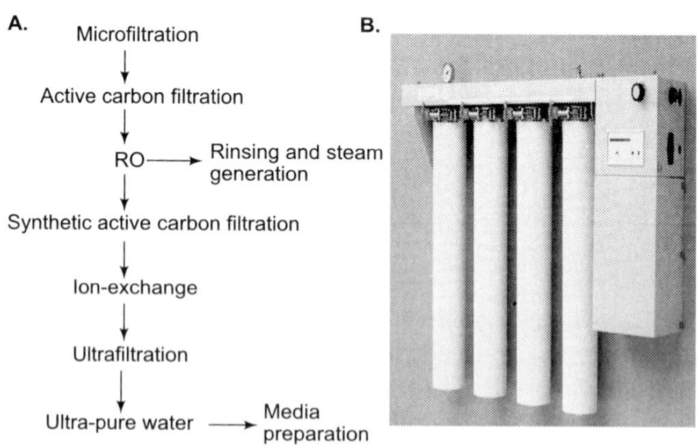

Figure 5.11 Ultra-pure water generator. **A.** general scheme for Millipore ultra - pure water generating system; **B.** ultra-pure water generator. [*Source*: Millipore Corporation, Bedford, Massachusetts, USA]

5.6.1 Maintenance of ultra-pure water generating system

Ultra-pure water generator has a system capacity, beyond which the performance of the filtration device is compromised. It is recommended to replace filter cartridges and membranes before exceed individual capacity. Nevertheless, schedule maintenance of the membrane filters and cartridges are important to generate certified grade of water. Inside of purification system, bacteria tend to grow on accumulated organic impurities in filters and later in water, resulting in pyrogen contamination even after fourth stage of purification. Ideally, the purification media should not be left (filled with water) in unused condition for more than 3 months. Regular cleaning of the system using different cleaning agents is recommended for optimum performance of the water purification system. It is also recommended to carry out routine microbiological and endotoxin tests on the product water.

5.7 BASIC EQUIPMENT

An exhaustive list of the cell culture equipment is shown in Table 5.3. The most common equipment used in animal cell culture and their maintenance are described below.

5.7.1 Centrifuge

Centrifuges are routinely used in animal cell culture as a part of subculturing of cells. Centrifuge generates aerosols when in operation, which increases the risk of infection or contamination. It is preferable to select centrifuge buckets, which can be sealed to avoid the formation of aerosol. Ideally, the centrifuge should have a window in the lid so that the condition of the materials can be observed without opening the lid. This will reduce the risk of the operator being exposed to hazardous materials, in case centrifuge tube is broken during operation. Care should be taken not to over-fill the tubes and to balance them accurately. These few simple steps will reduce the risk of aerosol formation. For convenient of handling, centrifuges are preferably located near the BSCs. Swing-out bucket refrigerated centrifuges are most popular in cell culture due to the nature of the applications. Centrifuge could be either table-top or floor-top model, but its site of installation should be such that accessibility for cleaning and maintenance is not difficult (Figure 5.12).

5.7.2 Inverted microscope

As the name suggests, an inverted microscope is upside down compared to a conventional up-right light microscope. In inverted microscope, the light source and the condenser are on the top above the stage pointing down. The objectives and turret are below the stage pointing up. The only characteristics that are similar to conventional microscope are: (a) a specimen is placed on top of the stage (generally kept in inverted position), (b) the binocular or trinocular tube is not upside down but in the standard position pointing at a conventional viewing angle. Inverted microscopes are useful for observing living cells or organisms at the bottom of a large container (e.g., tissue culture plates) under more natural conditions than on a glass slide, as is the case with a conventional microscope. Phase contrast microscope offers superior imaging of multi-well plates and cultures. Inverted microscope could be for bright field as well as in fluorescence field, and image can be for recorded through a digital camera. A popular model of inverted microscope (Olympus IX51) is shown in Figure 5.12.

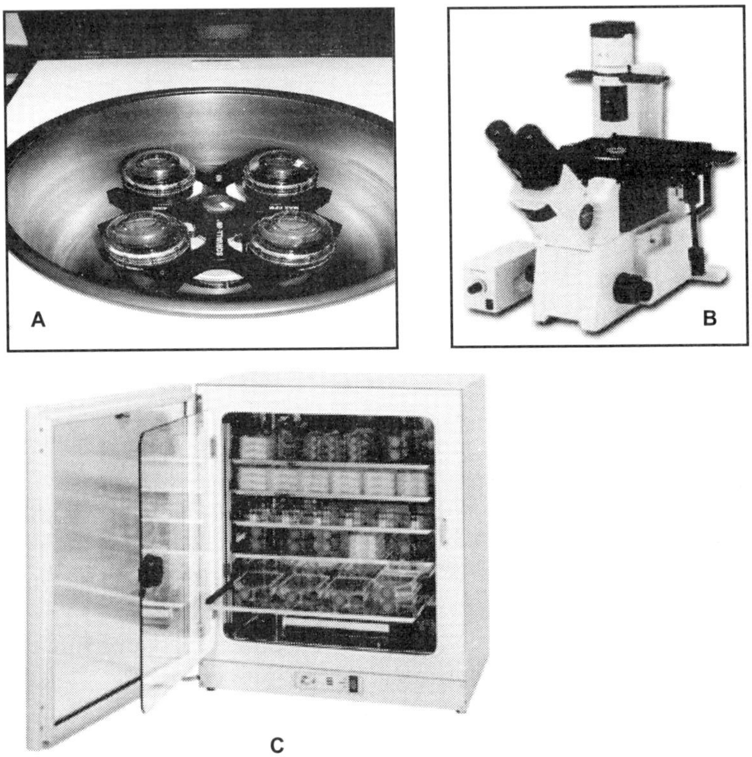

Figure 5.12 Common cell culture equipment. **A.** swing bucket refrigerated centrifuge (table top); **B.** Inverted phase contrast microscope; **C.** CO_2 incubator.

5.7.3 CO_2 incubator

Animal cells require a strictly controlled environment in which they can grow. CO_2 incubators are routinely used to provide the correct growth conditions such as temperature, humidity, and CO_2 levels in a controlled and stable manner (Figure 5.12). The incubators can be set to run at temperatures 5°C above the ambient temperature to 37°C or more, and can be set to provide CO_2 at the required levels (e.g., 5–10%). In a standard CO_2 incubator, the O_2 level in the air is assumed to be 20%. However, certain specialized cells require lower levels of O_2 hypoxic, which is controlled by flowing pure N_2 and O_2 in the incubator. The CO_2 level is controlled on the basis of two properties of CO_2 gas—infrared adsorption and thermal conductivity. Since, CO_2 incubators are one of the major sources of contamination in cell culture, various modified versions of the incubators are available now-a-days. Copper-coated incubators are available to reduce the risk of microbial contamination due to the inhibitory activity of copper metal. The inclusion of water bath treatment fluid in the incubator water may also reduce the risk of bacterial and fungal growth in the water trays. However, there is no substitute for routine cleaning of the incubators. *In situ* sterilizable incubators are available, where circulating hot air through the jacket ensures sterilization and temperature control. In general, CO_2 incubators are tabletop model, but floor-top models of different capacities are also available.

5.8 CULTURE VESSELS

Tissue culture plastic wares of wide capacities and different shapes are available from a number of companies. Plastic wares are generally available in three formats: Petri dish, flask, and well plate. Plastic wares are made of polycarbonate and available in pre-sterilized (γ-irradiated) form. The sizes or the capacities of tissue culture plastic wares are on the basis of available surface areas. For example, tissue culture flasks (T-flasks) are available in three different sizes (e.g., 25, 75, 150 cm^2). The photographs of commonly used plastic wares are shown in Figure 5.13. Some glass bottles, e.g., Roux bottles are often used in cell culture. It should be kept in mind that there is no special culture vessel available for growing anchorage dependent cells. Certain specialized cells (e.g., hepatocytes, neurons, endothelial cells, etc.) do not grow on ordinary plastic wares, hence the culture surface is coated with extracellular matrix components like collagen, gelatin, fibronectin, laminin, or poly-lysine.

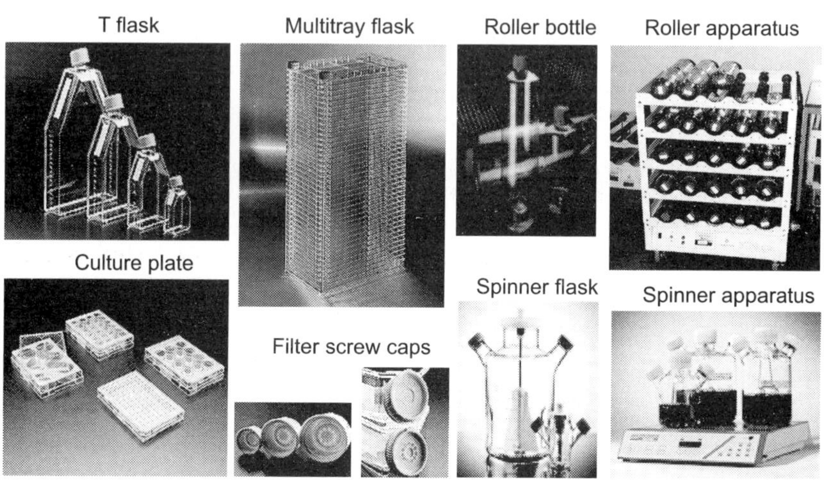

Figure 5.13 Cell culture plastic wares and culture vessels. [*Sources*: Nalge Nunc International, USA; Corning Life Sciences, USA]

For large-scale culture of both anchorage-dependent and independent cells, roller bottles are commonly used. Roller bottles are made up of either glass or plastic with improved optical properties to examine cells under inverted microscope. In roller bottles, cells grow on the inner surface and are continuously bathed in medium as the bottle rotates at low speed (0.5–5 rpm). The apparatus used to rotate the roller bottle is known as roller apparatus (Figure 5.13). A typical roller apparatus can accommodate 10–50 bottles of varying sizes (700 to 1750 cm^2 surface areas). Roller bottles are also available in corrugated surface, which increases growth area by several folds. Roller bottles are associated with one major problem of uneven growth of cells. Spinner flask apparatus are often used for homogeneous suspension culture or both anchorage-dependent and independent cells in which the cell suspension is slowly mixed (10–40 rpm) with the help of a magnetically driven agitator.

Nunc and Corning have introduced further large-scale cell culture system, commonly known as cell factory. It is multi-tray unit sealed together to give surface area for cell growth as high as 6000 cm^2. It is a static culture system; no rolling device is required like roller bottles (Figure 5.13). Culture startup and harvesting of cells from cell factory is relatively difficult than roller bottles and other plastic wares, primarily due to the nature of the connecting ports. The cell factory produces extremely high cell yields using little working space of the laboratory.

5.9 SUMMARY

1. GMP regulation ensures controlled manufacturing of biological/drugs, which are safe and effective for human use.
2. GMP includes design of the manufacturing facility and equipment, SOP, documentation and labeling, training of manpower, process control and validation, auditing and error management.
3. Animal cell culture facility is maintained in Class 10,000 clean room area.
4. Animal cell culture process is susceptible for contamination with bacteria, yeast, mould, and mycoplasma.
5. Animal cell culture (open manipulation) is carried out in Class IIA or IIB1 biosafety cabinet.
6. Air and cell culture media is sterilized in 0.1/0.22-μm membrane filters; asymmetric membranes are preferred for media sterilization.
7. Ultra-pure water is used for preparing cell culture media: water should be free from pyrogen, suspended matters and organisms. The permissible labels of salt and total organic carbon are 18 mΩ.cm and 10 ppb, respectively.

ABBREVIATIONS

ACDP	Advisory Committee on Dangerous Pathogens
ATCC	American type culture collection
BSC	Biosafety cabinet
BL	Biosafety level
cGMP	Current good manufacturing practice
SOP	Standard operating procedure
DAPI	4',6-diamidino-2-phenylindole
ELISA	Enzyme-linked immnunosorbent assay
ECACC	European Collection of Cell Cultures
FDA	Food and Drug Administration
HEPA	High efficiency particulate air
HPLC	High pressure liquid chromatography
HVAC	Heating-ventilation-air conditioning
HIV	Human immunodeficiency virus
LAL	Limulus amoebocyte lysate
PTFE	Polytetrafluroethylene
PVDF	Polyvinylidene fluoride
PES	Polyethersulfone
RO	Reverse osmosis
TOC	Total organic carbon

REFERENCES

1. Aiba S, Humphrey AE and Millis NF (1973). *Biochemical Engineering*, 2nd edition, p. 137. University of Tokyo Press, Japan.
2. Advisory Committee on Dangerous Pathogens (ACDP) guidelines, 1985.
3. Dorman RG (1964) In: *High-efficiency of air filtration*. (PAF White, SE Smith eds.) p. 67. Butterworth, London.
4. Harper GJ (1985) In: *Animal Cell Biotechnology* (RE Spier and JB Griffiths, eds.) vol. 1, p. 141. Academic Press, London.
5. Primary Containment for Biohazards: Selection, Installation and Use of Biological Safety Cabinets (2nd edition). U.S. Department of Health and Human Services, CDC-NIH, 2000.

SUGGESTION FOR FURTHER READING

1. Blosse P (2005) *Sterile Filtration of Gases*, Pall Life Sciences, Report-June.
2. Cardona M and Blosse P (2005) *Guidelines for selecting sterilizing-grade filters*. CEP. Pall Corporation, p. 34.
3. Chiu YH and Gueriguian JL (1991) *Drug Biotechnology Regulation- Scientific Basis and Practices*. Marcel Dekker, Inc., New York.
4. Freshney RI (1987) *Culture of animal cells. A manual of basic technique*. 2nd edition, p. 207. Alan R. Liss, Inc. New York.
5. Sinclair A and Monge M (2004) *BioProcess International* October (supplement) p. 26.
6. Pharmaceutical inspection convention- PIC/S. (2004). *Validation of aseptic processes*. PI 007- 2, July.
7. US Department of Health and Human Services, NIH (1986). *Guidelines for research involving recombinant DNA molecules*. Fed. Reg. 51 (88) May 7. FR Doc. 86-10120.
8. Wiling SH and Stoker JR (1997) *Good manufacturing practices for pharmaceuticals* by, 4th edition. Marcel Dekker, Inc. New York.

6

Environmental Factors and Cell Culture Methods

The commercial exploitation of animal cells initiated subsequently acquiring techniques for the cultivation of viruses. Besides viral vaccines, many therapeutic products and reagents are manufactured using animal cells. The major difficulties in animal cell culture-based processes are:

(i) stringent growth requirements by the cells;
(ii) requirement of solid support for growth;
(iii) culture is prone to microbial contamination; and
(iv) low yield of product.

This Chapter describes relation of animal cells with their environment, basic cell culture techniques, and different types of bioreactor used for the culture of animal cells.

6.1 ANIMAL CELLS AND THE CULTURE ENVIRONMENT

Dynamic interactions of animal cells with the environment take place at various levels. Some of them are discussed below:

1. ***Extracellular environment***: This includes interaction of a cell with other cells, with substratum (culture surface), and in response to physical (e.g., shear force, temperature, etc.) and chemical (e.g., pH, growth factors, nutrients, etc.) stimuli.
2. ***Cellular-molecular status***: This includes the type of cells, cytoskeleton structure, lipid bilayer, position of cell in cell cycle, cell signaling cascade, and cell detachment (in case of anchorage dependent cells).
3. ***Metabolic status*:** Uptake of nutrients and removal of toxic metabolites.

Animal cells either differentiate or multiply and hence require various signal inputs mediated by adjoining or distance cells. These signals regulate and maintain the expression of various genes required for cellular processes. The input cell signaling molecules (ligands) referred here are the soluble paracrine hormones or growth factors. The insoluble cell surface molecules (certain cell adhesion molecules) are also involved in cell-cell interaction process. Another important group of cell signaling molecules (e.g., integrin) are crucial for many primary cells to control differentiation and to maintain viability by interacting with the extracellular matrix (ECM) components produced by them. The adhesion molecules present on the cell surface interact with the counter ligands (e.g., ECM components) and help them to remain attached with the substratum. Some of the important cell adhesion molecules are heterodimers of integrin β1 subunit with various integrin α subunits, CD44, and LFA-1. The ECM components normally attached with these receptors are collagen (type I, IV, VI), fibronectin, laminin, heparin sulphate proteoglycans, etc.

The effect of both shear and temperature on cells needs better understanding. The flow pattern of fluid may alter cellular functions due to excessive shear force imposed on them. Animal cells are sensitive to shear force; they cannot withstand shear exerted by turbulent flow of the medium. The basal culture media is considered as low viscous Newtonian fluid. Addition of serum and/or polymers increases the viscosity of the medium, and thus minimizes the *Kolmogoroff length* to protect cells from turbulent shear. Cells are also exposed to shear stress when they collide with solid objects of the bioreactor. Another important physical factor that governs cell growth and its productivity is temperature. Animal cells are extremely sensitive to the fluctuation of culture temperature; temperature shock of few degrees above the optimum may result in the expression of heat shock proteins at the expense of desired protein products.

Biological are manufactured in both mortal (primary/cell strain) and immortal (cell line) animal cells. It is important to understand their cellular properties in terms of (a) whether the cells are anchorage-independent (suspension type) or anchorage-dependent, (b) whether contact inhibition or density dependent inhibition affects cell growth, (c) whether culture media require only serum supplementation, or other growth factor supplementation is necessary, and (d) physiological state of cells with reference to product formation. In view of practical bioprocess, the two most important physiological states of animal cells are growth and maintenance. Animal cells require two different culture conditions in growth and maintenance states. Most interestingly, the majority of therapeutic proteins are produced at very high specific rates in the maintenance state. It is needless to mention here that simplified media are required in the maintenance state. In maintenance state animal cells attain at highest density level where the population growth rate is found to be low and the cells gradually approach to G_1/G_0 state.

6.2 ENVIRONMENTAL PARAMETERS

The culture environment influence the growth and metabolism of animal cells. Although the magnitude of effect may vary from cell to cell, the responses are more or less similar.

6.2.1 Concentration of carbon source

The distribution of carbon source for synthesis of various cellular constituents is governed by the glucose consumption rate. It has been found that if the initial glucose concentration in the culture medium for rat hepatoma is maintained below 0.5 mM, about 50% of the carbon is incorporated in the synthesis of nucleic acids. At 10-fold increase in glucose concentration, 90% of it is converted into lactic acid (Renner *et al.*, 1972). Similar observation has been reported in case of FS-4 cells (Glacken *et al.*, 1986). Again, glucose and glutamine consumption by animal cells is reciprocally controlled by their respective concentrations. It means, if the glucose concentration is low, glutamine is rapidly consumed and vise-versa. It has been noted that specific oxygen consumption rate in mesenteric rat lymphocytes is increased by 75% in glutamine-supplemented culture (Nakashima *et al.* 1984). In batch culture, high initial glutamine concentration (up to 10 mM) leads to an increase in cell number (Butler and Spier, 1984). The specific glucose consumption rates in tumor cells differ from that in case of primary cells. Above certain level of glucose, its specific consumption rates is tumor cells is dramatically increased.

6.2.2 pH

The buffering systems used in animal cell culture media have been elaborated in Chapter 4. In general, the optimum pH of the culture media is 7.4, although many hybridoma lines grow well and secrete more monoclonal antibodies at pH 7.0 or below. The death rate of anoxic CHO cells is greatly increased by lowering pH from neutral range to 6.5 (Haveman and Hahn, 1981). In a closed culture vessel, maintaining of adequate headspace volume is most important, as the pH of the medium is adjusted by CO_2 of the headspace. In a closed culture system, typically 5 to 10-fold greater headspace than medium volume are recommended. As cells grow and generate CO_2 maintaining pH of the medium becomes simpler.

6.2.3 Dissolved oxygen concentration

Like any other aerobic organism, dissolved oxygen level is most critical in animal cell culture process. Most often, dissolved oxygen concentration becomes the growth-limiting factor in high cell density culture, primarily due to its poor solubility and low diffusion coefficient in medium. Oxygen is required for energy generation via oxidative phosphorylation. It also used in the synthesis of cellular components. Oxygen regulates the functional activities of a variety of cell types. For example, oxygen level increases erythropoiesis in bone marrow cells. The specific oxygen consumption rate varies from cell type to cell type (Tables 6.2 and 6.3). The optimum partial pressure of oxygen (pO_2) in a batch bioreactor for growth may vary with the cell type, but in general it is close to 80 mm Hg pressure (Thomas, 1986). Again, the optimum pO_2 for growth may not be the same as that for the product formation (Miller *et al.*, 1987). Further, high-level of oxygen is found to be toxic to the cells, and the oxidative damage results in DNA degradation, lipid peroxidation, polysaccharide depolymerization, and hydroxylation of aromatic compounds (Fee, 1981).

6.2.4 Waste metabolite concentration

Lactate and ammonia are two major metabolic waste products of animal cells. In a typical batch culture, depending upon the media composition, the lactate and ammonia concentrations may build up to 35 mM and 5 mM, respectively. The inhibitory effects of these metabolites differ from cell to cell. The growth inhibitory effect of ammonia is more pronounced than lactate.

6.2.5 Shear effects

Animal cells are more sensitive to shear due to lack of cell wall. Shear has effects on both growth and metabolism of animal cells. The mechanisms for hydrodynamic cell death in microcarrier cultures have been discussed in this Chapter. The extent of shear effect on cells depend on the culture system (bioreactor) used, cell type, and the nature of the culture (free suspension or microcarriers based). In general, free suspension cells are less prone to shear damage than adhered cells growing on microcarrier beads, which has been discussed later in this Chapter.

6.3 SUSPENSION VERSUS ADHERED CELL CULTURE

On the basis of the type of culture, animal cells are divided into two classes: adhered or anchorage-dependent cells (require solid surface to grow) and suspension or anchorage-independent cells (grow without solid surface). Except blood, all other tissue-originated primary cells do not grow in culture unless physical support is provided to them. Even for blood cells, at various stages of maturation they anchor with the ECM components or with the other cells; whether it is in the spleen or in lymph node, or in the bone marrow. It is believed that *oncogenic* transformation leads to the loss of anchorage dependency in many primary cells, although in most cases solid surface helps in growth of transformed cells. The merits and demerits of both the culture types are explained in Table 6.1. It may be noted that problems associated with the monolayer culture of adhered cells can be avoided by adopting microcarrier culture technique. Since suspension cells can be easily grown and maintained in culture, it has not been separately discussed.

Table 6.1 Merits and demerits of suspension and adhered cell culture

Properties	*Suspension cells*	*Adhered cells*	*Remarks*
1. Handling	Convenient	Difficult	Adhered cells require trypsinization to remove culture from solid surface
2. Automation and monitoring	Easy	Difficult	Automation and growth monitoring is difficult due to heterogeneous nature of adhered cells in culture
3. Scale up	Easy	Difficult	Volumetric scale up is difficult in case of adhered cells; require more space to grow
4. Medium change culture	Difficult	Convenient	Medium change is simple in adhered culture. In suspension culture any manipulation requires centrifugation to separate cells from medium
5. High cell density to	Difficult culture	Convenient	Adoptable in adhered culture due to easy perfusion of cells; washing of cells is a problem in free suspension culture
6. Operation cost	Less	Expensive	Cell growth support matrices are expensive
7. Biocompatibility	Less	More	Cells require adaptation to grow in suspension, which may reduce the product yield

6.4 ADHESION OF CELLS TO CULTURE SURFACE

Adhesion of cells to solid surface is considered a biphasic process. In the first phase, physical contact between a cell and the activated surface is taken place, which subsequently leads to the attachment of the cell. This is also known as phase of nucleation by which free cells tend to accumulate on the solid surface. The attached cells spread in the second phase, which is also known as the phase of extension. In this phase, major rearrangement of cytoskeleton elements occurs, which lead to gross change in cell shape and projection of surface receptors (Geiger *et al*., 1992). The initial attachment of cells results from electrostatic attraction and van der Waals forces between cells and the solid surface. Animal cells membrane possess a net negative charge, the extent of which is dependent on the physiological state of the cells. Thus, cells get attracted towards positively charged surfaces. The traditional glass and plastic culture surfaces are negatively charged, so for cells' attachment, binding of negative change with divalent cations (Ca^{+2}, Mg^{+2}) is most important. It has been shown that the attachment and spread of cells on growth surface depend on the density of the surface charge, rather than the polarity. Irrespective of the same polarity, the attachment proteins (e.g., fibronectin, laminin, etc.) and Ca^{+2}, Mg^{+2} ions of serum-supplemented media bridge the contacts between the cells and growth surface. Serum supplemented (10%) culture media contains enough fibronectin, which instantaneously adsorbs to activate the solid surface for attachment and spread of cells. Multivalent heparin sulphate proteoglycans also mediate the adhesion of cells to solid surface by co-ordinate binding to glycoproteins on the cell surface and the glycoprotein (fibronectin) adsorbed on the solid surface. Other serum proteins adsorbed on the solid surface are found to increase its wettability. The solid surface (plastic culture plate) used for the culture of cells are hydrophobic. So, increasing wettability (hydrophilicity) of the surface improves adhesion of cells. Serum contains cold insoluble globulin (CIG), which is an attachment glycoprotein that improves wettability of the surface. Certain cells secrete attachment glycoproteins, while they are growing. The attachment of cells to the solid surface is found to be specific in certain cases. Different materials (solid substrata) have been used as support culture of animal cells are alum-borosilicate glass, polystyrene, polycarbonate, cellulose acetate, polyethylene terepthalate, stainless steel, and titanium.

6.5 SOLID MICROCARRIERS

The concept of microcarrier culture originated from the principle of *unit culture* system. For the first time, the anchorage-dependent cells were cultivated in suspension by van Wezel (1967). DEAE Sephadex A-50 anion exchange beads (exchange capacity ~ 3.5 meq/gm) were used as solid support to grow the cells. Ten years later, Levine *et al.* (1977) showed significant improvement in *microcarrier* culture by using low charge density beads (< 2 meq/gm). Since then, wide varieties of microcarrier beads have been introduced to culture animal cells in suspension. Following criteria are used for evaluating a new microcarrier for anchorage-dependent cell culture process.

1. ***Surface properties***— An even surface with specific degree of curvature allows adhesion, spread and proliferation of cells. The materials used for manufacturing microcarriers should be nontoxic and biocompatible. The base material should be such that the growth of cells can be examined by standard microscopic observation.
2. ***Size and density***— A narrow size distribution is preferred so that the degree of confluence achieved in each microcarrier bead is approximately similar. Best growth of cells is obtained when size distribution varies between 100 and 230 μm. The density of microcarrier beads is maintained slightly higher than the culture medium, so that they can be easily separated by gravity settling in unstirred culture. The density of the microcarrier beads also should be sufficiently low to allow uniform suspension by gentle stirring. The optimum density varies between 1.03 to 1.045 g/cm^3.
3. ***Rigidity***— Microcarrier particle should be rigid enough to withstand collisions between impellers, vessel wall, and within particles.

On the basis of above criteria, three different Cytodex microcarrier beads (Pharmacia, Uppsala, Sweden) were developed, which differ in surface properties. The Cytodex-1 microcarrier is based on cross-linked dextran matrix, substituted with positively charged N,N-diethylaminoethyl (DEAE) groups. The charged groups are present throughout the matrix, on the surface as well as inside. The Cytodex-2 microcarrier is formed by substituting a cross-linked dextran matrix with only a surface layer of positively charged N,N,N-trimethyl-2-hydroxyaminopropyl groups. The total charge capacity in Cytodex-2 microcarriers is lower than Cytodex-1. The Cytodex-3 microcarrier is developed based on the aspect of biocompatibility for culturing specialized cells. It is uncharged and consists of a surface layer of denatured collagen covalently linked to the matrix of cross-linked dextran. The schematic presentation of the three alternate Cytodex microcarriers and their physical properties are shown in Figure 6.1 and Table 6.2, respectively. Besides Cytodex brand, many other types/brands of microcarriers are commercially available, some of the popular products are HyQSpheres (Hyclone), Superbeads (Flow Lab.), CellCarrier (Daymoon), CultiSpher G (Percell, Sweden), etc. The photomicrographs of different animal cells growing on some of these microcarriers are shown in Figure 6.2.

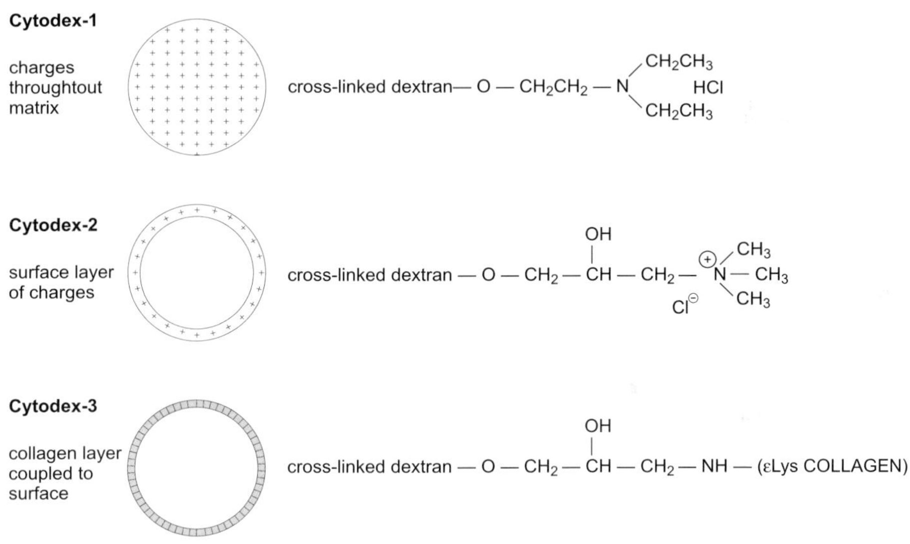

Figure 6.1 Surface chemistry of three alternate forms of Cytodex microcarriers

Table 6.2 Physical properties of Cytodex microcariers

Physical properties	*Cytodex-1*	*Cytodex-2*	*Cytodex-3*
Density (g/cm^3) in 0.9% NaCl	1.03	1.04	1.04
Size $d_{5\text{-}95}$ (μm), the range between the diameter at 5% and 95% of the volume of a sample microcarriers in 0.9% NaCl	131–220	114–198	133–215
Area (cm^2/g dry weight) in 0.9% NaCl	~6,000	~5,500	~4,600
Number of microcarriers/g dry weight	$\sim 6.8 \times 10^6$	$\sim 5.8 \times 10^6$	$\sim 4.0 \times 10^6$
Swelling index in 0.9% NaCl (ml/g dry weight)	18	15	14

Source: Microcarrier cell culture– principles and methods. Pharmacia, Uppsala, Sweden, 1986, p. 18.

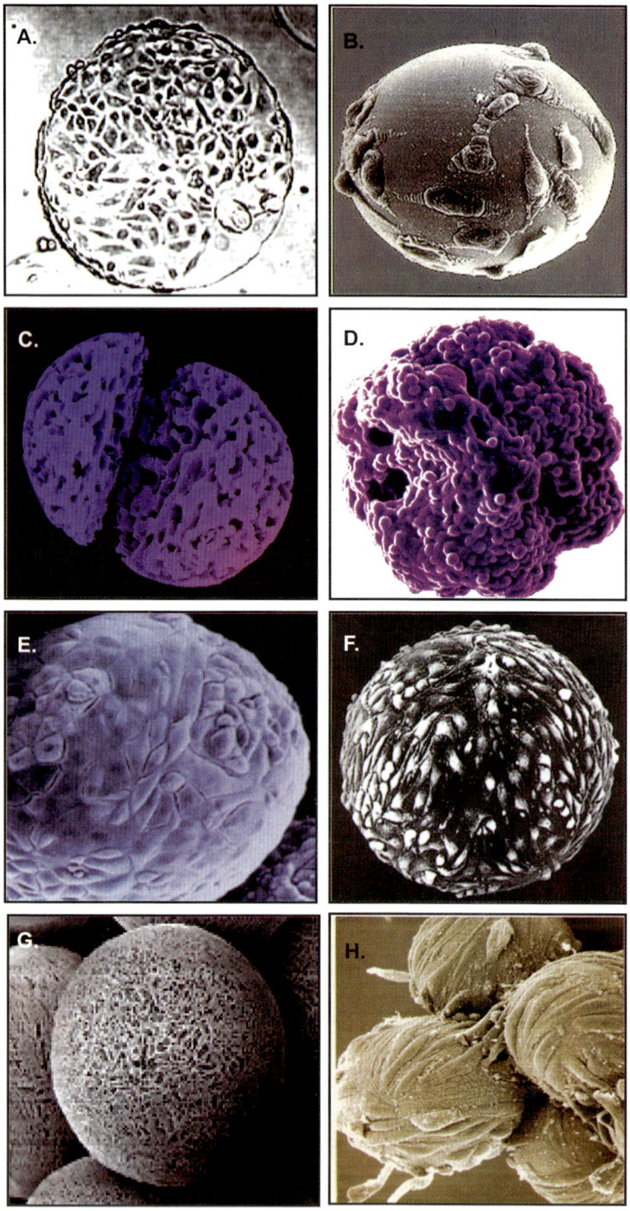

Figure 6.2 Growth of animal cells on microcarrier particles. A. Vero cells on Cytodex-1 microcarriers; B. Fibroblast cells on Cytodex-3 microcarriers (Pharmacia); C. Porous gelatin microcarrier CultiSpher; D. CHO cells on porous gelatin microcarriers [Reproduction by permission of Dr. Kjell Nilsson, Perall Biolytica, Sweden]; E. Epithelial cells on collagen-coated HyQSpheres microcarriers [Hyclone, Logan, UT]; F. Animal cells on Nunc Biosilon polystryene beads (Fisher Scientific); G. Porous biodegradable microcarriers [*Source*: JM MacDonald, BME, University of North Carolina at Chapel Hills]; H. EMS cells on CellCarrier: high silica glass coated cross-linked polystyrene microcarriers [Daymoon, CA, USA].

6.5.1 Kinetics of cell attachment on microcarrier beads

The attachment rate of animal cells in positively charged Cytodex-1 microcarriers increases with the increase in cell number. The kinetics of attachment at different cell numbers are shown in Figure 6.3A, where the profile of attachments in all cell numbers are almost identical. The $t_{1/2}$ value (time require to attach 50% free cells) decreases with increase in the cell number, indicating that faster attachment can be achieved by increasing cell numbers in the inoculums. The effect of available surface area on cell attachment is shown in Figure 6.3B. It can be seen that the number of cells attached to unit surface area is inversely proportional to the attachment surface at low available surface area. At surface area lower than 10 cm^2/ml, the physical contacts between each bead and the surrounding free cells is more, therefore more of cells are attached to the unit surface area of the microcarriers. At intermediate range (15–40 cm^2/ml), cell-to-bead ratios are just sufficient to maintain the same extent of cellular attachment to unit surface area. At very high available surface area (<50 cm^2/ml), the number of cells attached to unit surface area is reduced due to limitation of free cells. On the basis of the acceptance range of microcarrier density used in culture, the cell attachment rate is found to be independent of the surface area. Thus, the attachment of cells to microcarriers is a function of the free cells, and the rate equation for attachment can be simplified to fit a *unimolecular* type first order reaction in which the reactant can be presented as free cells that are converted into attached cells on the microcarriers surface. In addition, following assumptions are made: (i) cells do not grow during attachment, (ii) each cell has the same affinity for attachment on the surface, and (iii) no cell clump is present. The rate equation may be written as:

$$dN/dt = kN_F \qquad 6.1$$

$$\text{or, } = k\,(N_{F0}-N_A) \qquad 6.2$$

Where, N_A, N_{F0} and N_F are number of attached cells ($t = t_1$), free cells ($t = 0$), and free cells ($t = t_1$), respectively. The above Eq. (6.2) can be integrated between limits $t = 0$ where $N_A = 0$, and $t = t$ where $N_A = N_{Ff}$, then

$$-\ln\,[(N_{F0} - N_{Ff})/\,N_{F0}] = kt \qquad 6.3$$

The initial attachment rate constant (k) can be calculated by plotting left hand side of Eq. (6.3) versus time (t). A typical plot of attachment of Vero cells to Cytodex-1 microcarriers is shown in Figure 6.3C. The initial value of k calculated from the slope is 0.07 min^{-1}. The *biphasic attachment* is seen in medium supplemented with 10% serum, where the attachment rate in the second phase reduces to 0.011 min^{-1}. This transition can be evidenced at about 80-85% attachment of cells (Mukhopadhyay *et al.*, 1993). It is important to note that faster and monophasic attachment can be observed in serum free medium.

The scattering of cells on microcarrier beads follows *Poisson distribution.* Uniform distribution of cells is most desirable to improve population doubling of cells during the course of the culture. Non-uniform distribution of cells on

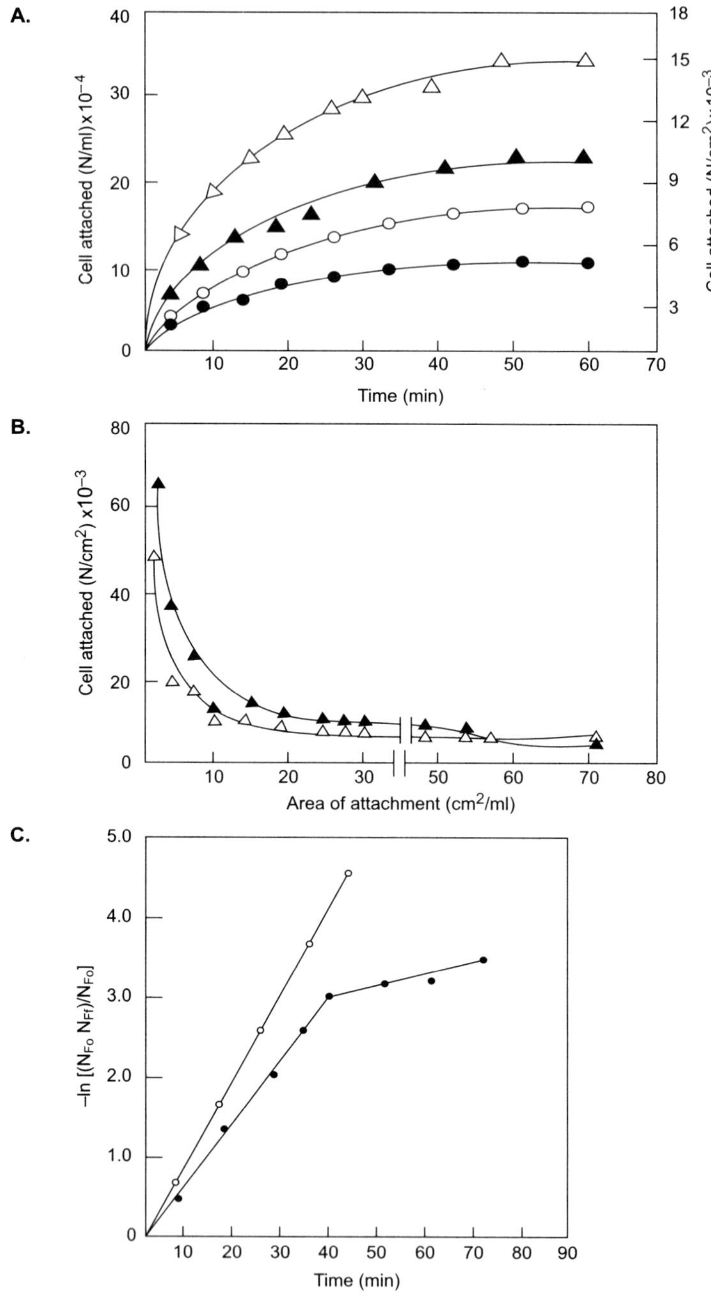

Figure 6.3 Attachment kinetics of Vero cells on Cytodex-1 microcarriers. **A.** Effect of cell numbers on kinetics of attachment; (•): cell-to-bead ratio = 5; (o): cell-to-bead ratio = 7; (▲): cell-to-bead ratio = 10; (△): cell-to-bead ratio = 15. **B.** Effect of available surface area on attachment; (△): After 30 min; (▲): After 60 min. **C.** Determination of attachment rate constant. (o): control, DMEM only; (•): DMEM containing 10% FCS. [Taken from Ref. 12]

microcarriers may occur due to size distribution of beads, inadequate mixing during attachment and low cell number in the inoculums. The number of the unoccupied (*cell free beads*) microcarriers may increase at low inoculum densities, which may affect population doubling. Again, more cells tend to attach the slow moving smaller beads than the fast moving larger beads.

6.5.2 Effect of microcarrier density on attachment and growth rate of cells

The *attachment rate constant* (k) of animal cells on Cytodex-1 microcarriers depends on its density. Initially, k improves with the increase in microcarrier density; later it declines (Figure 6.4A). At lower levels of microcarriers, the interacting

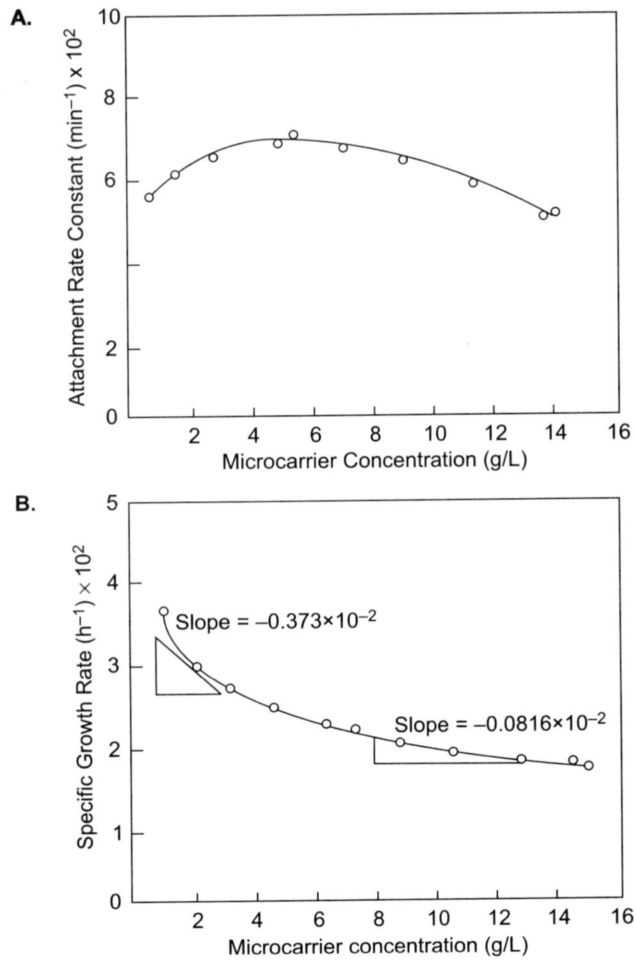

Figure 6.4 Influence of microcarrier density on the growth of Vero cell. **A.** Attachment rate constant at different densities of microcarriers; **B.** Specific growth rate at different densities of microcarriers. [Taken from Ref. 12]

species (cells and beads) are relatively far apart from each other, which minimize the chances of physical contacts between them, thus lower the attachment rate. At intermediate densities, the probability of physical contact between cells and the beads is increased, and hence the ‘k’ is also increased. The decrease of ‘k’ with an increase in microcarrier densities can be explained by exceeding detachment rate over the attachment rate due to inter-bead collisions. The specific growth rate (μ) decreases with an increase in microcarrier densities, but it changes in two distinct phases (Figure 6.4B). The absolute value of ‘μ’ is more sensitive to the change in the microcarrier densities at lower range. The apparent decrease in growth rate is believed to be due to the increase in frequency of *collision* in *turbulent region*, specifically at higher densities of microcarriers.

6.6 HYDRODYNAMIC CELL DEATH

The *hydrodynamic death* of cells, attached on microcarriers, may be due to many reasons. Notable among them are turbulence eddies, microcarrier collisions, and collisions with solid objects. The size of the turbulent eddies plays an important role in hydrodynamic damage of cells on microcarriers surface. If the size of the eddies are relatively larger than the diameter of the beads, the entrained microcarrier would be able to rotate to reduce the *torques* and forces on its surface. As a result, the attached cells will not experience with the forces generated by large eddies. The rotation of a microcarrier is restricted in case of small size of eddies (lower than the diameter of the beads), as a result cells are exposed to shear force generated by small eddies. In consequence of that, the cells on the microcarriers surface are instantaneously damaged. The *turbulent eddies* in the *viscous dissipation regime* are in between the size of a free cell and a microcarrier bead, therefore the free suspended cells are not affected by this force. In the viscous dissipation regime, the size of the smallest eddy is calculated by Kolmogorov length scale (L). Thus

$$L = [n^3/e]^{1/4} \qquad 6.4$$

Where, ν is the *kinematic viscosity* (cm^2/s), and ε is the power dissipation in unit mass of fluid (cm^2/s^3). The size of the smallest eddy decreases with the increase in power dissipation or decrease in kinematic viscosity. In a typical stirred tank bioreactor, the power input ‘P’ through the impeller is determined by direct measurement of motor power consumption, or is calculated from the impeller parameters, as follows:

$$P = N_p \rho n^3 d_i^5, \qquad 6.5$$

$$\varepsilon = P/\rho V \qquad 6.6$$

Where, ‘N_p’ is the dimensionless impeller *power number*, ‘ρ’ is the fluid density, ‘n’ is the impeller rotational rate (rps), ‘d_i’ is the impeller diameter, and ‘V’ is the medium volume in which the energy dissipation takes place.

Cell death may occur through interactions between two microcarriers on which they are adhered. The severity due to collision-induced cell death increases with an increase in microcarrier loading and the agitation speed. It has been shown that the specific growth rate of Vero cells is reduced with an increase in microcarrier loading (Figure 6.4B). In a gentle agitated system, microcarriers follow fluid streamlines. The relative motion between two neighboring microcarriers will arise from the action of turbulent eddies. The collision between two microcarrier beads can occur in two ways. First, the two neighboring microcarriers may be influenced independently by two small turbulent eddies, giving them independent velocities in the orientation that they collide each other (Figure 6.5A). Second, the collisions may occur in large eddies by a shear mechanism. Beads on different *streamlines* will have different velocities; if the distance between two streamlines is less than one bead diameter, those beads having different velocities may collide with each other (Figure 6.5B). The volumetric frequency of bead-to-bead collisions (F_c) is found to be of second order in microcarrier density (M_d), and thus,

$$F_c = KM_d^{\,2} \qquad 6.7$$

where, K is a collision frequency constant.

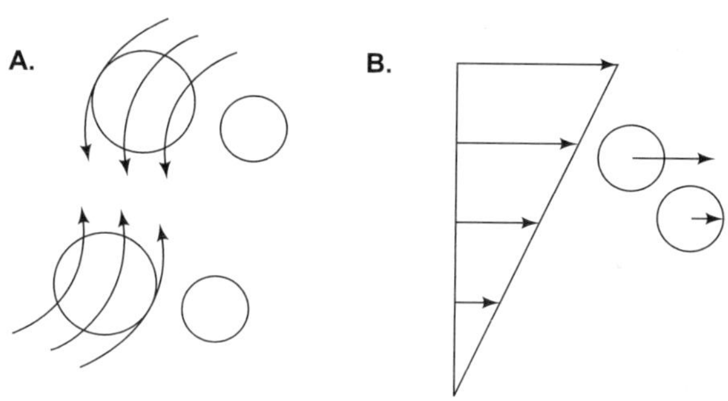

Figure 6.5 Hydrodynamic cell death due to bead-bead collisions. **A.** Small eddy collision causes direct impact; **B.** Shear-based collision is pronounced in large eddies. [*Source*: Animal Cell Biotechnology, Spier & Griffiths, eds. Vol 4, p. 71., AP]

The number of cell death from each collision will depend on the cells in unit surface area of the bead, and the effective surface area exposed to each collision between two colliding beads (Croughan *et al.*, 1988).

The third kind of interaction responsible for cell death is collisions of microcarrier beads with the solid objects of the bioreactor (e.g., wall, impeller blade, pH and dissolved oxygen electrodes, etc.). Most important bioreactor design variables are impeller diameter and the rotation speed. The growth of animal cells on microcarriers has been correlated with *integrated shear factor* (ISF) as follows:

$$ISF = 2\pi N d_i/(d_t - d_i) \qquad 6.8$$

Where, $2\pi Nd_i$ is impeller tip speed, 'N' is impeller speed, 'd_i' impeller diameter, and 'd_t' bioreactor diameter. Thus, ISF increases with an increase in 'N' and 'd_i'. It has been shown that the maximum growth of chick embryo fibroblasts (Sinskey *et al.*, 1981) and FS-4 (Hu, 1983) on microcarrier beads is sharply reduced above ISF of 90 s^{-1} and 25 s^{-1}, respectively.

6.7 OXYGEN SUPPLY IN CELL CULTURE MEDIUM

Cell culture medium can be aerated singly or by combination of methods: surface aeration, sparging of air/oxygen, *bubble free aeration, medium perfusion*, increasing *partial pressure* of oxygen in air inlet, and increasing overhead pressure in the bioreactor. The demand for oxygen is found to be high in case of transformed cells as compared to the normal cells; again it may change from cell to cell. Therefore, a particular method of oxygen supply may not be sufficient for all cell types. Again, the demand for oxygen is increased with the cell number. Cell culture in T flasks and in roller bottles relies on oxygen supply through surface aeration. However, due to low oxygen solubility in serum-supplemented culture medium (< 6 mg/l), its' limited diffusion rate (in static culture) or inadequate surface aeration (in low mixing culture) may result in oxygen limitation in high cell density culture. Oxygen requirements for numerous important cell types at different cellular densities are shown in Table 6.3. It is clear that closed culture system is not suitable for any of the following cells, and continuous supply of oxygen is necessary. Three basic questions need to be addressed to select oxygenation system for animal cell bioreactors, these are (a) damage of cells at air-liquid interface in bubble aeration, (b) ability of media to deliver oxygen to the cells, and (c) ability of various systems to transfer adequate amount of oxygen to the cells. Therefore, efforts have been made to supply oxygen to the cells in harmless manner. Table 6.4 describes alternate oxygenation methods and their capacities in a pilot scale animal cell culture bioreactor.

Table 6.3 Oxygen requirement at different cell densities

Cell type	*Respiration rate ($\mu gO_2/10^6$ cells.h)*	*Oxygen demand at different cell densities (mgO_2/l.h)*		
		10^6/ml	5×10^6/ml	10^7/ml
BHK–21	6.4	6.4	32	64
WI38	16	16	80	160
Hybridoma	9.6	9.6	48	96
HeLa	12.8	12.8	64	128

[Taken from Ref. 15]

Table 6.4 Alternate aeration methods of a 40 L stirred bioreactor (30 L working volume)

Aeration method	*Oxygen supply (mg/l.h)*	*Number of cells supported for growth (×10⁶/l)*
	Air supply: 600 ml/l.h, 40 rpm	
Surface aeration	0.5	0.08
Direct sparging	4.6	0.76
Spin filter sparging	3.0	0.40
Perfusion (1 volume/h)	12.6	2.10
Perfusion (1 volume/h) + Spin filter sparging	15.9	2.65
	Oxygen supply: 600 ml/l.h, 80 rpm	
Spin filter sparging	51.0	8.50
Perfusion (1 volume/h) + Spin filter sparging	92.0	15.00

Oxygen consumption rate: 2–6 μg/10^6 cell.h. [Taken from Ref. 5]

6.7.1 Oxygenation by surface aeration

Surface aeration is the simplest way for *oxygenation* of cell culture media, but it is less efficient in case of animal cells having high oxygen uptake rate, or in case of high cell density culture. Oxygen is transferred from headspace to the medium through air-medium interface. Assuming that (i) oxygen transfer follows *Fick's law of diffusion*, (ii) dissolved oxygen concentration in medium is almost zero, and (iii) headspace air is fully saturated with oxygen, the maximum rate of *oxygen diffusion* to the culture by surface aeration will be 1.5 μg/cm^2.h (Spier and Griffiths, 1984). In a typical 5L bioreactor of aspect ratio (height-to-diameter ratio) 2.0, the maximum oxygen transfer through the *gas-liquid interface* will be about 1 mg/h. For BHK-21 cells, the oxygen consumption per liter culture (cell density: 1×10^6/ml) is about 6.4 mg/l (Table 6.3). Therefore, surface aeration alone is not enough to provide adequate quality of oxygen to the culture. The limitation of oxygen transfer through surface aeration is critical with increase in the volume of the medium, as the liquid-air surface area (A) to medium volume (V) ratio reduces with increase in charging volume in the bioreactor. The diffusion of oxygen from headspace can be improved by agitating medium surface, providing a floating stirrer to improve surface agitation, or providing partial pressure of oxygen in the gas phase above the normal level.

6.7.2 Oxygenation by direct sparging of air

Sparging of air through the medium is most efficient method of oxygenation. However, direct sparging of air may be detrimental to the cells due to cell-bubble interactions at the surface of *bubble disengagement.* Two possible mechanisms that may cause harm to the animal cells by air bubbles (Handa-Corrigan *et al.*, 1989) are: (a) damage due to rapid oscillations caused by *bursting bubbles* in medium containing antifoam, and (b) damage due to *entrainment* of cells in draining liquid films in low protein containing medium. The damage of cells is found more pronounced in case of low serum containing or defined low-protein serum-free medium. The extent of damage can be minimized by using large air bubbles (low surface energy than smaller bubbles), low airflow rate, and incorporating *surface-active agents* in the medium (0.02% Pluronic F-68 or 10% serum). Both serum and Pluronic F-68 stabilize the cell-liquid interface by forming highly condensed interfacial structure.

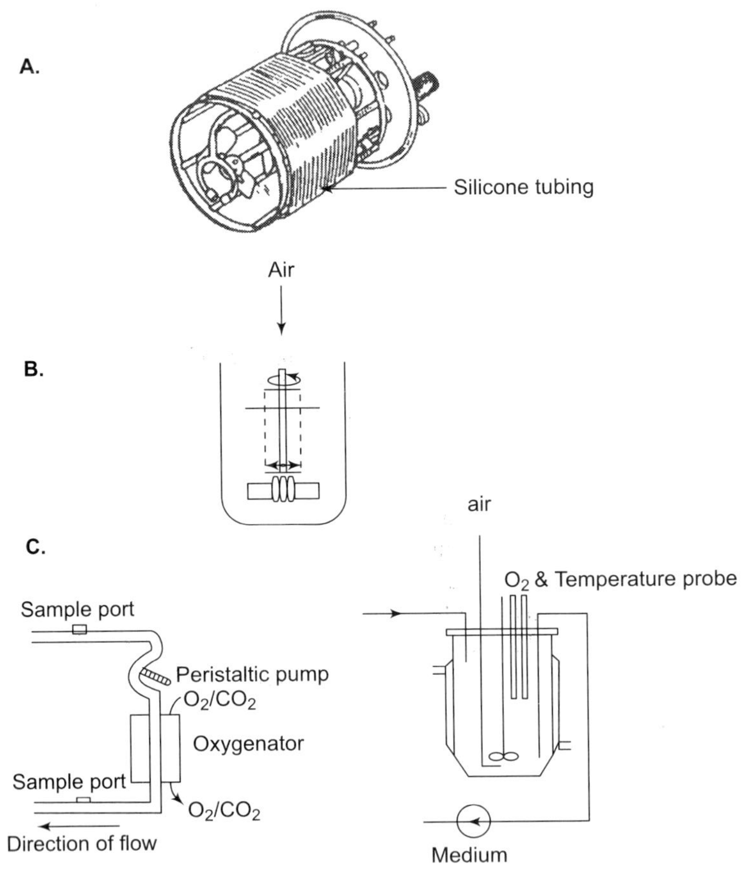

Figure 6.6 Safe oxygenation of mammalian cell culture medium. **A.** Silicon tubing for bubble-free aeration; **B.** Sparging cage; **C.** Indirect oxygenation systems.

In order to avoid cell damage by direct sparging of air, two novel ways of media aeration devices have been developed, these are (a) *bubble-free aeration*, and (b) *cage aeration*. By using oxygen-permeable tubing, it is possible to carry out effective oxygen transfer without producing air bubbles in the bioreactor. The rate of oxygen diffusion through the membrane is directly proportional to the air pressure applied to the tubing inlet. Thus, by increasing length of the tubing (increasing the gas-medium interface) and partial pressure, it is possible to improve oxygen transfer (Figure 6.6A). Thin-walled silicon tubing or porous polypropylene tubing has been commercially used for bubble-free aeration system. A similar technology was used in 150L commercial scale bioreactor (Diessel GmbH & Co., Hildesheim, Germany); however, membrane-based aeration system is found inconvenient to use. The major problems in membrane-based aeration system are associated with cleaning and replacement of tubing, and scale-up of the system. The scale-up problem is due to the requirement of unmanageable length of tubing to cater oxygen demand in pilot scale bioreactor.

The sparging cage aeration is based on the principle that the aeration is carried-out inside an enclosed stainless steel wire cage, which is primarily cell free. The cage is partially submerged, and the air bubbles are disrupted as the foam leaves the cage into the headspace of the bioreactor (Figure 6.6B). Several modified versions of cage aeration device have been reported, but all of them were developed for similar objective 'avoiding damage to the cells due to direct sparging'.

6.7.3 Oxygenation by medium perfusion

The best method of oxygenation in animal cell culture bioreactor is medium perfusion. Despite the fact that the solubility of oxygen in serum containing media is about 25% lower than the MilliQ water, medium perfusion is the safest and most efficient method of oxygenation (Table 6.4). In a typical batch culture, perfusion technique is not directly used for oxygenation. In an open-loop or a close-loop perfusion system (continuous/ intermittent), the culture is fed with oxygenated medium, whereas the effluent medium is reoxygenated in a separate oxygenation chamber (Figure 6.6C). The medium is reoxygenated either by gas-exchange through a permeable membrane or by direct sparging of air/oxygen.

6.7.4 Oxygenation by increasing partial pressure in the headspace

It is one of the most popular methods used in the fermentation industries. The principle of increasing oxygen level in medium is based on *Henry's law*, that raising of gas-phase partial pressure increases the solubility of a gas in the liquid phase. Dissolved oxygen concentration can be increased by simply enriching oxygen level in the headspace using oxygen-enriched air or by elevating backpressure of the bioreactor. Backpressure increases both the solubility of oxygen as well as its diffusion rate. This method may not be applicable in all cell culture due to oxygen toxicity and sensitivity of cells to hydrostatic pressure.

6.8 TYPES AND OPERATION OF CELL CULTURE BIOREACTOR

Like microbial cultures, animal cell culture processes are also carried out either in *batch* or in *continuous* mode. However, the batch culture is more popular in which fixed volume of medium is inoculated with cells. In batch culture, no fresh medium is added after inoculation, and thus the nutrients are gradually exhausted with the growth of the cells. At the same time toxic metabolic end products are accumulated in the culture medium. Typical bioreactors used for batch culture and the characteristic growth kinetics are shown in Figures 6.7A and 6.7B. These culture systems can be used for both suspension and anchorage-dependent cells. In batch culture, medium volume in maintained constant and it is less contamination prone. The batch culture is started with typical inoculum densities of 0.5 to 2×10^5 cells/ml, which can reach up to 1 to 3×10^6 cells/ml at the end of the culture, depending upon the cell type. As mentioned before, in batch culture, the environment is continuously changed due to many reasons like, exhaustion of nutrients, accumulation of toxic metabolites, and limitation of growth surface (in case of anchorage-dependent cells). All these factors can lead to growth inhibition of the cells.

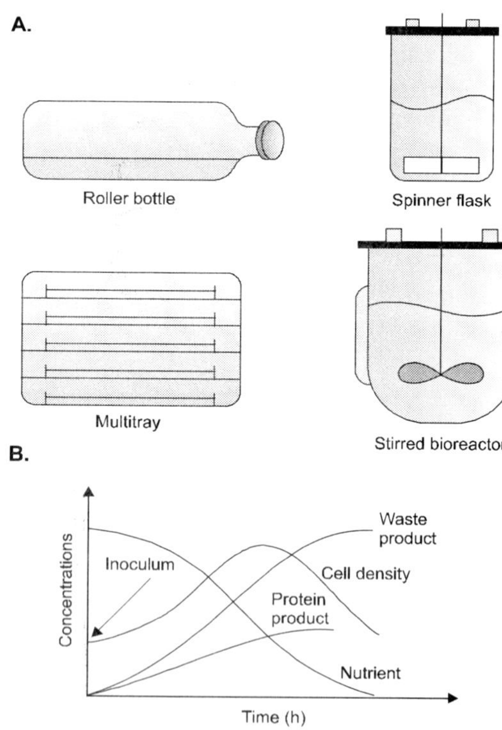

Figure 6.7 Animal cell bioreactors and batch kinetic profiles. **A.** Bioreactors used for batch mode of operation; **B.** Kinetics of cell growth, nutrient consumption, waste product accumulation, and product formation in a batch type bioreactor.

Batch bioreactor

In a stirred batch bioreactor, the unsteady state mass balance in terms of cell number is

$$d(NV)/dt = \mu NV - k_d NV \quad 6.9$$

For constant volume medium, V = constant, Eq. (6.9) becomes:

$$dN/dt = (\mu - k_d)N \quad 6.10$$

where, V is medium volume, N is cell number (here the cell mass is replaced with the cell number), 't' is batch culture time, μ is specific growth rate, and k_d is a *specific death constant.* If we consider that μ is approximately constant during the batch growth and equal to μ_{max} and k_d is also a constant, then we can integrate Eq. (6.10) between initial (t = 0) and final batch time (t = t):

$$N_t = N_0 e^{(\mu_{max} - k_d)\,t} \quad 6.11$$

Assuming that there is no limitation of nutrients or any kind of inhibition due to hydrodynamic effect and physico-chemical factors, the death rate of cells can be considered negligible low as compare to the growth rate, i.e. $k_d << \mu_{max,}$ Eq. (6.11) reduces to:

$$N_t = N_0 e^{\mu_{max} \cdot t} \quad 6.12$$

where, 'N_t' cell number at time t = t.

The simplified equation to determine cell number in terms of the consumption of *growth limiting substance* can be written as:

$$N_t = N_0 + Y_S (S_0 - S_t) \quad 6.13$$

where, Y_S is the yield of cells in unit substrate (glucose) consumption, S_0 and S_t are the substrate concentrations at t = 0 and t = t, respectively. The Eq. (6.13) can be used in cases where the batch growth of cells is based on single limiting nutrient. However, in cell culture media glucose is not the only limiting nutrient, other constituents of the medium may act as limiting substrate, so for each such limiting nutrient there will be a Eqs. like 6.13.

In batch culture, continuous changing of culture environment enforces alteration in cellular metabolism and growth, which will adversely affect on the productivity. The synthesis of most of the animal cell derived products (e.g., monoclonal antibodies) is considered *non-growth associated,* and maximum product concentration is achieved in the extended stationary culture. Again, the specific growth and product formation rates are much low in animal cells. Therefore, typical batch culture process is not reommended to improve productivity of animal cells derived products. Except viral vaccines, continuous culture technique may be adopted for all other animal cell culture processes. In continuous cell culture process, in addition to continuous supply of nutrients, removal of toxic metabolic wastes in also taken place. The continuous cell culture process endorse several advantages over the batch culture, these are:

1. ***Allows economic production***: In a batch culture, the production of monoclonal antibodies from hybridomas is limited by the lifetime of the culture. However, in continuous culture the product is obtainable for infinite periol of time due to unlimited life of hybridoma, which offers economic advantage.

2. ***Offers simple downstream processing***: In a typical continuous process, cells are retained inside the bioreactor and the production phase is segregated from the growth phase. Thus, downstream processing of the final products become relatively simple, as a few cells contaminate in the effluent, the product is less contaminated with serum and cellular proteins, and the concentration of the product in the effluent is increased.

3. ***Prevents unnecessary product loss***: In batch culture there is a chance of degradation of products (temperature or enzymes mediated), and suppression of product formation by feedback inhibition. In continuous culture, rapid isolation of products minimizes such loss.

Fed-batch bioreactor

In batch culture, the periodic additions of fresh medium, with or without removal of spent medium, can realize some of the benefits of continuous culture. This type of culture is technically known as *fed-batch* and *semi-continuous*. In fed-batch culture, fresh medium is continuously or intermittently added, hence it is a variable volume process. On the other hand, the semi-continuous culture is a constant volume process, where a known volume of the culture is replaced with equal volume of fresh medium. Inspite of the fresh medium added, these types of culture also experiences a fluctuating culture environment. The culture is labor intensive and prone to contamination and residual concentrations of toxic metabolites cannot be reduced below the inhibition levels (Figures 6.8A and 6.8B). In stirred fed-batch bioreactor, the unsteady state mass balance in terms of cell number is

$$d(NV)/dt = FN_0 + mNV - k_dNV \qquad 6.14$$

where, F is volumetric flow rate and N_0 is cell number in the feed. In batch culture V is constant, whereas, in fed-batch culture, V is a function of time, so it cannot be replaced. Thus, Eq. (6.14) can be rewritten as:

A.

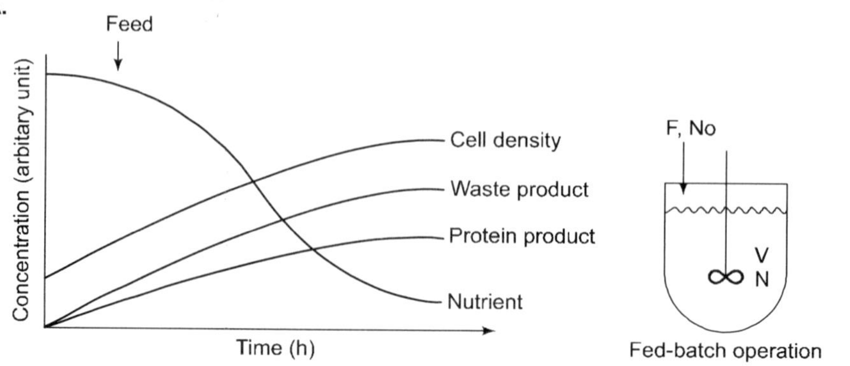

(*Contd.*)

B.

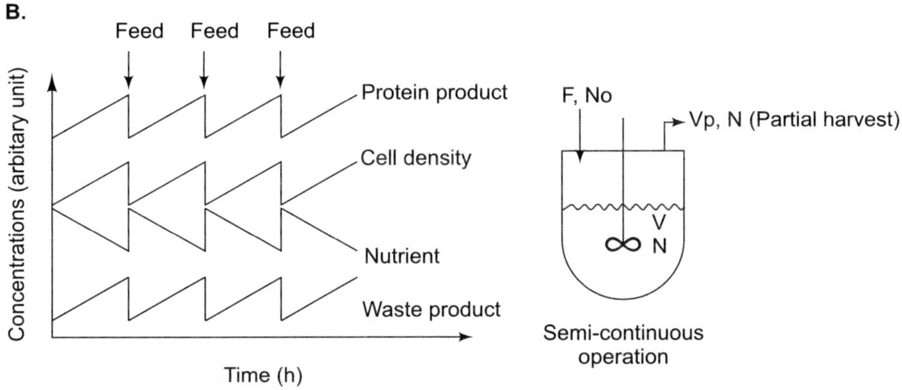

Figure 6.8 Schematic diagram. **A.** Fed-batch system with kinetics of nutrient consumption, etc; **B.** Semi-continuous culture system with kinetics of nutrient consumption, etc.

$$N\ (dV/dt) + V\ (dN/dt) = FN_0 + (\mu - k_d)\ NV \quad 6.15$$

or,

$$NF + V\ (dN/dt) = FN_0 + (\mu - k_d)\ NV \quad 6.16$$

where, unsteady-state mass-balance equation for fed-batch bioreactor is

$$dV/dt = F$$

Dividing Eq. (6.16) by V and rearranging the different functions, we get

$$dN/dt = (F/V)\ N_0 + (\mu - k_d - F/V)\ N \quad 6.17$$

As dilution rate D = F/V, Eq. (6.17) can be written as

$$dN/dt = D.N_0 + (\mu - k_d - D)\ N \quad 6.18$$

Eq. (6.18) can be simplified as

$$dN/dt = (\mu - D)\ N \quad 6.19$$

as, $N_0 = 0$, and $k_d << \mu$.

Continuous flow bioreactor

Continuous-flow or *chemostat culture* is a homogeneous open system with a fixed culture volume process. The bioreactor is fed with fresh medium in a constant rate, and spent medium containing cells and products are withdrawn from the system at the same flow rate. The flow of medium and kinetic profiles for nutrient consumption, cell growth, etc. of a continuous flow stirred bioreactor are shown in Figures 6.9A and 6.9B, respectively. In a continuous culture process, the flow rate is set at a particular value and the rate of growth of the cells is automatically adjusted to this flow rate. Thus, system steady state is reached at each medium flow rate, and all the state variables (e.g., cell density, nutrient concentrations, product concentration, etc.) become independent of the time. The material balance for the cells in chemostat culture can be written as

$$FN_0 - FN_x + \mu N_x V - k_d N_x V = V[dN_x/dt] \quad 6.20$$

where, F is medium flow rate, N_0 is inlet cell density, N_x is steadysate cell density, V is the working volume, and dN_x/dt represents the change of cell number in the bioreactor with time, and $k_d << \mu$.

At steady state, $dN_x/dt = 0$, thus Eq. (6.20) becomes

$$(N_x - N_0) / \mu N_x = V/F = \tau_m \quad 6.21$$

where, 'τ_m' is the mean residence time of the medium.
Since, the inlet stream is free from cells ($N_0 = 0$), Eq. 6.21 becomes

$$\tau_m = N_x / \mu N_x = 1/\mu = 1/D \quad 6.22$$

where, 'D' is *dilution rate* and equal to specific growth rate of cells (μ) at steady state, and reciprocal of the *residence time* (τ_m). Thus, in continuous culture system, the cellular growth is determined by the medium flow rate. The continuous culture process assures high 'μ' value as compared to the batch culture and low residual concentrations of metabolic wastes. Inspite of this, it remains a less efficient culture system, as cell density cannot reach to a high level because of continuous removal of cells with the spent medium. Change of steady state cell density with an increase in D is shown in Figure 6.9C. Initially, steady

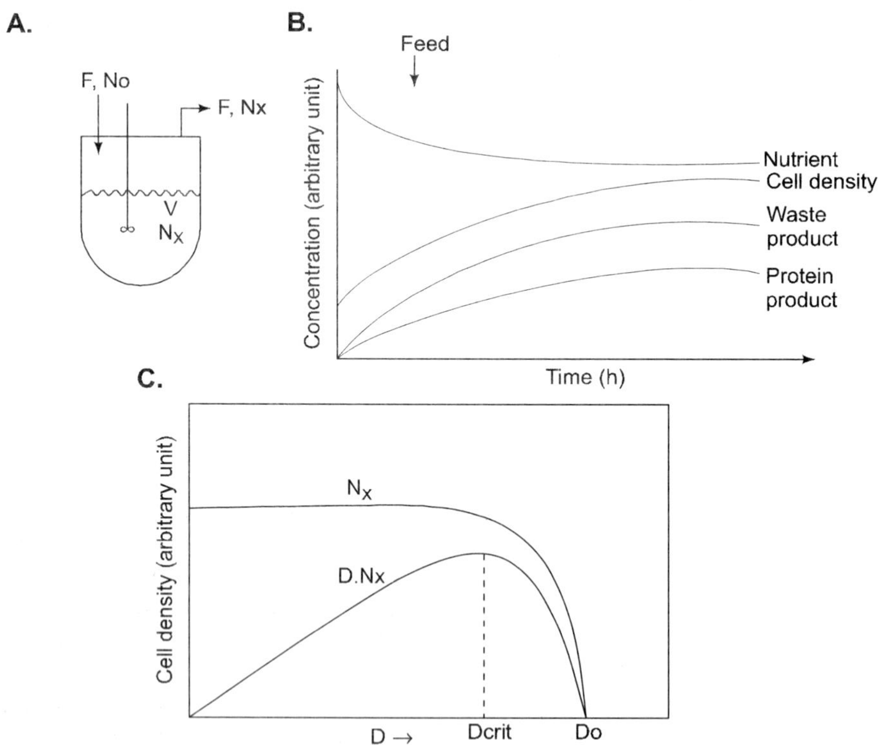

Figure 6.9 **A.** Schematic view of chemostat culture system, **B.** Steady state levels of limiting nutrient, cell density, etc; **C.** Change in cell density with dilution rate.

state cell density (N_x) remains constant, as specific growth rate (μ) increases with D. Beyond a certain value of dilution rate, further increase of D results in the decrease of N_x, which is known as *critical dilution rate* (D_{crit}). At D_{crit}, the specific growth rate of cells is found maximum (μ_{max}). Further increase in D ($>D_{crit}$) is associated with sudden decline in N_x, which finally approaches 'zero' corresponding to complete *washout point* dilution rate (D_o). This problem is more serious in case of continuous animal cell culture, since 'μ' does not significantly change with 'D'. Therefore, high flow rate (high dilution rate) continuous culture is not suitable for animal cell culture processes.

Continuous perfusion bioreactor

One way to improve cellular productivity ($D.N_x$) in bioreactor is to recycle the cells by separating them from the spent medium, which is known as *cell recycling* bioreactor (Figure 6.10A). The same principle has been used to develop continuous *perfusion* bioreactor; the only difference is that instead of separating cells outside the bioreactor, they are separated within. This change has been introduced to avoid contamination and shear damage to the cells. The perfusion culture system is a homogeneous closed system with a fixed culture volume process that allows cell retention in the bioreactor, and at the same time cells are perfused at a high rate with fresh medium (Figure 6.10B). In perfusion bioreactor, the products and the metabolites are continuously removed from the culture. The perfusion culture attains a steady state when cells stop multiplying (confluent state) or multiply at a low rate to replace the dead cells. The perfusion culture is analogous to a recycling bioreactor with 100% cells recycling, and a chemostat with cell immobilized system. At a steady state, the mass balance equation for suspended cells and adhered cells can be written as below, assuming that in Eq. (6.20) the feeding medium contains no cells, and $k_d << \mu$.

$$- FN_S + \mu N_S V + \eta_T \mu N_A V = 0 \qquad 6.23$$

where, N_S and N_A are suspended and adhered cells, respectively. Due to diffusion limitation, the specific growth rate of multi-layer adhered cells on microcarriers is reduced to $\eta_T\mu$ in place of μ for suspended cells, where, η_T is the total effectiveness factor to account restrictions of internal and external mass-transfer (oxygen and other nutrients) on growth. In case of suspension culture and monolayer adhered cells on solid microcarriers ($\eta_T \mu N_A V = 0$), the above Eq. (6.23) can be written as,

$$- FN_S + \mu N_S V = 0 \qquad 6.24$$

Rearranging and dividing by V, Eq. (6.24) gives

$$D_P.N_S = \mu N_S \qquad 6.25$$

where, D_P is the dilution rate in perfusion culture. Similarly, in case of multilayer adhered cells on microcarriers, Eq. (6.24) reduces to

$$D_P.N_S = \eta_T \mu N_A \qquad 6.26$$

It is important to note that both Eqs. (6.25 and 6.26) reduce to Eq. (6.22) for a chemostat culture similar to μ ($\eta_T\mu$) = D_P, as $N_X = N_S$ or N_A. The significant difference in the relationship between dilution rate and growth rate in continuous (chemostat) culture and continuous perfusion culture is that in case of former the growth rate is controlled by the dilution rate, and D approaches to μ_{max}, $N_X \rightarrow 0$. On the other hand, in continuous perfusion culture, D at steady state can be greater than μ. Accordingly, the dilution rate is no longer limited to μ_{max}, and perfusion bioreactor can be operated at D considerably higher than D_{crit} without washout. Thus, in continuous perfusion bioreactor it is possible to minimize the toxic effects of metabolites as compared to other modes of bioreactor operations and at the same time, specific growth rate and final cell number can be maintained high (Fig. 6.10C).

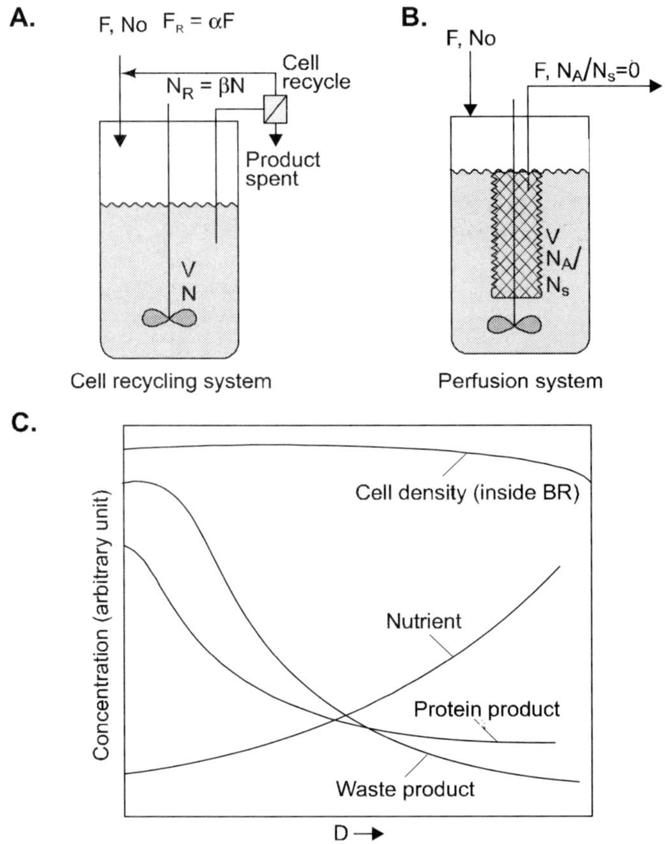

Figure 6.10 **A**: Schematic view of recycle bioreactor, **B**. Schematic view of perfusion bioreactor; **C**. Change in cell density with 'dilution rate' in perfusion bioreactor.

6.9 CELL RETENTION SYSTEMS IN PERFUSION BIOREACTOR

In perfusion culture, cell retention can be achieved either by (a) *homogeneously distributed*, or (b) *heterogeneously distributed* system. In homogeneously distributed systems, culture medium moves around the free suspended cells or cells growing on microcarriers or cells entrapped in capsules. This ensures excellent mass transfer characteristics. The retention of cells in these perfusion systems may be achieved either by gravity settling (internal or external) mesh filter, or high-density microcarriers (Figure 6.11A). Cells growing on microcarriers or encapsulated cells can be retained in the bioreactor on the basis of above methods. However, for free suspension culture gravity settling and mesh filtration methods are two useful options for separating cells. Cells settling become easy when they grow on microcarrier beads. The density of microcarrier beads ($\rho = 1.04$ g/ml) is higher than water, so carriers' free zone can be easily created in a homogeneous perfusion bioreactor. Titanium impregnated microporous microcarriers have been developed by Percell Biolytica, Sweden, for growing cells in suspension culture. Due to high specific gravity ($\rho = 1.6$ g/ml), even in fluidized bed or expanded bed (closed-loop) bioreactor, beads-free zone is formed at the exit of the bioreactor, which offers simple harvest of cell-free medium (Figure 6.11B). Static mesh-filter based perfusion system has always been a difficult device for

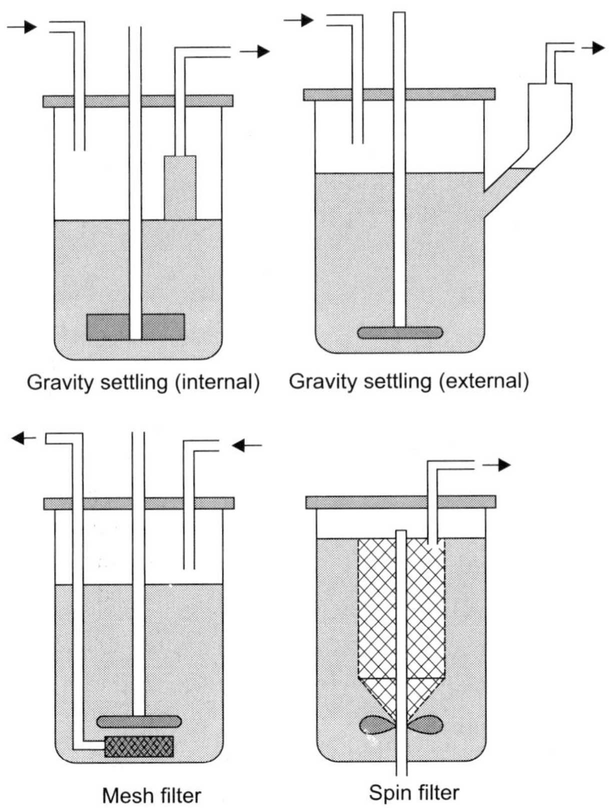

Figure 6.11A Cells retention devices in homogeneously distributed perfusion systems (Gravity settling- internal, Gravity settling- external, Mesh filter, Spin filter).

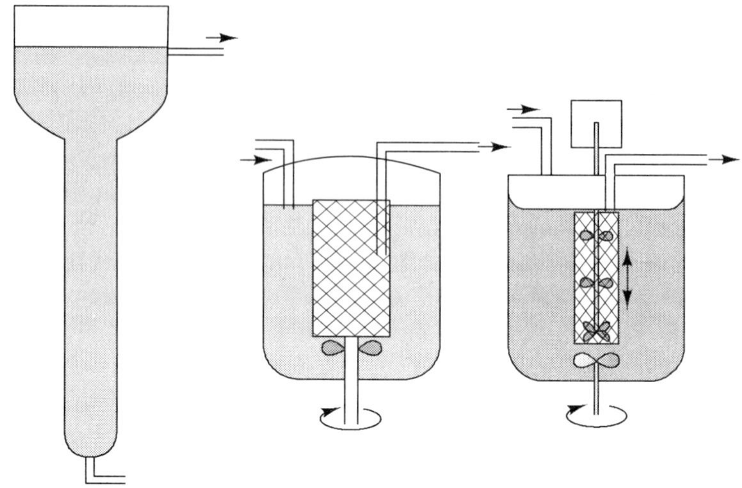

Figure 6.11B Cells retention devices in homogeneously distributed perfusion systems (Fluidized bed, Rotating wire cage, Chemcell™).

scale-up due to blockage in continuous operation. Spin filter based perfusion system has been introduced to minimize blockage of filter mesh. As spin filter rotates, a boundary layer of liquid is created around the filter mesh, which prevents attachment of cells and subsequent filter clogging. The filter is made of stainless steel mesh having a pore size (~3 μm) smaller than a cell, but which is not free from fouling in long duration culture. The spin filter also has the problem of limited perfusion rate (maximum up to 2 culture volume/day) due to small pore size filters. To avoid these problems, rotating wire cage bioreactors have been developed for hybridoma culture. The mesh size of rotating wire cage is 53 μm, which allow 95% retention of free hybridoma cells in the wire cage bioreactor with 3-culture volume/day perfusion rate (Figure 6.11B). Spin filter and rotating cage filter devices are suitable for perfusion of cell growing on microcarriers. Another useful cell retention device has been reported in the literature for perfusion of microcarriers based cell culture system. The trade name of such device is Chemcell™ (Chemap AG, Switzerland), which has been used up to 100 L scale. The Chemcell is based on 50 μm stainless steel mesh static cylinder, inside of which is a cell-free zone through which medium is sucked-out (Figure 6.12B). A set of vibrating disc (Vibro mixer™) moves vertically up and down to deliver medium from cell-free zone to the culture, and draws cell-free medium from culture to inside of the cylinder. Continuous movement of medium across the wire mesh prevents clogging.

Heterogeneously distributed perfusion systems (Figure 6.12) are used to achieve cells at much higher densities that are otherwise difficult to obtain in homogeneous systems. In these systems, cells are retained on the basis of (a) *immurement*—retaining cells within a compartment, which allows easy flow of medium, and (b) *entrapment*—immobilizing cells on porous substrate.

Shear sensitive cells are conveniently grown at very high densities in these bioreactors. Hollow fiber bioreactor is popular among all heterogeneously distributed perfusion systems. It is made up of thousand of hydrophilic porous fibers. Cells are grown and retained on the extracapillary space and medium is supplied via the lumen of the fibers (Figure 6.12A). Hollow-fiber system is particularly suitable for the cells normally grown in suspension (e.g., hybridomas). Most hybridomas do not attach on the tissue culture plates owing a small volume of cytoplasm than nucleus, so they cannot flatten out to make multiple tight attachments as anchorage-dependent cells can do. Flat membrane bioreactor is an alternative to hollow-fiber system. In this system, animal cells are immobilized between two different sheets of flat membranes forming three compartments (Figure 6.12B). The ultrafiltration membranes between medium and cell compartments have a molecular weight cut-off allowing nutrients and oxygen diffusion to the cells but no product diffusion to the medium compartment. The microfiltration membrane between the cell and the product compartments has pore size of about 0.2-0.5 μm, which allows product but no cell leakage to the product compartment. In this configuration, cells can be continuously fed with medium, and product harvest intermittently when a proper concentration is achieved. Membroferm system is the trade name of the membrane bioreactor, marketed by MBR Bioreactor (Wetzikon, Switzerland).

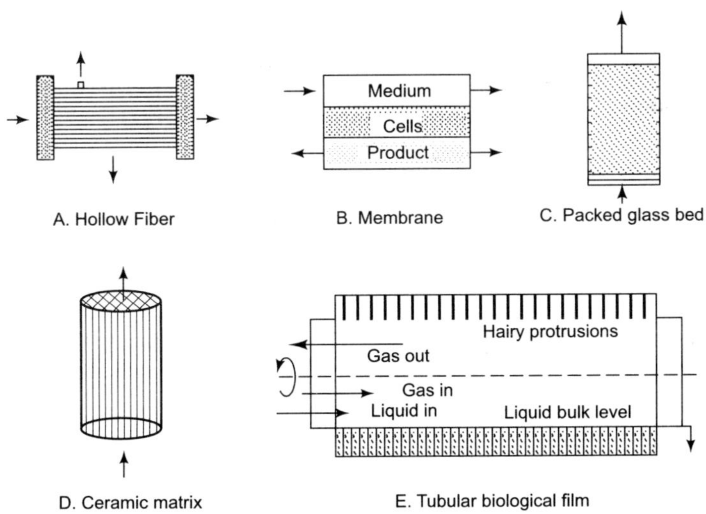

Figure 6.12 Cells retention devices in heterogeneously distributed perfusion systems (Hollow fiber, Membrane, Packed glass bed, Ceramic matrix, Tubular biological film).

Packed glass beads perfusion system is suitable for small-scale production with anchorage dependent cells. The bed is packed with 3-5 mm glass beads, through which medium is continuously perfused (Figure 6.12C). Ceramic matrix based system has quite a different approach for *high-cell density* perfusion culture. Animal cells are immobilized by adsorption to a honeycombed ceramic matrix. The commercial name of such system is Opticell. It consists of many square channels running the length of a cylinder, providing rigid support for cell adsorption (Figure 6.12D). The ceramic matrices are of two types— one possessing smooth and the other porous surface. The smooth surface ceramic is generally used for culture of cells that need to harvest by trypsinization. The porous matrix is used to trap anchorage-dependent or independent cells. This type of core is generally used for the production of secreted biomolecules. The ceramic cartridge is perfused with medium in a close-loop via an oxygenator.

The tubular biological film bioreactor is based on the principle of roller bottles for continuous perfusion culture. The surface of the cylinder is modified to have cell carrier matrix in the form of hairy protrusions on which cell can be attached in case of anchorage dependent, or kept immobilized in a stagnant liquid film if they are suspension cells (Figure 6.12E). The vessel rotates at a speed comparable to that of roller bottle and can be continuously or intermittently perfused with fresh medium and mixture of gases.

6.10 SUMMARY

1. Both physical and chemical environment can modulate growth and productivity of animal cells.
2. Culture of suspension cells is more convenient than adhered cells. However, using microcarrier based technique many problems associated with adhered cells can be resolved.
3. Attachment of free cells to solid microcarriers follows first order kinetics. The rate of attachment and growth reduces with an increase in microcarrier density.
4. Cell growing on microcarriers are affected by three kinds of hydrodynamic forces: turbulence eddies, microcarrier collisions and collisions with solid objects.
5. In animal cell bioreactor, oxygen is supplied to the culture by combination of different methods: air/oxygen sparging through medium, surface aeration, and medium perfusion, etc.
6. Continuous perfusion system maintains high cell density, and at the same time minimizes the accumulation of metabolic byproducts.
7. In perfusion bioreactors, cells are retained by sedimentation, spin filtering, attaching on high-density microcarriers, entrapment, hollow fiber/flat membrane filtration, etc.

ABBREVIATIONS

BHK	Baby hamster kidney
CIG	Cold insoluble globulin
ECM	Extracellular matrix
DEAE	N,N-diethylaminoethyl
ISF	Integrated shear factor
LFA	Leukocyte functional antigen

REFERENCES

1. Butler M and Spier RE (1984) *J Biotechnol.* 1, 187.
2. Croughan MS, Hamel J-FP and Wang DIC (1988) *Biotechnol. Bioeng.* 32, 975.
3. Fee JA (1981) In: *Oxygen and Life* Royal Society of Chemistry (Special publication) no. 39, p. 77.
4. Geiger B, Salomon D, Takeichi M and Hynes HO (1992) *J Cell Sci* 103, 934.
5. Griffiths JB and Looby D (1987) Proceedings 8th. ESACT meeting, "Israel" Butterworths, Guildford.
6. Glacken MW, Fleischaker RJ and Sinskey AJ (1986) *Biotechnol. Bioeng.* 28, 1376.
7. Hu W-S (1983) Ph.D. Thesis, MIT Cambridge, MA, USA.
8. Honda-Corrigan A, Emery AN and Spier RE (1989) *Enzyme Microb. Technol.* 11, 230.
9. Haveman J and Hahn GM (1981) *J Cell Physiol.* 107, 237.
10. Levine DW, Wang JS, Wang DIC and Thilly WG (1977) *Somatic Cell Genetics* 3, 149.
11. Miller WM, Wilke CR and Blanch HW (1987) *J Cell Physiol* 132, 524.
12. Mukhopadhyay A, Mukhopadhyay SN and Talwar GP (1993) *J Chem. Tech. Biotechnol.* 56, 369.
13. Nakashima RA, Paggi MG and Pedersen PL (1984) *Cancer Res.* 44, 5702.
14. Renner ED, Plagemann PGW and Bernlohr RW (1972) *J Biol. Chem.* 247, 5767
15. Spier RE and Griffiths B (1984) *Dev. Bio. Stand.* 55, 81.
16. Sinksey AJ, Fleischaker RJ, Tyo MA, Giard DJ and Wang DIC (1881) *Ann. NY Acad. Sci* 369, 47.
17. Thomas JN (1986) In: *Mammalian Cell Technology* (Thilly WG, ed.) p. 109. Butterworths, Boston.
18. van Wezel AL (1967) *Nature(London)* 216, 64.

SUGGESTION FOR FURTHER READING

1. Cherry RS and Papoutsakis ET (1990) In: *Animal Cell Biotechnology* (RE Spier and JB Griffiths eds.) vol. 4. p. 71. Academic Press, London.
2. Griffiths JB (1988) In: *Animal Cell Biotechnology* (RE Spier and JB Griffiths, eds.) vol. 3. p. 179. Academic Press, London.
3. De Bruyne NA (1988) In: *Animal Cell Biotechnology* (RE Spier and JB Griffiths, eds.) vol. 3. p. 141. Academic Press, London.

4. Croughan MS and Wang DIC (1991) In: Animal Cell Bioreactors (CS Ho and DIC Wang, eds.) p. 213. Butterworth-Heinemann, Boston.
5. Prokop A (1991) In: Animal Cell Bioreactors (CS Ho and DIC Wang, eds.) p. 21. Butterworth-Heinemann, Boston.
6. Shintani Y, I wamoto K and Kitanot K (1988) *Appl. Microbiol. Biotechnol.* 27, 533.
7. Doran PM. (1995). *Bioprocess Engineering Principles*. Academic Press, London.

7

Separation of Viable Cells

Typical animal cell culture is based on one particular cell type (pure culture), whether it is used for the production of recombinant proteins, monoclonal antibodies, or viral vaccines. A single cell-type confers great advantage in terms of studying cellular properties, as compared to mixed with other cell types. In many studies, and even in clinical applications, more than one type of cells are used. Again, in many clinical application, one specific class of cells is separated (e.g., depletion of T cells from bone marrow for allogenic transplantation) from the heterogeneous population of bone marrow cells. Since peripheral blood and bone marrow are two major sources of clinically useful cells, this Chapter deals with the various techniques for separation of cells. Blood or bone marrow cells can be separated on the basis of two basic principles: (i) physical (density and size) properties, and (ii) biological properties (affinity of cell surface receptors to specific antibodies), which are elaborated below.

7.1 PHYSICAL PROPERTIES

7.1.1 Density gradient centrifugation

On the basis of buoyant density (Table 7.1), peripheral blood or bone marrow cells can be fractionated into 3-major fractions in a single-step or continuous-density gradient centrifugation. Ficoll and Percoll, the two density gradient media, are commonly used for isolating low-density *mononuclear cells* (MNCs) from high-density erythrocytes/ granulocytes/dead cells present in blood or bone marrow (Boyum, 1968). The success of this separation technique depends on two main parameters—the density of the gradient medium and the applied *g*-force on the cells. For example, high-density erythrocytes/granulocytes/dead cells can be separated (pelleted) from light-density MNCs and platelets, if subjected to *g*-force that exceeds buoyant densities of these cells in Ficoll-Hypaque solution (ρ = 1.077 g/ml). On the other hand, owing to low-density, MNCs and platelets do not penetrate dense Ficoll-Hypaque layer and collected at the interface and aqueous phase, respectively. The main disadvantage of density gradient separation technique is that the product, in this case MNCs, is often contaminated with platelets and erythrocytes.

Table 7.1 Densities of different blood cells

Cell type	*Density (ρ) in g/ml*
T cells	1.077
B cells	1.077
Monocytes	1.064
Granulocytes	1.09
NK cells	1.06
Erythrocytes	1.115

Source: Dr. Peter Van Vlasselaer, Activated Cell therapy, Mountain View, CA, USA.

The single-step gradient technique to isolate MNCs has been described below:

1. Take 10 to 15 ml of Ficoll-Hypaque in a 50 ml conical tube, and carefully layer 20 ml of 1:2 diluted (in PBS) bone marrow cells on top of the Ficoll cushion.
2. Centrifuge the tube in a properly balanced swing-out centrifuge bucket at 600 *x g*, at 20°C for 30 min, without brake.
3. Collect the buffy-coat layer (cells from the interface) consisting of most of the MNCs, and wash the cells 3 times in the culture medium.

The positions of different cell fractions after centrifugation are shown in Figure 7-1.

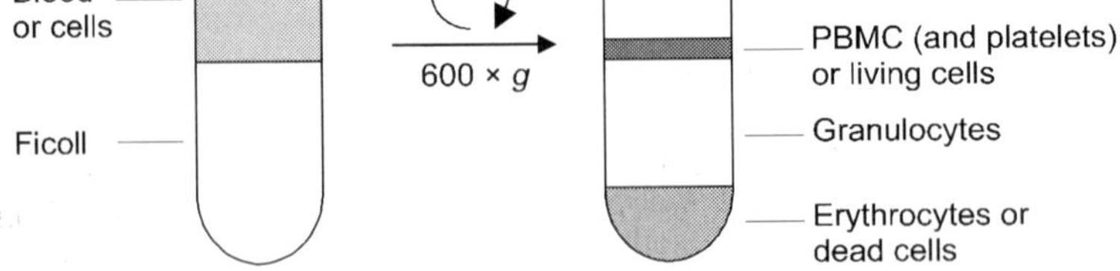

Figure 7.1 One-step density gradient separation of blood cells. Mixture of cells can be separated on the basis of their densities. High-density erythrocytes are collected at the bottom of the tube, whereas lighter density mononuclear cells are retained at the interface between Ficoll-Hypaque and aqueous medium.

7.1.2 Centrifugal elutriation

It is also known as counter flow centrifugation elutriation, or counter-streaming centrifugation. *Centrifugal elutriation* is based on the principle of differences in sedimentation velocity of cells to be separated (Figdor, 1997). A suspension of cells is loaded in the elutriation chamber (Figure 7.2A), which rotates on its axis and the medium is pumped into the chamber in a direction opposite to the *g*-force (Figure 7.2A). The chamber is designed such that at the entry it diverges, and at the exit it converges into a small tube. The cells enter into the chamber stay at points where the outward-directed centrifugal force is balanced with the inward-directed fluid dynamics, maintained by appropriate rotor speed and flow rate of liquid. Therefore, the cells differ in *sedimentation velocity* (volume and density) tend to distribute themselves on different horizontal zones (*equilibrium zones*), depending upon their physical properties at a given chamber rotor speed and the flow rate of the medium. In other words, cells are distributed according to their size from larger to smaller at entry and exit, respectively. Immediately after loading, cells will redistribute according to sedimentation velocity (Figure 7.2B). Thus, the cells can be fractionated either by increasing the flow rate of the medium or decreasing *g*-force by reducing rotating speed (Figure 7.2B).

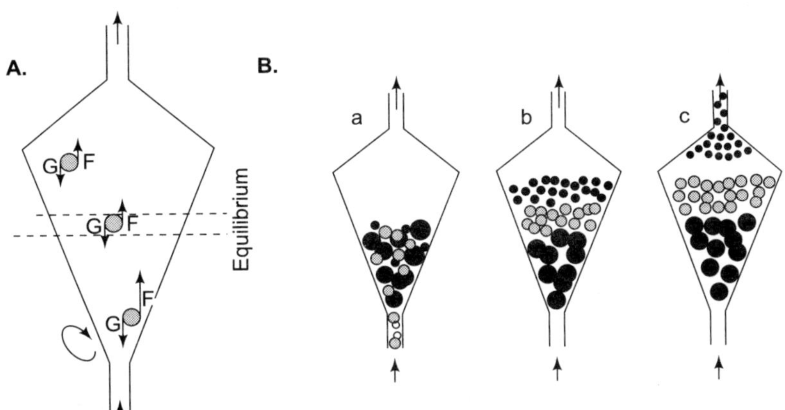

Figure 7.2 Centrifugal elutriation. A. Entered cells are experienced with two opposing forces: G-force due to centrifugation and F-force due to flow rate of the liquid. B. Just after loading (a) cells are rapidly oriented on the basis of sedimentation velocity, (b) finally fractionated, and (c) according to the size when medium flow rate is increased or rotor G-force is decreased. [Adapted from Ref. 2]

The separation efficiency in a centrifugation elutriation is influenced by several parameters. Besides the differences in sedimentation velocity, flow rate, rotor speed, and sample volume are the determinants of the performance of separation process. The other parameters that might affect the elutriation process are the temperature of the medium and its density and shape of the elutriation chamber. The most popular application of centrifugal elutriation system is the isolation of progenitor hematopoietic cells from cytokine mobilized peripheral blood.

7.2 BIOLOGICAL PROPERTIES

Each mammalian cell, specifically bone marrow or immune cells, expresses unique cell surface molecules (markers), based on which they are distinguished from each other. High affinity monoclonal antibodies specific to these surface markers are now available, by which cells can be identified and separated. The binding affinity of these antibodies towards specific markers is the basic biological property by which a purified fraction of cell may be obtained from a heterogeneous population of cells.

7.2.1 Panning

The *panning* is the simplest and less accurate method to deplete undesired cells or to select target cells from heterogeneous mixture of cells. Antibodies specific to the target cell(s), to be depleted or selected, are coated onto the polystyrene plastic dishes. Properly diluted cell suspension is applied on the coated surface. The cells are incubated at 4°C for 45-60 min, before unbound cells are removed from the dish by washing (Figure 7.3). It is a simple technique for separating target cells and the method is compromised with the purity of the target cells.

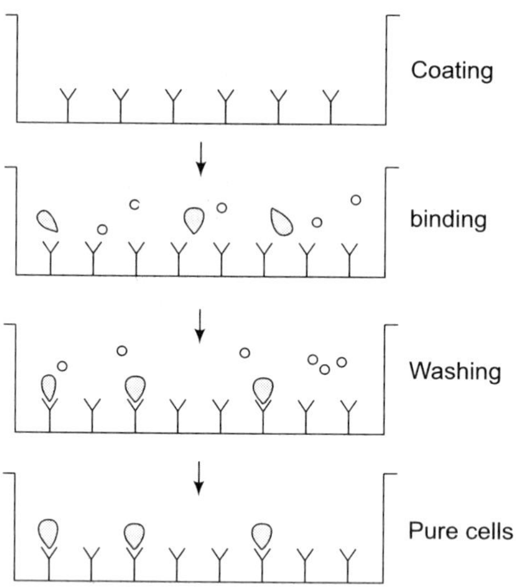

Figure 7.3 Panning of target cells using cell surface specific antibody. (Y): antibodies; (◊): cells expressing antigen specific to the antibodies coated on the plastic plate.

7.2.2 Magnetic cell sorting

Selective cell separation requires a method to identify the target cells and a method to separate them. In *immunomagnetic* separation, identification is achieved by the use of a monoclonal antibody directed to a specific cell-surface molecule(s), which may or may not be biologically exclusive to the target cells (Kantor, 1997). Once the target cells are identified, it is possible to link them with solid phase to separate cells by using different physical processes. The general principle for magnetic cell sorting technique is explained diagrammatically in Figure 7.4. A suspension of viable cells is incubated with specific monoclonal antibody targeted to a particular cell population, covalently coupled with the magnetic microbeads (50-100 nm diameter). The mixture is exposed to strong magnetic field, where the target cells attached with the magnetic microbeads are attracted to the magnetic field, leaving other cells free in the suspension, which can be easily separated out by decanting. Alternately, the cell suspension is passed over a column containing a ferromagnetic matrix placed in a strong magnetic field. The matrix creates a magnetic field gradient to filter-out magnetically labeled cells, while unlabeled cells flow through the column. The target cells are then removed by washing the column with a buffer in absence of magnetic field (Figure 7.4). The purity of the target cells obtained in magnetic cell sorting is very high (> 90%). There are many operating variables that determine the purity of the target cells. These are specificity of the antibody to the target antigen, ratio of cells to the microbeads, time of incubation of cells with the beads, operating temperature, nature of the column and the gradient of magnetic field developed in the ferromagnetic matrix, etc.

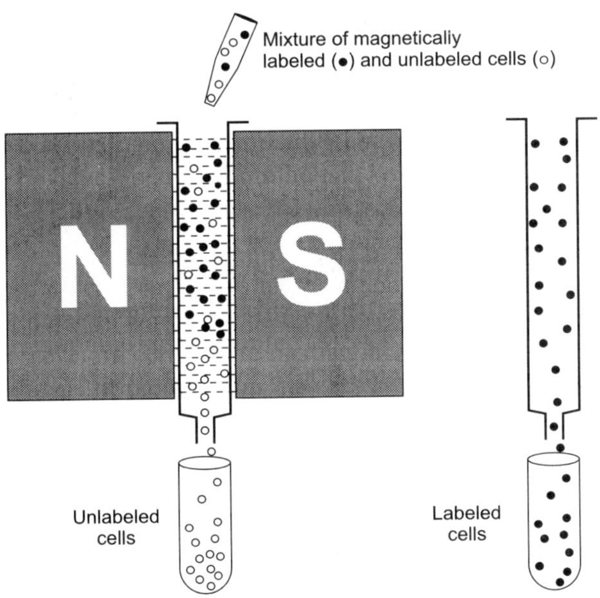

Figure 7.4 Isolation of magnetically labeled cells. In the first step, the magnetically labeled target cells are captured in the column by magnetic field. The target cells are separated in the second step by detaching the column from the magnetic field, followed by washing with buffer.

A number of paramagnetic solid phases have been used for cell separation. They are widely varied in their size, shape, magnetic properties, and also in method for collection. Microbeads can be easily collected in a magnetic field generated by small magnet. Paramagnetic nanoparticles and magnetic colloids are too small to be collected by above mechanism, and high gradient magnetic field is used for their collection. The advantage of magnetic nanoparticles is two fold—they remain dispersed due to Brownian motion, and are supermagnetic (Gee, 1997). The magnetic core is small enough for the particles to retain any residual magnetism, when keep them away from the magnetic field.

Immunomagnetic separation can be indirect or direct. In the indirect separation, cells are first treated with the monoclonal antibody or biotinylated-monoclonal antibody, and subsequently reacted with the beads coated with anti-I_GG antibody or avidin (Figure 7.5A). One major advantage of indirect separation is that a mixture of monoclonal antibody can be used to identify the target cells, followed by incubation with a single capture reagent. This process is useful when the densities of target antigens is changed in the cells. Direct separation is based on the incubation of target cells with microbeads that are directly coated with target monoclonal antibody (Figure 7.5B). Direct separation has the advantage of being single-step procedure that eliminates extensive washing, and thus minimizes the time of separation. However, direct procedure may show binding complication with the target antigen due to steric effect, since both the active molecules (cells and antibody conjugated magnetic beads) are present on the solid phase.

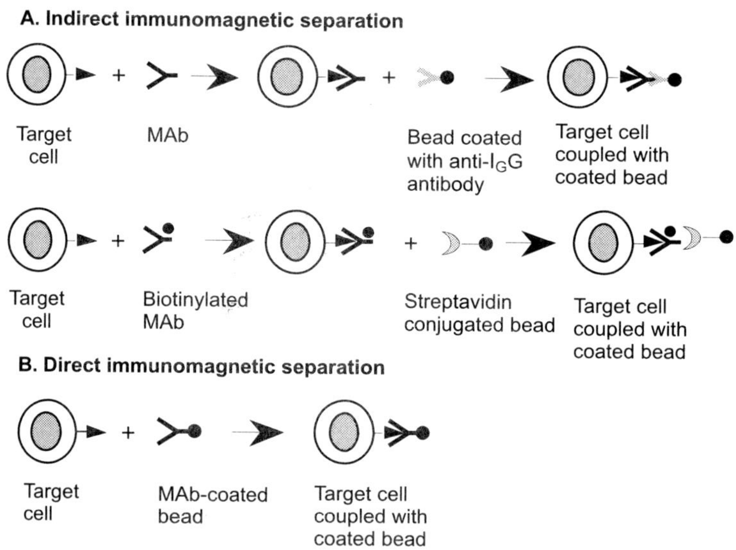

Figure 7.5 Indirect and direct immunomagnetic separation of cells. Target antibody ≻; Target antibody conjugated with biotin (≻•;) Target antibody conjugated magnetic beads (≻•;) Avidin/Streptavidin conjugated magnetic sbead (⟩–•).

The magnetic separation process has two basic modes: *positive* and *negative selection*. In positive selection, target cells are captured in the microbeads, which are later recovered from the column (Figure 7.6A). It has the advantage that a single specific monoclonal antibody for the target cell can be used to separate cells, but the main disadvantage is that the microbeads remain attached with the target cells for long period of time. In negative selection, the aim is same, but first the above process is followed to remove all nontarget cells, leaving microbeads free target cell(s) population in the effluent (Figure 7.6B). Negative selection is the method of choice when the type of the contaminating cells are familiar (i.e., respective antibodies are available), and selected cells are to be used for *in vivo* studies. The detailed procedure for positive selection of hematopoietic progenitor stem cells ($CD34^+$) from cord blood, counterstained with CD133 antibody is shown below:

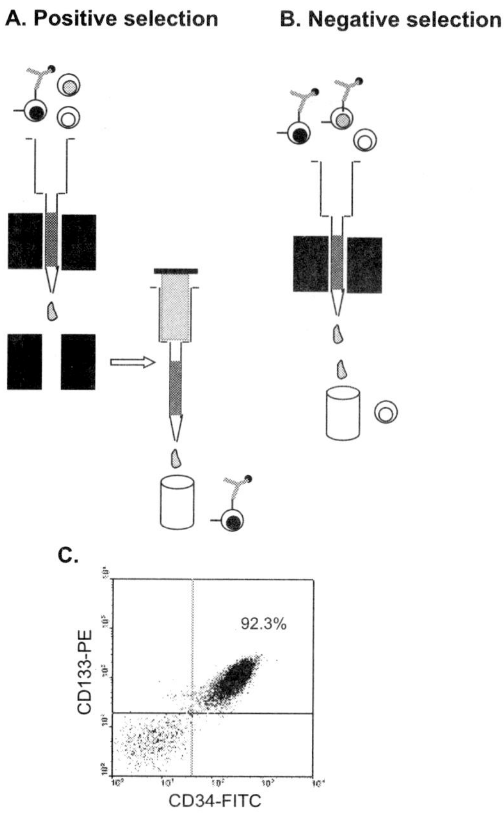

Figure 7.6 Model of purifying cells by using immunomagnetic separation technique. **A.** Positive selection; **B.** Negative selection of cells using immuno-magnetic beads. **C.** Positive selection of $CD34^+$ cells, counterstained with CD133 is shown in the dot-plot.

1. Take 50×10^6 cells (single-cell suspension) in a tube and wash the pellet in labeling buffer (degasses PBS pH 7.2, containing 2 mM EDTA).
2. Resuspend cell pellet and label them with biotinylated antibody (CD34-biotin) for 10-15 min at 4-8°C (10 μl antibody/10^7 cells/ 100 μl suspension).

3. Wash the cells to remove unbound primary antibody by adding 5 ml of labeling buffer and centrifuge at 300 x*g* for 10 min. Repeat the washing step one more time.
4. Suspend cell pellet in 450 μl labeling buffer, add 50 μl of Streptavidin-microbeads, mix the suspension gently, and incubate at 4-8°C for 15 min.
5. Wash the cell pellet in 5 ml of labeling buffer and resuspend pellet in 500 μl of separating buffer (degasses PBS pH 7.2, containing 0.5% BSA and 2 mM EDTA).
6. Apply cell suspension on LS column (previously equilibrated with separation buffer) placed on magnet stand. Collect the unlabeled cells for analysis, and wash the column with 3 × 3 ml of separation buffer.
7. Remove the column from the magnet stand and place on a collection tube. Apply 5 ml separation buffer on the column, and immediately flush out magnetically labeled cells by firmly applying pressure on the plunger.
8. Pellet down the cells, label one aliquot of cells with CD133-PE and CD34-FITC antibodies antibody, and analyze by flow-cytometry to check the purity of the cells. The dot-plot of the analysis is shown in Figure 7.6C.

7.2.3 Fluorescence-activated cell sorter (FACS)

It is the most powerful technique for separating cells of specific type from a heterogeneous mixture of cells. The principle for the separation of cells is based on optical and fluorescent properties of the cells, and the machine used for this purpose is known as *flow-cytometer*. This technique is predominantly used to identify fluorescence intensity of the cells labeled with fluorescent-conjugated antibodies, which are morphologically/structurally identical. For example, different subsets (differentiated populations) of lymphocytes are morphologically uniform—small round cells with a dense nucleus and little cytoplasm. However, they are distinct in terms of the expression of specific characteristic cell-surface proteins. Flow cytometer detects individual cell that passes in a stream of fluid through a laser beam. Each time a cell passes, the laser beam deflects the light from the detector, and this interruption of the laser signal is recorded. If some of these cells are labeled with specific monoclonal antibodies tagged with fluorescent dyes, they will experience excitation by the laser beam and emit light that is again recorded by a second detection system (photomultiplier tubes) located at a right angle to the laser beam. The simplest form of the instrument counts cell number and records the level of fluorescence intensities by the individual cells (Figure 7.7A). By using a standard program, computer generates plots of cell number versus fluorescence intensity in the histograms, or as dot plots showing the corrected data for two parameters. In case of antibody-conjugated fluorochromes, the intensity of signals may vary 3 to 4 orders of magnitude. Thus, logarithmic scales are used to represent the data. Dot-plot of lymphocytes, co-stained with CD8-FITC (green fluorescence) and CD4-PE (red fluorescence) is shown in Figure 7.7B.

The sorting of cells is carried out based on the principle of droplet sorting. After the cells pass through sensing device, the fluid stream is broken into small droplets. Some droplets are empty and others contain cells. The droplets containing cells are given electrical charge and are physically separated from uncharged droplets by an electrostatic deflection. The formation of droplets, charging and finally the deflection of the droplets is shown in Figure 7.7C. Once the cell passes through the analyzing zone (laser excitation followed by detection of fluorescence emission), droplets are formed by the action of vibrating transducer on the liquid. If the cells analyzed in the laser beam are to be sorted, an electrical charge is applied to the stream. The stream is charged as the drop breaks off and separated from the stream. The charge on the sample stream is then turned off and the following droplet is not charged. The droplets pass through an electric field formed between two plates across which high voltage is applied. Charged drops are deflected in the electric field toward a collection tube. Most sorters allow drops to be charged either positively or negatively, which allows drops to be deflected by two plates of opposite charge. Thus, simultaneously two different populations of cells can be sorted based on the applied charge on the dropets.

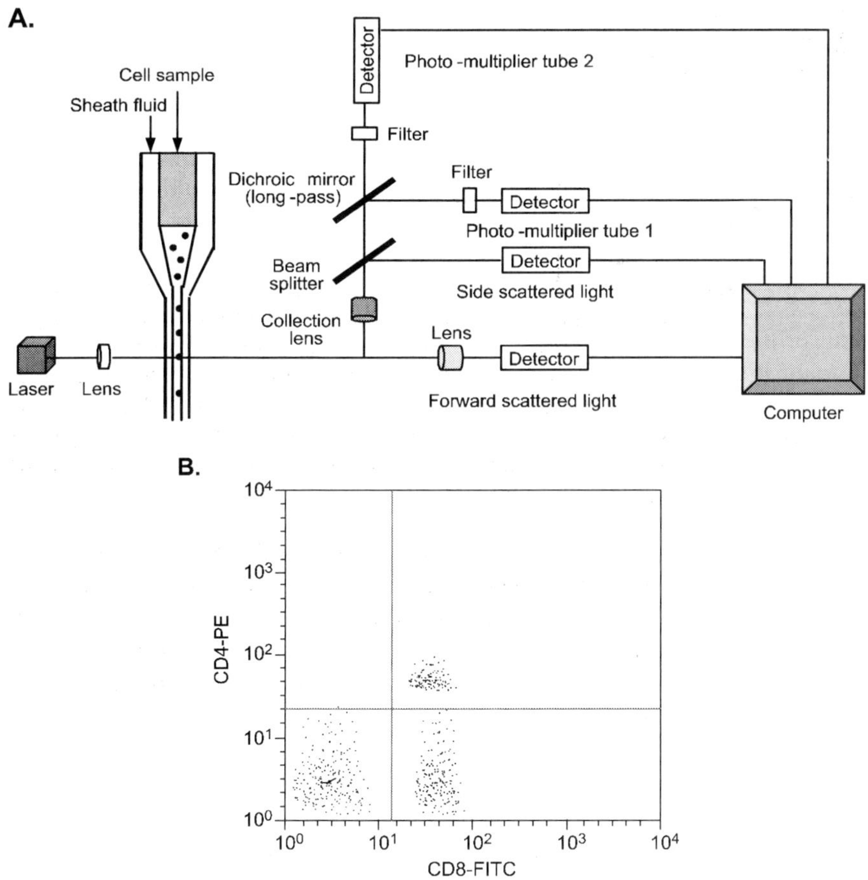

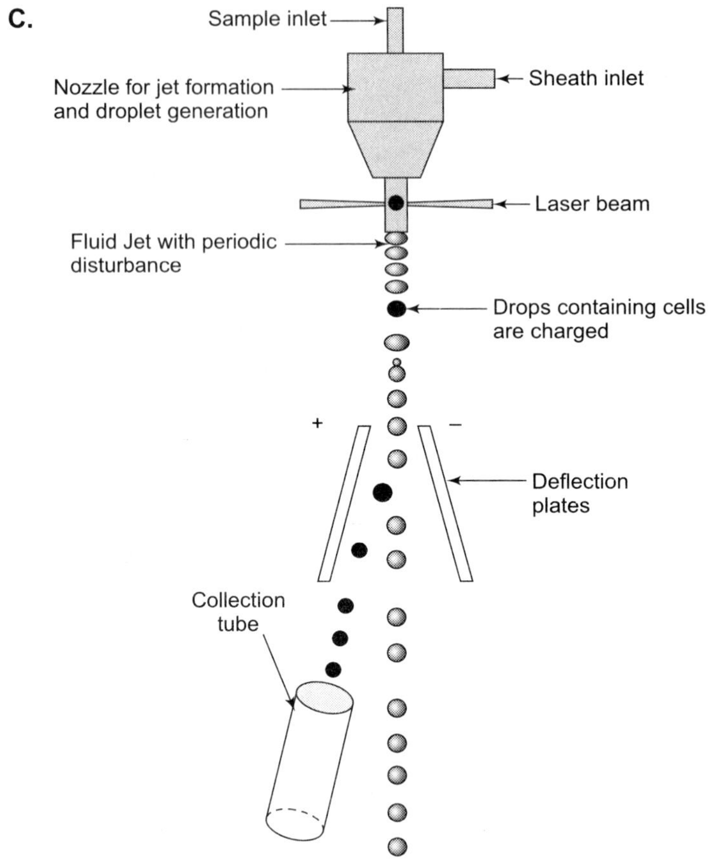

Figure 7.7 **A.** Optical system of a simplified three parameters flow cytometer. The diagram shows, in schematic form, the arrangement of mirrors, optical filters, and detector photomultiplier tube (PMT). Single-cell suspension is forced through the nozzle of the machine under pressure. The cells, confined to the axis at the resultant fluid stream by concentric sheath of cell-free fluid, pass through a laser beam, which is focused into the stream; **B**. Dot-plot of flow cytometry analysis of lymphocytes, **C.** Schematic representation of the essential features of fluorescence activated cell sorter.

The major limitations of FACS technique are that the method relies on costly equipment, and the sorting process requires more time compared to the magnetic cell sorting technique. In a standard instrument, sorting of viable cells can be performed at analysis rates of 10,000 cells/sec with greater than 98% purity. The purity of cells and sorting rates are mutually exclusive, high sorting rate compromises with the purity of the cells.

7.3 SUMMARY

1. Mammalian cells are separated based on (i) physical, and (ii) surface properties.
2. The purity of cell in centrifugal elutriation apparatus depends on its sedimentation velocity, rotation speed, flow rate, and loading of sample.
3. Magnetic cell sorting and FACS depends on the cell surface antigens and the specific antibodies.

ABBREVIATIONS

FACS: Fluorescence-activated cell sorter
MNC: Mononuclear cell
PBS: Phosphate buffered saline

REFERENCES

1. Boyum A (1968) *Scand J Clin Invest* 21, 77.
2. Figdor CG, Preijers F, *et al.* (1997). In: *Cell separation methods and application* (D Recktenwald, A Rasbruch, eds.), p. 43. Marcel Dekker, Inc. New York.
3. Gee AP (1997) In: *Cell separation methods and application* (D Recktenwald and A Rasbruch, eds.) p. 175. Marcel Dekker, Inc. New York.
4. Horan PK, Wheeless LL Jr. (1977) *Science* 198:149.
5. Kantor AB, Gibbons I, *et al.* (1997). In: *Cell separation methods and application* (D Recktenwald, A Rasbruch, eds.) p. 153. Marcel Dekker, Inc. New York.
6. Melamed MR., Lindmo T., Mendelsohn ML, eds. (1990) Flow cytometry and sorting, 2nd ed. New York, Wiley-Liss.
7. Khurana S, Bhaskar S, Mukhopadhaye A (2007) In: Advanced techniques in soil microbiology (A vermaa, Rolf Oelmuller, eds.), p. 92, Springer-Verlog, Berlin.

SUGGESTION FOR FURTHER READING

1. Funderud S, Kjell N, *et al.* (1987) In: *Lymphocytes- A practical approach* (Klaus, GGB, ed.) p. 55-64. IRL Press, Oxford.

Section III

Therapeutic Proteins and Bioengineering Aspects

8

Stable Gene Expression in Mammalian Cells and Monoclonal Antibody

Each animal cell acts as a factory for the synthesis of complex proteins, whether they are membrane associated or secretory in the case of recombinant proteins. During past 20 years, a considerable extent of scientific and technical advancement have been made in the production of therapeutic proteins, which resulted into more than 100-fold increase in the yield of many products. The improvement in yield has been possible due to better understanding of cellular metabolism (by media design/*metabolic engineering*), mammalian gene expression (by vector design, selection of high yielding clone, and *host cell engineering*), and high cell density culture (by *process engineering*). This chapter deals with stable gene expression in mammalian cells and development in therapeutic monoclonal antibodies.

8.1 PATHWAY FOR RECOMBINANT PROTEIN PRODUCTION IN MAMMALIAN CELLS

The regulatory events for the synthesis of recombinant therapeutic proteins in mammalian cells are shown in Figure 8.1. The regulatory events are broadly divided into four groups: (i) design of expression vector; (ii) integration of plasmid in the specific location of the chromosome; (iii) stability of mRNA and its processing; and (iv) translation of protein, its post-translation modification and secretion (Barnes and Dickson, 2006). Some of these aspects have been discussed below.

8.1.1 Mammalian expression vector

The term *vector*, in molecular biology, indicates the vehicle by which naked DNA or collection of nucleotide sequences is introduced into a host, which could be a single cell or tissue of a live organism. The function of mammalian expression vector is to ensure correct and efficient expression of cloned gene and the corresponding protein. The expression of mammalian proteins has been carried out either in transient or stable expression system. In transient expression, the vector sequences are not integrated into the chromosome but are maintained extrachromosomally for short duration (maximum 5-day). The transfected cells produce low quantity of protein, and its quantity primarily depends on the *transfection* efficiency. The genes of the interest are delivered to the cells (commonly used cell lines are HEK-293, COS, and BHK) through plasmid or virus vectors. In the production of therapeutic proteins, stable transfection technologies are used, where plasmid sequence is integrated into the chromosome of the host cells. Thus, the stable expression of proteins for a longer duration is possible. The present discussion has been limited to the mammalian stable expression systems for the production of therapeutic proteins.

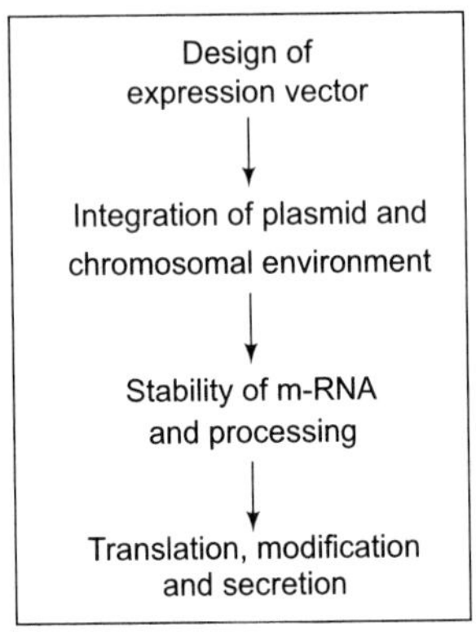

Figure 8.1 Regulatory events for the expression of recombinant proteins in mammalian cells.

The mammalian expression vector contains a minimum of six structural components for promoting efficient expression of heterologous genes. These are as follows:

(1) **Backbone shuttle sequences:** These allow the movement of plasmids from one organism to another. For more efficient propagation of mammalian expression vectors, bacterial replication and selection sequences are used as the backbone of the plasmid (e.g., pBR322 replication origin).

(2) **Appropriate translational initiation sequence:** The secondary structure of 5′ untranslated region (UTR) of many genes can inhibit the expression in a recombinant system. Again, mammalian cDNAs often contain multiple ATG codons within this region, only one of which can act as the appropriate translational initiation signal for authentic foreign protein. It is essential that the vector contains no additional ATG codons in the 5′ leader, otherwise this would give aberrant translation initiation.

(3) **Intron:** Splicing of primary mRNA transcripts is a common event in mammalian system. Addition of an intron to the 5′ UTR of a vector can significantly improve the expression level.

(4) **Polyadenylation signal:** Polyadenylation of mature mRNA is found to be critical for its own stability.

(5) **Mammalian promoter/enhancer:** Promoters direct transcription in a defined direction, and the enhancer elements augment transcription from a promoter. The promoter consists of specific sequences where RNA polymerase II initiates transcription. The most commonly used enhancers/promoters in mammalian expression vectors are human cytomegalovirus (hCMV), semian virus (SV40), and mouse metallothionein (mMT).

(6) **Marker gene:** Used to select stably transfected cells and to amplify the gene.

The expression vectors used for the production of recombinant proteins have been discussed latter; however, the general structural feature of *monocistronic expression vector* is shown in Figure 8.2. The choice of an expression system for the production of recombinant therapeutic proteins depends on many factors, these include cell type and growth characteristics, expression levels, nature of application (therapeutic vs diagnostic) of the protein, as well as regulatory issues and economy.

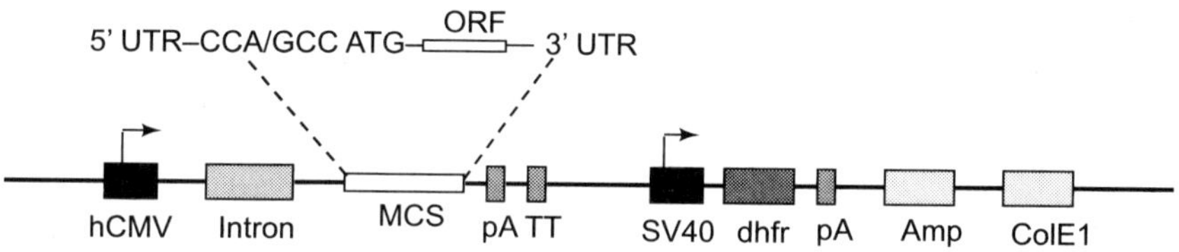

Figure 8.2 Structural components of a mammalian expression vector. The ColE1 origin of replication and the ampicillin-resistance gene allow vector propagation in bacteria. The *dhfr* gene allows both selection and gene amplification. hCMV, human cytomegalovirus enhancer/promoter; MCS, multiple cloning site; pA, polyadenylation signal; TT, transcription terminator; SV40, large tumor (T) antigen; UTR, untranslated region; ORF, open reading frame. The optimal Kozak translational initiation sequence (CCA/GCC) before initiating ATG of cDNA gene expression has been shown. [Taken from Makrides, 1999]

8.1.2 DNA delivery and its chromosomal integration

Nonviral gene transfer is the preferred approach for establishing stable clone for manufacturing therapeutic proteins. Various methods for delivering gene into mammalian cells have been discussed in the literature. Among them, calcium phosphate precipitation, lipofection, electroporation, and optoporation are more popular (Norton and Pachuk, 2003). It is important to remembere that the efficiency of transfection depends on the method of DNA delivery and the cell line used. The recombinant gene and the selection (marker) gene are preferably to be present in the same vector, but they can be present in two different vectors. The efficiency of stable transfection is improved if the plasmids are linearized before transfection. However, supercoiled plasmid DNA molecules are naturally linearized in the nucleus, with the help of endo- and exo-nucleases, before integration into the chromosome (Finn *et al.*, 1989).

The integration of DNA in the chromosome is a random event; the site of integration plays a major role in the transcription rate of the transgene. Two types of chromatin are found in mammalian genome: heterochromatin and euchromatin. The heterochromatin is highly condensed and silenced region of the chromatin, whereas the euchromatin is more open and active. Genes normally active in a euchromatin region will typically be silenced, often showing a variegating *phenotype*, when placed adjacent to or within a heterochromatin region by *chromosomal rearrangement* or *transposition*. The molecular basis of gene expression within the defined chromatin area is controlled by *epigenetic modifications* of DNA (methylation) and *histones* (acetylation, methylation, phosphorylation and ubiquitination). Due to random integration process of foreign gene, there is a fair amount of chance that the transgene will get positioned in the heterochromatin region (also known as *negative positioning*), resulting into little or no expression of target protein due to *gene silencing*. Even transgene integrated in the euchromatin may be inactivated due to neighboring heterochromatin. Two most important approaches can be adopted to prevent transgene from the effect of negative positioning. These are either by physically preventing the surrounding chromosomal environment from influencing transcription of the transgene or alter the overall epigenetic environment of the surrounding chromosomal DNA (Barnes, 2006). An episomal replicating vector has been described, which contains the SV40 origin of replication and the scaffold/matrix-attached region (S/MAR) from the human interferon-β gene. S/MARs are DNA sequences associated with chromosomal origins of bi-directional replication. The vector replicates at low copy number in CHO cells and is stably maintained for more than 100 generations without selection (Piechaczek *et al.*, 1999). Students are advised to consult references to understand how to prevent the transgene from negative positioning (Kwaks and Otte, 2006; Kwaks *et al.*, 2003; Heng *et al.*, 2004).

8.1.3 mRNA stability, translation, and secretion of protein

The stability of mRNA is most critical for the steady production of recombinant proteins. Some factors that control mRNA expression may vary with the culture environment. If the expression of mRNA is higher than the translation and proteins secretion rates, the expression system becomes unstable and unbalanced. Another example of unstable expression system is due to high translation rate, every protein molecule does not get proper chance for modification and secretion. This may lead to the aggregation of protein and its subsequent degradation. In case of a complex protein (e.g., antibodies), if *synchronization* between mRNA synthesis, protein translation, and secretion is failed, it may result in the synthesis of one of the components in greater proportion. For example, it has been found that in case of CHO and NSO cells, light chain mRNA and intracellular accumulation of light chain protein are found to be more than heavy chain. Thus, these cells produce a mixture of light chain and a complete antibody molecule. A similar result was observed in case of human chorionic gonadotrophin (hCG) synthesis by the chorionic and many tumor cells, in which αhCG is always synthesized in much higher proportion than the β-counterpart.

8.2 SELECTION AND GENE AMPLIFICATION

Stably transfected cells produce *heterologous* proteins on a continuous basis. As the gene transfer efficiency is much lesser than 100%, it is of paramount importance to select the clone that expresses gene of interest at highest level. Selection ensures stable integration of the heterologous gene in the chromosome of the host cells. Many selection systems used in practice are based on metabolism of purine and pyrimidine (Figure 8.3). The biosynthetic pathways for purine and pyrimidine are independent, as two different enzyme systems are involved for their metabolism. In animal cells, if one pathway is defective, alternative pathway remains active to continue metabolism. If a specific drug is used to block the alternate pathway, cells start dying unless the mutant gene of the defective pathway is replaced with the wild-type gene.

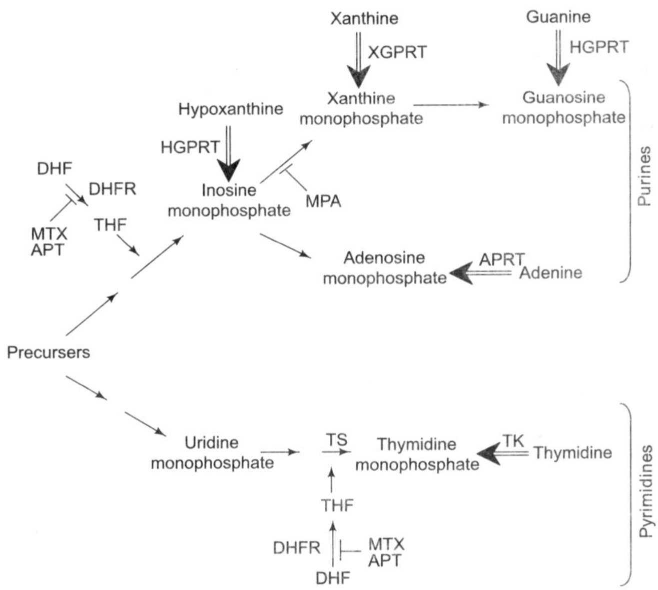

Figure 8.3 The biosynthetic pathways for metabolism of purine and pyrimidine by mammalian cells, and the enzyme systems involved. The pathways showing inhibitors of key enzymes, and the intermediate precursors. For example, MTX is the inhibitor of DHFR enzyme.

Eukaryotic chromosome is replicated once per division and this appears to be due to block in re-initiation of replication. However, in some cases, the chromosomes are repeatedly replicated without cell division, such as DNA virus infected cells and salivary gland cells of certain insects. Most interestingly, it has been found that a certain portion of chromosome in a number of cell types replicates several times without cell division, which is termed as *DNA amplification* (Stark and Wahl, 1984). The major function of localized amplification of chromosome is to produce larger quantities of gene product than could be achieved by single copy gene. It has been observed that the treatment of cells with methotrexate (MTX) results in the appearance of a few resistant cells, many of which contain multiple copies of a region of DNA containing the gene encoding dihydrofolate reductase (*dhfr*)—the enzyme that is inhibited by MTX. This DNA may be amplified from several hundred to thousand folds and is often present as small extrachromosomal elements, known as *double minute chromosomes*. Therefore, selection and amplification of genes based on dihydrofolate reductase (*dhfr*) has become more popular in vector-based gene amplification. Glutamine synthetase (*gs*) and hypoxanthineguanine phosphoribosyl transferase (*hgprt*) are also used for selection and gene amplification. The important consideration in choosing an agent for selection is the degree of its selectivity or stringency. The clone expressing highest level of target gene can be selected in the presence of a more stringent agent, which automatically reduces the workload for screening of the clone.

8.2.1 Dihydrofolate reductase (*dhfr*) amplification

This system of selection of gene is most attractive, as a number of dhfr-deficient CHO cell lines are available, which can not grow without nucleosides in the culture medium. In DHFR system, expression of a recombinant protein can be augmented by exposing cells to methotrexate (MTX), a drug that blocks the activity of DHFR (Gandor *et al.*, 1995). In some cases, the selected clones retain the amplified vector sequences for extended period even in the absence of MTX selection pressure. Usually, the concentration of MTX varies between 50 to 500 nM during selection of the transformed clone. The typical cell types used for transfection of genes are the dhfr$^-$ varients of CHO cell line such as CHO-K1, DUKX-B11, DG44. The selection process requires exposure of the transfected cells to MTX for 2-3 weeks. By that time most of the cells die, but a few cells become resistant to the selection pressure even at highest concentration of MTX. The surviving cells expressing DHFR, in the presence of increasing concentrations of MTX, may contain several hundred to few thousand copies of the integrated plasmid embedded in the chromosomes (Wurm, 1986). The incorporation pattern of *c-myc* and *dhfr* sequences has been studied in CHO cells in the presence and absence of MTX selection pressure. Quantitative analysis of hybridization patterns showed substantial heterogeneity in cells growing in MTX-containing medium. On an average, 3-fold more integration of *c-myc* sequences and protein expression has been found in the presence of MTX than in its absence (Pallavicini *et al.*, 1990). The specific productivity of the recombinant protein produced by the selected cells has been found to augment by 10 to 20-fold, as compared with the unamplified cells (Zettlmeissl, 1988). The level of co-expressed proteins not only depends on the copy number of the gene, but is eventually limited to the capacity of the secretion apparatus of the host cells. In case of t-PA, the maximum secretion rate from CHO cells is reported to be 50 pg/cell.day (Kaufman, 1985). Different DHFR mammalian expression vectors have been developed for the production of recombinant proteins, however, most of them are proprietary in nature. In one such vector, pSVM.*dhfr* (Figure 8.4), the expression of *dhfr* gene is under the control of weak mouse mammary tumor virus (mMTV) promoter. In this case, the highest concentration of MTX requirement for amplification of gene is much lower (20 nM) than that has been used (500 nM) in strong SV40 early promoter-based expression vector, pSV2.*dhfr*.

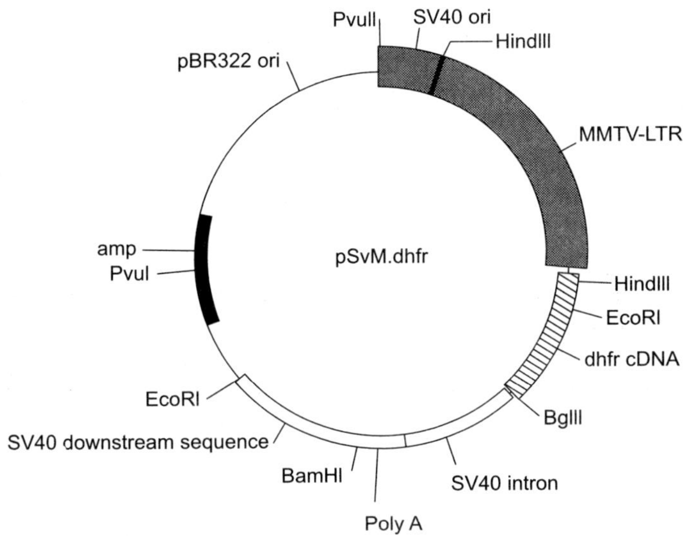

Figure 8.4 Amplifiable vector (pSVM.*dhfr*) of size 6.5 kb for dhfr⁻CHO cells. [*Source:* Celltech Therapeutics Ltd]

8.2.2 Glutamine synthetase (*gs*) amplification

Beside DHFR, a novel amplification system based on enzyme glutamine synthetase has been developed. Glutamine is a key metabolite in a number of biosynthetic and catabolic pathways. It is an essential component of cell culture media, and may also be produced from glutamate and ammonia by the means of GS. Certain cell lines (e.g., CHO) express sufficient amount of GS to survive in an glutamine-free medium. Specific and irreversible inhibition of GS can be mediated by methionine sulphoximine (MSX). A transfected GS gene in a mammalian expression vector may act as a dominant marker in cells if it confers protection for concentration of MSX, which is just sufficient to kill the wild type cells and the natural MSX-resistant varients. The rodent myeloma do not express sufficient GS to survive in glutamine-free medium. Thus, transfected GS gene can be considered as a selectable marker, provided the cells are cultured in glutamine-free medium. In this case, amplification of vector copy can be achieved by using gradual higher concentrations of MSX in the selection medium. For example, at concentrations of 10 to 100 μM MSX, resistant clones can be identified in selected NSO cells that amplified the desired transgene complex containing GS gene and the desired gene of interest (Bebbington *et al.*, 1992; Barnes *et al.*, 2000).

There are many advantages of using GS-based amplification system as compared with the DHFR system:

(1) Initial transfectants have higher production rates than those obtained in DHFR, so high level expression is possible in the first round of gene amplification.
(2) It is compatable with many cell types.
(3) MSX is not carcinogenic.
(4) As cells utilize ammonia to produce glutamine, the toxic metabolic end product (ammonia) is catabolized to form important nitrogen and carbon source (glutamine).

Since the exogenous glutamine by-passes the GS pathway, the medium to be used in culturing transformed cells by this system should be glutamine and serum free (or dialyzed serum). Such culture medium extend multiple advantages like less accumulation of ammonia and less media protein load to be handled in the purification steps. On commercial scale, the recombinant protein has been produced upto 200 mg/l in GS-based expression system (Field *et al.*, 1989). The expression vector used for CHO-K1 cell line is pEE14 (9.4 kb). This vector (Figure 8.5) produces high copy numbers than other GS expression vectors in transfected cells. The vector contains a GS minigene (lacking many restriction sites), a single intron of the GS gene, and polyadenylation signals from the GS gene (GS poly A-1 and GS poly A-2) expressed from SV40 late promoter (SV40L). The GS minigene and SV40L have been inserted into the *Bgl*II site of pEE6hCMV-B. NSO is the preferred cell line for transfection with GS vector system, as it is highly tranfectable and the yield of glutamine-independent clones is practically negligible (1 in 10^8 cells) (Galfre and Milstein, 1981).

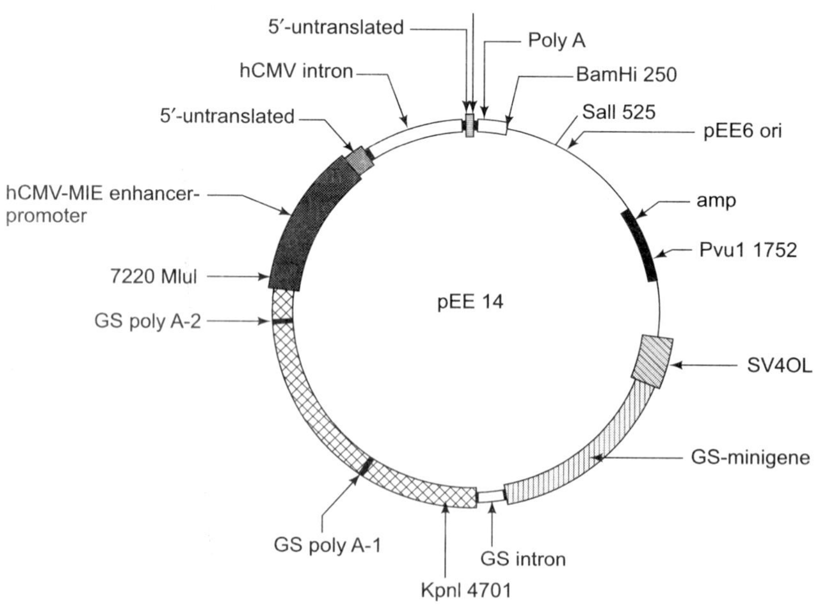

Figure 8.5 Expression vector (pEE14) of size 9.4 kb for CHO cells containing a GS-selectable marker. [*Source:* Celltech Therapeutics Ltd.]

8.3 MONOCLONAL ANTIBODIES (Mab)

Antibodies are the principal mode of defense against extracellular bacteria and exotoxins. They can intercept viruses before entry into the host cells. Antibodies are a mixture of closely related immune proteins, with slight but important structural and functional differences. They are produced to recognize and bind to foreign invaders, singling them out for elimination by the body's immune defenses. Human, mice and rats have the ability to make antibodies that have capacity to recognize (by binding to) virtually any antigenic determinant (*epitope*) and to discriminate between even similar epitopes. These make monoclonal antibodies the most attractive to target many molecules found in mammals such as receptors or other molecules found on the surface of normal cells, molecules specifically expressed on the surface of cancer cells, etc. In 1975, Milstein and Köhler described the process of cell fusion, the 'hybridoma technology', where the B cells conferred antibody production capability, while the myeloma cells enabled hybridomas to divide indefinitely and grow well in cell-culture. A single clone that produces desired antibody at high titer can be selected for large-scale culture for the production of monoclonal antibody. Many critical steps are involved for generating hybridomas and producing monoclonal antibodies.

8.3.1 Basic structure of antibody

Antibody is defined as *immunoglobulin* capable of binding specifically with the antigen that induced its production in a susceptible animal. Antibodies are produced by the plasma cell in response to the invasion of foreign molecules in a host. Immunoglobulins are glycoproteins, in higher mammals they are present in five distinct classes (e.g., IgG, IgA, IgM, IgD, and IgE). Each class of antibody differs in size, charge, amino acid composition, and carbohydrate content. IgG and IgA have different subclasses having wind functions.

Immunoglobulins have a basic unit of two light chains and two heavy chains hold together by disulfide bonds. The light chain has a molecular weight of 25 kDa, whereas in heavy chain molecular weight varies from 50 to 77 kDa depending upon the classes of immunoglobulins (on the basis of reducing SDS PAGE bands). The chains are folded into discrete regions called domains; light chains have two domains, and four to five domains are present in heavy chain. The N-terminal half of both light (L) and heavy (H) chains shows much sequence variability and is known as V_L and V_H domain, and the C-terminal half is known as constant region (C_L and C_H). The basic structure of IgG_1 subtype is shown in Figure 8.6. The constant region (C_H) is extended to the remaining part of the H chain. The portion of the antibody molecule that binds to antigen is formed by a small stretch of amino acids in the V regions of both H and L chain. Papain digestion of immunoglobulin generates two Fab fragments and one Fc fragment from each IgG molecules. Fab fragments are antigen-binding site, whereas Fc fragment is responsible for binding with antibody receptors present in the cell. Different fragments of immunoglobulin molecules are linked together by several disulphide bonds from H to H chains, H to L chains, and L to L chains.

Normal immune system generates an immunoglobulin for every possible immunogen that the system may encounter during lifetime. Instead of making completely separate immunoglobulin for each immunogen, the immune system is capable of rapidly changing and modifying the variable regions of immunoglobulins to suit a particular antigen. The N-terminal 110 amino acid sequences of the H and L chains constitute the variable regions of the immunoglobulin molecule. Certain sequences in the variable region show remarkable diversity and are termed *hypervariable regions* or *complementarity determining regions* (CDR). The CDR regions are the antigen-binding sites of an immunoglobulin.

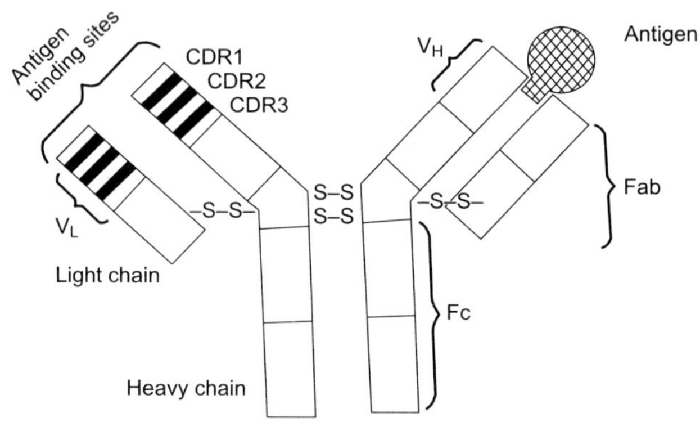

Figure 8.6 The basic structure of antibody (IgG). It consists of discrete domains like Fab, Fc, V_H, and V_L. The heavy and the light chains are joined together by disulfide bonds. The complementary determining regions (CDRs) represent the antigen binding domains.

8.3.2 Antibody response to vaccination (immunization)

First dose of vaccine or immunogen induces a primary antibody response within 5 to 7 days of its exposure. In primary response, low affinity IgM is synthesized, however, within few days to weeks (depending upon the nature of the immunogen), some antibody-producing B cells switch to make antibodies of the same antigen specificity but of an IgG, IgA, or IgE isotype with the help of activated T cells. When a second dose of vaccine (booster dose) is administered 2-4 weeks after the first dose, the memory B cells rapidly divide and begin to produce large quantities of antibody within very short period of time (Figure 8.7A). Secondary response generates mostly IgG isotype, as opposed to IgM. The activation of B cells occurs at various stages. In the final stage, B cells triggers a series of metabolic events, resulting in gene activation and expression of MHC Class II molecules, co-stimulatory molecules CD80, CD86, CD40, and receptors for interleukin-2, 4, 5, and 6 (IL-2, 4, 5, 6). Immunogens bound to the surface of immunoglobulins of the B cells are internalized and processed, and presented on cell surface as a complex with MHC Class II molecules. The B cells, recognized by activated CD4 helper T cells (Th2 type), and CD28, CD40 ligand, T cell receptor (TCR) bind with corresponding receptors/ligands on B cells (Figure 8.7B). During the T-B cell interaction, T cells secrete different cytokines (IL-4, IL-5, and IL-6) that have significant effects on B cells. Collectively, these cytokines enable activated B cells to proliferate and stimulate the activated B cells to synthesize and secrete antibodies and promote the differentiation of B cells into antibody-secreting plasma cells (Figure 8.7C). The Th2 cells also help B cells to undergo affinity maturation through somatic hypermutation (point mutations into the gene sequences encoding CDR) and to switch antibody isotype, that is, IgG, IgA, or IgE. A set of activated B cells, however, remains unstimulated with IL-6 and constitutes as the *memory B cells*. The memory B cells provide faster antibody response to the same immunogen in the future (Figure 8.7C).

Activation of B cells in response to T-dependent antigens has been briefly discussed in this section. However, students are advised to consult any textbook on Immunology to understand antigen processing and presentation, and the activation of T cells.

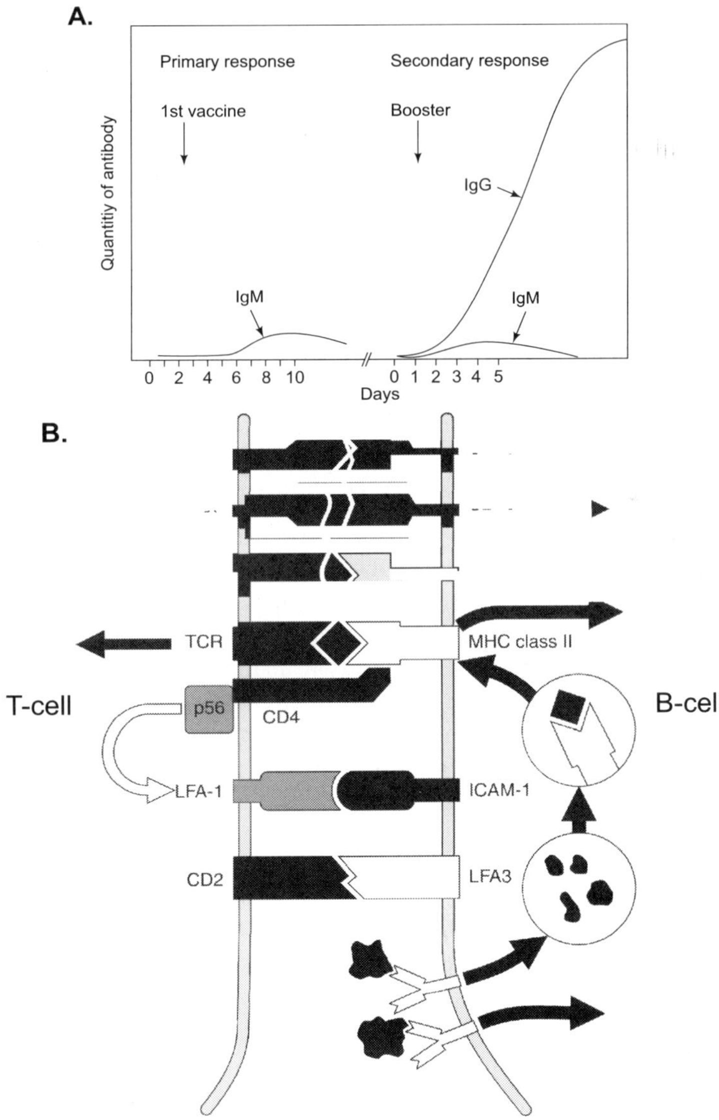

(*Contd.*)

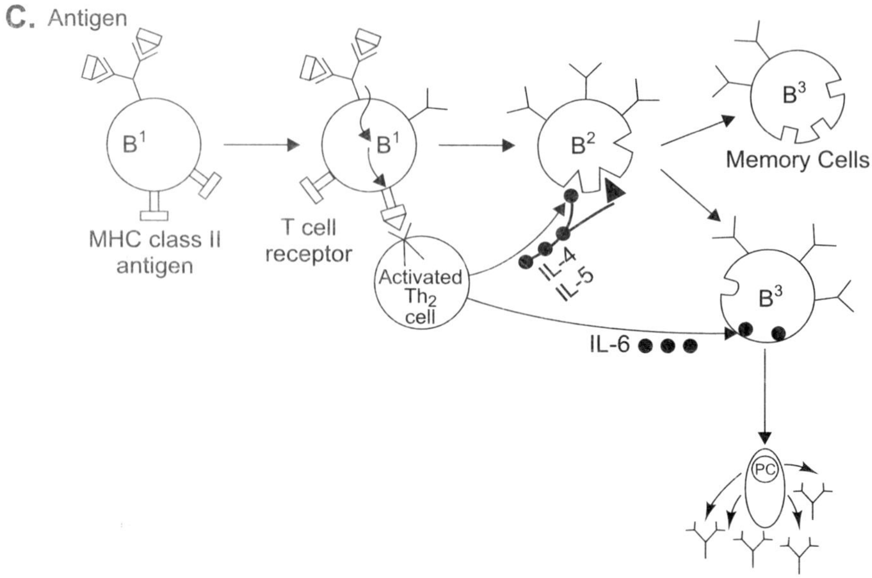

Figure 8.7 T cell mediated antibody response. **A.** Primary antibody response is obtained after injection of vaccine (immunogen), where IgM synthesis is predominated. In second dose (first booster dose), IgG antibody titer is predominated over any isotype; **B.** T cell - B cell interaction through specific receptors and cognate ligands; **C.** Different stages for the activation of B cells.

8.3.3 Therapeutic use of antibodies

Antibodies are raised by body's immune system to recognize and bind foreign invaders in order to neutralize their pathological effects on the host. This is possible only when our immune system is active, and not compromised in any respect. Synthesis of antibodies through our immune system (as mentioned above), to control over invading object, is the result of *active immunization*. It has been observed that antibodies produced in one person are effective when injected into another individual. Establishing immunity in an individual by injecting antibodies is known as *passive immunization*. In an *immunocompromised* individual, passive immunization by injecting immunoglobulins prepared from other individual known to have high levels of antibodies to the infectious agent has been found quite useful. For example, immunosuppressed allograft recipients may need cytomegalovirus (CMV) immune serum after inadvertent exposure to CMV. Passive vaccination is also used when a patient cannot make sufficient quantities of antibodies within short time to protect against diseases. For example, in a contaminated wound, *Clostridium tetani* can produce lethal quantities of toxins long before an unvaccinated individual could generate neutralizing antibodies against the toxin. In such a case, the patient is also given injection of anti-tetanus immunoglobulin together with vaccination against future exposure. Another indication for passive vaccination is snakebite, which requires the use of antivenin. Since individual's immune system through natural exposure is rarely available, it is necessary to use antisera prepared in other host to block the binding of toxins to the cell receptors. The theory of passive immunization made possible for the development of new era of therapeutic applications of monoclonal antibodies (Mab) in wide areas diseases like transplantation rejection, cancer, rheumatoid arthritis, etc. A partial list of FDA approved therapeutic antibodies for clinical applications, as well as antibodies under various stages of development is shown in Table 8.1.

Table 8.1: Therapeutic antibodies in the present and future use

Trade name	*Company*	*Application*
Herceptin	Genentech	Breast cancer
Remicade	Centocor, Abbott, Schering-Plough, Johnson & Johnson	Crohn's disease, Rheumatoid arthritis
Simulect	Novartis	Transplantation rejection
Zenapax	PDL/Roche	Transplantation rejection
Rituxan	IDEC/Genentech	Non-Hodgkin's lymphoma
Panorex	Glaxo Wellcome	Colon carcinoma
Synagis	Medimmune	RSV infection
OKT3	Johnson & Johnson	Transplantation rejection
Medarex, MDX44	Centeon	Psoriasis
ABX-EGF	Abgenix	Blocks EGF-anticancer
BEC2	ImClone Systems/Merck	Anti-idiotypic for small cell lung cancer and melanoma
TNT	Techniclone	Malignant glioma
CDP 835 CDP 835	Celltech/Schering Plough	Severe asthma
IDEC-151/ BB-217969	IDEC-SKB	Rheumatoid arthritis
M195	PDL	Acute myeloid leukemia E25
Genentech/Novartis	IgE - allergy	
LDP-01	LeukoSite	Inflammation following stroke and transplantation
LDP-02	LeukoSite	Crohn's disease, inflammatory bowles disease, ulcerative colitis
LDP-03 (campath)	LeukoSite	Chronic lymphocytic leukemia
C225	ImClone Systems/Merck	Blocks EGF (HER-1) receptor for cancer treatment

8.3.4 Mouse-mouse hybridoma

As the name suggests, it is a hybrid of antibody secreting specialized cell and an infinitely growing myeloma cell. Application of *hybridoma technology* has resulted in the generation of stable lines secreting structurally and functionally similar immunoglobulin molecules, which have both diagnostic and therapeutic applications. Monoclonal antibodies specific for human antigens are normally produced by mouse-mouse hybridoma, which is also produced by rat-rat hybridoma. The basic techniques used in the generation of mouse-mouse hybridoma are described in Figure 8.8. The initial step of making a therapeutic antibody is to generate a rodent hybrid, which can produce high affinity Mab specific for human antigen. The steps involved in generating hybridoma are elaborated below:

8.3.4.1 Immunization of mouse

Mouse (Balb/c) is given primary and booster immunization in a gap of 3-4 weeks with 5-20 μg pure human antigen with appropriate *adjuvant*. When the antibody level in sera is significantly increased as compared to the pre-bleed titer (titer before primary immunization), the mouse is aseptically sacrificed to collect the spleen. The spleen is teased with the help of a blunt-head sterile forceps or between two sterile glass slides to obtain single cell suspension. The erythrocyte depleted spleen cells are suspended at a density of 10×10^6 cells/ml in RPMI-1640 containing 10% FCS.

8.3.4.2 Fusion of immunized spleen cells with myeloma cells

Immunized spleen cells do not grow in culture, thus their antibody secreting capacity depends on their viability. In order to obtain a clone that can produce antibody for longer period of time, the immunized spleen cells are fused with myeloma (tumor) cells. Thus, the generated hybrid cells (hybridoma) will have characteristics of infinite life span and antibody producing ability. The criterion for selection of myeloma cells for fusion is that they should not secrete immunoglobulin light chains, otherwise these will combine with the heavy chain of monoclonal antibody to produce undesirable hybrid molecules. Myeloma cell line, commonly used as fusion partner, derived from sensitized spleen cells of Balb/c mice is NSO. This myeloma cell is sensitive to hypoxanthine, aminopterin, and thymidine (HAT). The actively growing NSO cells are fused with sensitized spleen cells by incubating in the presence of 50% PEG (M.W. 4000) solution. The adjacent plasma membranes of spleen and NSO cell fuse together making a hybrid cell, though the success rate is extremely low. The fusion mixture is cultured in the presence of HAT medium, initially in T-flask and later in 96-well plate. Since the myeloma cells do not synthesize hypoxanthine-guanine-phosphoribosyltransferase (HGPRT) in unfused state, they cannot survive. The unfused spleen cells also die due to their limited life span. Only the fused cells (hybridomas) are able to grow indefinitely because the spleen cell supplies HGPRT and the myeloma is immortal.

8.3.4.3 Selection and cloning of hybridoma

During culture in 96-well plate, some hybridoma clones may show positive for the antibody. It is important to reclone the hybridoma as soon as possible to avoid potential loss of the positive clone due to overgrowth of non-secreting cells (Khurana et al, 2007). To ensure that the antibody is monoclonal, cloning should be done 2-3 times before selecting the final clone of hybridoma. Cloning can be accomplished by either growing hybridoma in soft agar or by *limiting dilution technique.*

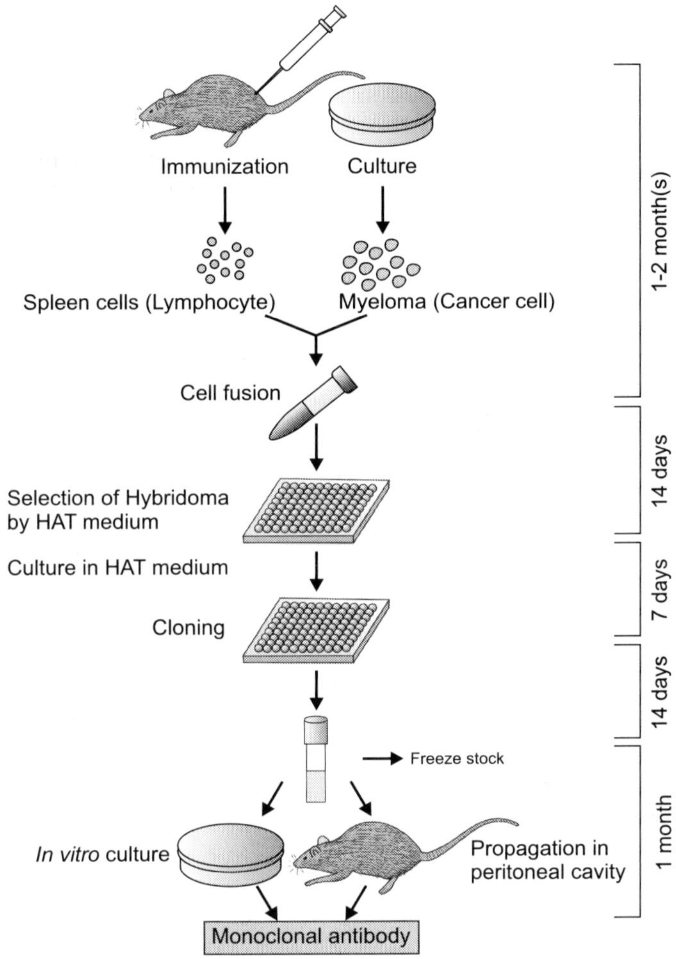

Figure 8.8 The principle steps involve for generating mouse-mouse hybridomas and monoclonal antibody.

8.3.4.4 Expansion of clone and production of antibody

Right clone of hybridoma (on the basis of yield of antibody and its specificity) is expanded to prepare LN_2 stock. Two techniques can be adopted to produce monoclonal antibody:

(a) The highly concentrated (5-10 mg/ml) antibody is obtained in ascitic fluid of peritoneal cavity of a mouse (monoclonal antibodies produced in ascitic fluid have been banned in USA and Europe).

(b) It can be produced by cell culture technique, however, the yield of antibodies by this technique is several folds lower than that produced in the ascitic fluid.

8.4 DEVELOPMENT OF HUMAN Mab

In earlier time, two different techniques, cell hybridization and viral transformation, used to follow for generating human hybridomas. In cell hybridization, two different hybrids are used: mouse-human and human-human. The problem of making mouse-human hybrids using mouse myeloma with sensitized human lymphocytes is instability of the clone. Human chromosomes preferentially segregate, thus extensive sub-cloning is necessary to obtain a stable hybrid. In human-human hybrids, the main limitation was the lack of suitable human myeloma cell lines. Scientists have tried to use B-cell lymphoblastoid line as a fusion partner with sensitized human lymphocytes, but the yield of secretory antibody was extremely low.

In all sense, mouse-mouse hybridoma is a much steady and safer method for making Mab. However, the antibodies synthesized by mouse systems are of limited use, as therapeutic agents, as human immune systems recognize such antibodies as foreign protein (Figure 8.9A). As a consequence, the immune systems trigger defense reaction against the antibodies by producing *human anti-murine antibody* (HAMA) response. The immune response causes a quick elimination of the foreign antibody, essentially the therapeutic value of antibodies after an initial treatment. To overcome the problem of HAMA response, alterate approaches have been attempted to modify nonhuman antibodies: by *chimerizing* with human antibodies and humanizing the mouse antibodies. Chimeric antibodies have a greater circulating half-life *in vivo*, *i.e.*, a longer duration of therapy that can vary from patient to patient (65-300 h). The circulating half-life of chimeric antibodies is significantly higher than murine antibodies (19-40 h), but much lower than human antibodies (300-500 h). They also have the advantage of repeated administration in chronic cases and greater intensity at lower doses than mouse antibodies.

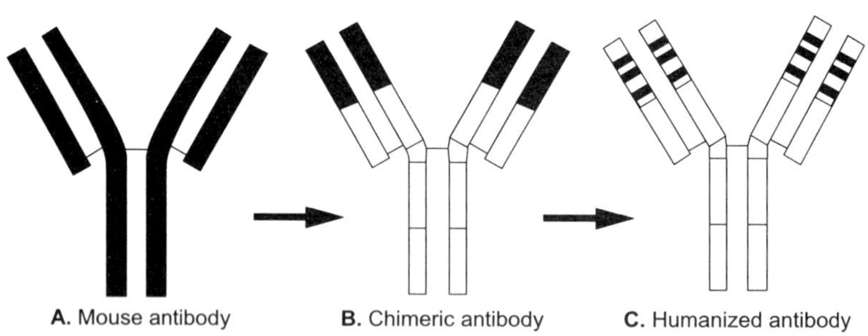

Figure 8.9 Stages for the development of therapeutic antibodies for human. **A.** Highly specific Mab (produced by mouse-mouse hybridoma) that recognizes target human antigen; **B.** Chimeric antibody, where constant domain of both the heavy and light chains are derived from the human antibody and the variable region is entirely murine; **C.** Humanized reshaped antibody, in which only CDRs on the human frame is replaced with that of murine antibody.

Recombinant molecular biology technique has been used to create chimeric antibodies. Constructs of gene fusion between DNA sequences encoding mouse antibody variable light (V_L) and heavy (V_H) chain domains and human antibody light chain and heavy chain of constant domains permit expression of chimeric antibodies (Morrison, 1985). These chimeric antibodies (Figure 8.9B) contain a large number of murine amino acid (33%) sequences and thus are also immunogenic to humans. Patients exposed to these chimeric antibodies produce *human-anti-chimera antibodies* (HACA). HACA is directed against the murine V region and can also be directed against the novel *V-region/C-region* (constant region) junctions present in recombinant chimeric antibodies.

To overcome immunogenicity in chimeric antibodies, murine monoclonal antibodies have been humanized for therapeutic applications. This has been achieved by CDRs grafting, in which murine CDR loops are grafted onto human framework regions. As CDRs take part in antigen binding, the specificity of murine antibody is achieved in humanized antibodies at the same time, immunogenicity of chimeric antibodies can be minimized. The humanized Mab contains only 10% sequence of mouse and 90% human antibody (Figure 8.9C). It contains certain amino acid substitutions from the mouse Mab into the framework of the humanized Mab to retain the correct shape and binding affinity for the target antigen (www.patentsonline.com).

8.4.1 Designing chimeric antibodies

Chimeric antibodies, which have largely (about 67%) human antibody like composition, are generated at high level by recombinant DNA technology. It is important to remember that chimeric antibodies are formed from the mouse-mouse hybridoma clones producing highly specific antibodies for humans. The basic steps for making a human-mouse chimera (e.g., for anti-TNFα antibody) are described below:

1. As a first step, a mouse-mouse hybridoma clone is generated, which produces high affinity monoclonal antibody against human TNFα. It is important to know the sequence of V_L and V_H chain regions of the genes.
2. Total RNA is extracted from this high yielding clone to convert them into cDNA.
3. V_L and V_H chains are amplified by RT-PCR method.
4. Amplified genes are cloned into a suitable mammalian expression vector. Most of the vectors used for antibody expression in mammalian cells are evolved from pSV2 series (Gorman *et al.*, 1991). New universal expression vectors have been proposed, which can be used to generate stable clones for consistent expression of high level of secreted recombinant antibodies (Xiong *et al.*, 2005). The heavy chain expression vector pYR-GCEVH (containing human constant region gene, Cγ) is designed in such a way that gene of variable region (V_H) can be introduced into it with the help of BamHI and NotI sites (Figure 8.10A). Similarly, in the light chain expression vector pYR-GCEVL (containing human constant region gene, Cλ or Cκ) gene of the variable region (V_L) is inserted as below (Figure 8.10B).

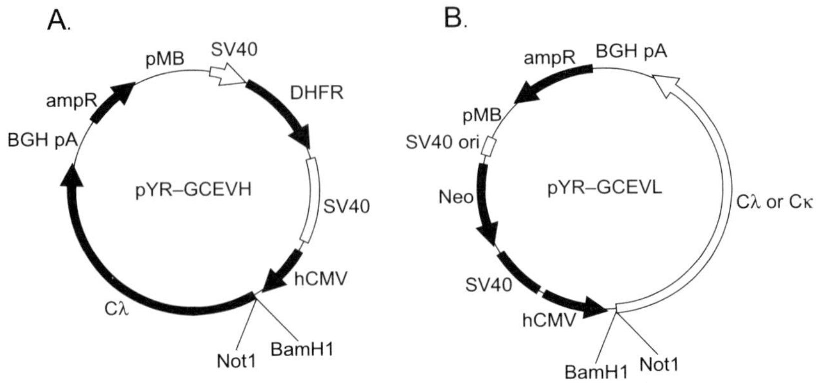

Figure 8.10 Expression vectors for human-mouse chimeric antibody. **A.** Heavy chain expression vector (pYR-GCEVH); **B.** Light chain expression vector (pYR-GCEVL). [Taken from Ref. 19]

5. Cos-7 cells are transiently transfected with the above two expression vectors. ELISA and other techniques are followed to authenticate human anti-TNFα antibodies resent in the culture supernatant.
6. Stable transfection is carried out in dhfr$^-$CHO cells using Lipofectamine transfection reagent or equivalent. These two vectors contain the hCMV promoter, BGH polyadenylation site, C region, and origin of replication. The pYR-GCEVH contains the weakened DHFR gene, while pYR-GCEVL contains the Neo gene. Co-transfection with these plasmids, followed by culture in GHT (glycine, hypoxanthine, thymidine) free DMEM containing G418 will help to select DHFR$^+$ and Neo$^+$ clone (Xiong *et al.*, 2005). The expression of transgenes in the transfected cells may be amplified by stepwise increment of MTX concentration in the growth medium.

8.4.2 CDR-grafting for humanized antibody

In its simplest form, the method used for CDR-grafting consists of three CDRs of the heavy chain and the three CDRs of the light chain taken from a rodent Mab onto a framework of a human antibody (Ketteleborough, *et al.*, 1991). Thus grafting of CDRs in place of entire variable domains of rodent antibody results humanized antibodies, less immunogeneic than chimeric antibodies. A basic outline of CDR grafting technique has been described in Figure 8.11. It is assumed that the CDRs of an antibody do not carry any species-specific function, thus such a reshaped human antibody is identical with the immunoglobulins found in human. However, the successful design of a high-affinity CDR-grafted antibody essentially requires the key murine residues be substituted into the human acceptor framework to preserve the CDR conformations (reviewed in Winter and Harris, 1993). Although progress has been made towards identifying such framework residues either experimentally (Foote and Winter, 1992) or by computational methods (Queen *et al.,* 1989), this process is generally unique for each reshaped antibody.

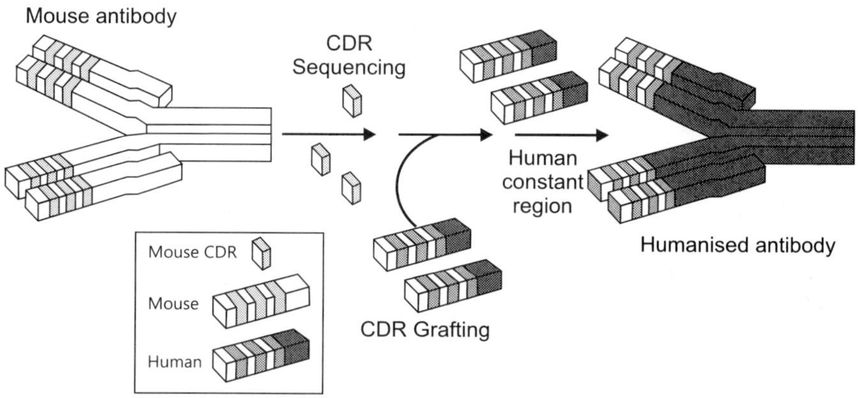

Figure 8.11 The basic technique of CDR grafting. The Mouse CDR sequences are grafted on human V_H and V_L chain frameworks. The CDR grafted regions are fused with human constant region to obtain humanized antibody.

The CDRs grafting technique for humanizing antibodies relies on two assumptions: (i) antigen binding sites are limited in the CDRs, and no other variable regions are involved in antigen binding, (ii) the frameworks of heavy and light chain variable regions serve as scaffold to support CDRs for facilitating antigen binding (Gussov and Seemann, 1981). Three basic steps are involved in humanizing a murine antibody:

***Step*-1**: Sequence the murine antibody and utilize this information. Select a known human framework whose amino acid sequence has the maximum homology to the framework regions in the original murine Mab; thus human framework most likely accept the murine CDRs without any conformational change.

***Step*-2:** Identify '*key contact*' amino acids in the framework region of the murine antibody using molecular modeling process (Vasquez and Payne, 1993). The 'key contact' amino acids are those amino acids present in the *murine framework* that influence the shape of CDRs in such a way that they can able to bind with the target antigen. In humanizing antibodies, murine CDRs with 'key contact' amino acids are transferred in human framework.

***Step*-3:** The grafting is achieved by using oligonucleotides to replace the CDRs of an appropriate human immunoglobulin gene with the CDRs (with 'key contact' amino acids) from a rodent Mab with desired specificity. For that, six synthetic oligonucleotides are constructed to include each murine CDR sequence. Each specific murine oligonucleotide is flanked with about 12 nucleotides cluster complementary to the *flanking regions* within the human framework to be used to host the murine CDRs.

Humanized variable heavy and light genes are cloned into separate expression vectors that contain high level expression of promoters, selection markers, and immunoglobulin constant-region genes. A suitable host cell line is co-transfected with both expression vectors, which is followed by selection in antibiotic containing medium. High-level expression of the *humanized antibodies* can be achieved by sub-cloning, followed by selection using gene amplification protocol using DHFR (in CHO cells) or GS system (in CHO / NSO cells).

8.5 SUMMARY

1. Positioning of transgene in chromatin is most crucial for stable and high level expression of target protein. Transgenes are targeted for embedding in the euchromatin region.
2. Therapeutic proteins are produced using nonviral expression system that can give rise to stable clone, and gene amplification is possible.
3. Murine monoclonal antibodies elicit immune-response (HAMA) in human subject, which reduces the efficacy of the antibodies. Chimeric or humanized antibodies contain lesser proportion of murine amino acid sequence, thus the chance for the formation of anti-antibodies is reduced.
4. A therapeutic monoclonal antibody is not possible to design unless highly specific mouse monoclonal antibody against human antigen is available and the gene sequences in the variable regions or CDRs are known.

ABBREVIATIONS

IgG, IgA, IgM, IgD, and IgE: Immunoglobulin G, A, etc.
cDNA: complementary DNA
C_H: Constant heavy chain
CMV: Cytomegalovirus
CDR: Complementarity determining regions
DHFR: Dihydrofolate reductase
GS: Glutamine synthetase
HGPRT: Hypoxanthine guanine phosphoribosyl transferase
hCG: Human chorionic gonadotrophin
HAMA: Human anti-murine antibody
HACA: Human anti-chimera antibodies
HAT: Hypoxanthine, aminopterin, thymidine
IL-4/5/6: Interleukin-4/5/6
MTX: Methotrexate
mRNA: messenger RNA
mMTV: Mouse mammary tumor virus
MSX: Methionine sulphoximine
Mab: Monoclonal antibody
PEG: Polyethylene glycol
SV40: Semian virus 40

mMT: Mouse metallothionein

S/MAR: Scaffold/matrix-attached region

TNFα: Tumor necrosis factor α

TCR: T cell receptor

UTR: Untranslated region

V_L & V_H: Variable light and heavy chain

REFERENCES

1. Barnes LM, Bentley CM, Dickson AJ (2000). *Cytotechnology* 32, 109.
2. Barnes LM, Dickson AJ (2006). *Curr. Opin. in Biotechnol.* 17, 381.
3. Bebbington CR, *et al.* (1992). *Bio/Technology* 10, 169.
4. Field R, Cockett M, Froud SJ (1989). In: Adv. In animal cell biology and technology for bioprocesses (RE Spier, JB Griffiths, J Stephenne, PJ Crooy, eds.). Butterworyhs & Company Publishers Ltd. UK, pp 195.
5. Finn GK, Kurz BW, Cheng RZ, Shmookler RJ (1989). *Mol. Cell. Biol.* 9, 4009.
6. Gandor C, Leist C, Fiechter A *et al.* (1995) *FEBS Lett.* 377, 270.
7. Galfre G, Milstein C (1981). In: *Methods in enzymology* (Langone JJ and Vulnakis H, eds.), vol. 73, p. 3-46, Academic Press.
8. Gorman, S D, Clark, M R, Routledge, E G, Cobbold, S P and Waldmann, H (1991). Proc. Natl. Acad. Sci. USA 88, 4181.
9. Gussov D and Seemann G. (1981). In: Methods in Enzymol. (JJ Langone, ed.) vol. 203. Academic Press, San Diego, California. p. 99.
10. Hua Xiong, Yuliang Ran, Jinliang Xing, Xiangmin Yang, Yu Li and Zhinan Chen. *Journal of Biochem. Mol. Biol.* vol. 38, no. 4, July 2005, p. 414.
11. Kaufman RJ, Wasley LC, Spiliotes AJ, *et al.* (1985) *Mol. Cell. Biol.* 5, 1730.
12. Kettleborough CA, Saldanha J, Heath VJ, Morrison CJ, Bending MM (1991) Protein Eng. 4, 773.
13. Morrison, S. L. (1985). *Science* 229, 1202.
14. Norton PA, Pachuk CJ (2003). In: *Gene transfer and expression in mammalian cells* (SC Makrides, ed.). Elsevier Science BV, Amsterdam, p. 265.
15. Pallavicini MG, DeTeresa PS, Rosette C, Gray JW, Wurm FM (1990). *Mol. Cell. Biol.* 10, 401.
16. Piechaczek C, Fetzer C, Baiker A, Bode J, Lipps HJ (1999). *Nucleic Acids Res.* 27, 426.
17. Stark GR, Wahl GM (1984). Annu. Rev. Biochem. 53, 447.
18. Vasquez M, Payne PW. (1993). *Chem Design Automation News.* November/ December.
19. Wurm FM, Gwin KA, Kingston RE (1986). *Proc. Natl. Acad. Sci., USA* 83, 5414.
20. Xiong A, Ran Y, Xing J, Yang X *et al.*, (2005) *J Biochem. Mol. Biol.* 38, 414.
21. Zettlmeissl G, Wirth M, Hauser HJ, Kupper HA (1988). *Gene* 73, 419.

SUGGESTION FOR FURTHER READING

1. Datar RV, Cartwright T, Rosen C-G (1993). *Bio/Technology* 11, 349.
2. Heng HHQ, Goetze S, Ye CJ, Liu G, Stevens JB *et al.* (2004). *J Cell Sci.* 117, 999.
3. Kwaks THJ, Otte AP (2006). *Trends Biotechnol* 24, 137.
4. Kwaks THJ, Barnett P, Hemrika W, Siersma T *et al.* (2003). *Nat. Biotechnol.* 21, 553.
5. Marino MH (1989). *BioPharm* 2, 18.
6. Makrides SC (1999). *Pro. Exp. Puri.* 17, 183.
7. Wurm F, Bernard A (1999). *Curr. Opion. in Biotechnol.* 10, 156.
8. Khurana S, Bhaskar S, Mukhopadhyay A. (2007): In: Advanced Techniques in Soil Microbiology (A Verma, Ralf Oelmullor, eds.); p. 92, Springer-Verlag, Berlin.

9

Large-scale Animal Cell Culture for the Production of Therapeutic Proteins

The biology and engineering revolution has lead us to identify and subsequently the production of large number of therapeutic proteins and cell-culture based viral vaccines. Despite the early enthusiasm that some of these protein products could be produced in bacteria and yeasts, a number of functional associated problems have forbidden the use of microorganisms as a host for the production of many therapeutic proteins. These are:

(a) Many therapeutic proteins are high molecular weight and complex glycosylated molecules, and the biological activity depends on the native conformation of the proteins. These extensive posttranslation modifications of protein are either completely absent or limited in the microbial systems.

(b) The heterologous proteins are usually retained in the cytoplasm of microorganisms as inclusion bodies, which require disintegration of cells to recover the protein, its solubilization, and subsequent refolding.

(c) Animal viruses can replicate only in animal host cells.

Despite relatively low product concentration, long batch-time, and expensive manufacturing costs, animal cells have been found indispensable for the bioprocessing of many complex therapeutic proteins. As a consequence of this in past 15 years, a multi-fold increase in the productivity of cell culture-based products has been registered by a combination of improved culture technique, vector systems, purification techniques, and engineering design of the production processess.

9.1 DEVELOPMENT OF LARGE-SCALE ANIMAL CELL CULTURE PROCESS

The impetus for large-scale animal culture process had been started for the need of cell culture-based viral vaccines for mass-scale vaccination program. Later, the advent of recombinant DNA and hybridoma technology have made large-scale animal cell culture process successful for the production of various therapeutic products. Figure 9.1 describes chronological development of cell culture process and corresponding improvement of rabies vaccine. Rabies was known for over 4000 years in Mesopotamia civilization as an incurable disease. For the first time, in 1885, Louis Pasteur was successful in growing rabies virus in the spinal cord of rabbit. The rabies vaccine produced in this way was an unpurified and inactive form of virus. The vaccine had many side effects, and 15 to 20 injections through intraperitoneal route were necessary for a patient. Later, the rabies vaccine was available in many forms, but the first-generation vaccine was produced in 1955, which was having fewer side effects than Pasteur's vaccine. The rabies viruses were produced either in suckling mouse brain or in duck embryos. Kissling introduced cell culture-based second-generation rabies vaccine in 1958. This vaccine was in purified form, more effective and costlier than earlier forms of the vaccine. With the development of cell culture techniques and identification of different host cells for the production of rabies viruses, the most improved form of the vaccine was introduced in 1964. The vaccine was produced in human diploid cells (MRC-5), and 2-3 doses of vaccine were enough to develop immunity. In 1983, two more purified rabies vaccines were

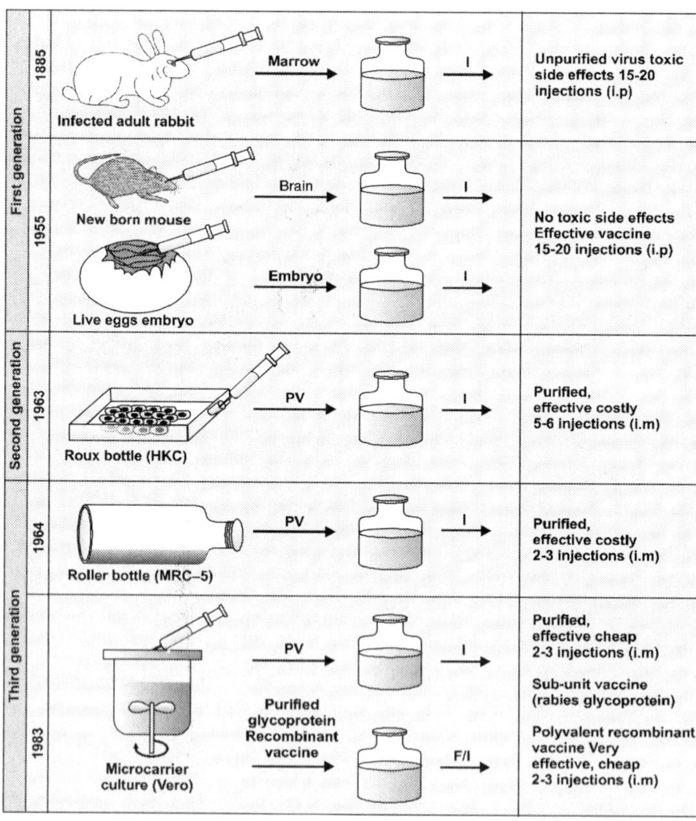

Figure 9.1 Chronological development of animal cell culture process in rabies vaccine production. I: inactivated; PV: Purified virus; HKC; Hamster kidney cell; i.p: Interaperitonial; i.m: Intramasculer; F: Formulated.

introduced—one produced in purified chick embryo cells and other in Vero cells. The manufacturing technology followed in these two forms of the vaccine was different, however, the vaccine was equally effective and cheaper than human diploid cells-based vaccine. Presently, the rabies vaccine is either produced in MRC-5 or in Vero cells.

9.2 ANIMAL CELL CULTURE-BASED PRODUCTS AND MARKET POTENTIAL

Based on primary and secondary market research, global market value of recombinant protein therapeutics in 2003 was estimated as US$ 32 billion, out of which 60-70%

Table 9.1 Market share of cell culture-based therapeutic products

Product	*Company*	*World-wide yearly sales year $ (million)*		*Indication*
1. Erythroprotein	Amgen, Roche, Johnson & Johnson	2003	9466	Stimulates the production of blood red cells, anemia related to chronic renal failure in dialysis patients.
2. Interferon β	Serono, Biogen, Amgen	2003	5473	Multiple sclerosis
3. Factor VIII & IX	Baxter Healthcare Corporation	1995	600	Blood clotting factor for hemophilia
4. Tissue plasminogen activator (t-PA)	Genentech Inc.	1995	300	Treatment of thrombolytic blood clots in heart-attacked patients
5. Herceptin against HER-2 (humanized)	Genentech Inc.	1999	120	Breast cancer
6. ReoPro against GPIIb, IIIa (chimeric)	Centocor, Lilly	1999	445	High risk angioplasty
7. Rituxan IDEC against CD20 (chimeric)	Genentech Inc.	1999	245	Non-Hodgkin's lymphoma
8. Synagis against RSV F protein (humanized)	Medimmune	1999	228	Respiratory Syncytial Virus infection
9. Remicade against TNFα (chimeric), Humira (human)	Centocor, Abbott, Schering-Plough, Johnson & Johnson	2004	2200	Rheumatoid arthritis, Crohn's disease
10. Avastin (humanized)	Genentech Inc.	2004	800	Colorectal cancer
11. Erbitux (chimeric)	ImClone, Bristol	2004	656	Colorectal cancer, head and neck cancer

was shared by the cell culture derived products (Pavlou and Reichert, 2004; Wurm, 2004). During past few years, the market grew at an average of 15% per annum, which was expected to continue for another 3-4 years. The market of cell culture-based recombinant therapeutic proteins has been divided into several segments: hematology, oncology, infectious diseases, arthritis, inflammation and immune disorders, central neural system, and others. Hematology segment (e.g., EPO, factor VIII and IX) accounts for more than 50% market share, followed by oncology segment. Oncology segment is dominated by therapeutic monoclonal antibodies. At present, 17 monoclonal antibodies, comprising of three murine, five chimeric, eight humanized, and one human antibody are available for clinical use. About 180 different antibodies shared by the cell culture derived products (Pavlou and Reichert, 2004a; Wurm, 2004). Therapeutic monoclonal antibody market is the fastest growing portfolio among various biotechnology products. The estimated global market in 2002 was US$ 5.4 billion, which is likely to grow up to US$ 16.7 billion in 2008 (Pavlou and Reichert, 2004b). On the basis of available information, the market shares of animal cell derived major recombinant therapeutic protein products are shown in Table 9.1.

Another important segment of animal cell culture-based product is viral vaccines, such as polio, rabies, rubella, measles, mumps, hepatitis A, chickenpox, smallpox, etc. Many of these vaccines are available for clinical applications; others are still in the developmental stages. The global market value of existing cell culture derived vaccines is much lower than the recombinant therapeutic products, and their growth has almost attained a plateau.

9.3 CELL-LINE DEPENDENT VARIATION OF PRODUCT QUALITY IN LARGE-SCALE MANUFACTURING PROCESSES

The economic production of therapeutic proteins requires proper screening of host cell by which authentic product can be synthesized in large quantities. In order to meet the requirement for the production of therapeutic protein, host cell should have the following characteristics:

(i) Gene of interest is highly expressed, and the protein product is efficiently secreted from the cells.
(ii) Appropriate post-translational modification (e.g., disulphide bonding, glycosylation, etc.) of heterologous protein is possible (similar to natural products) to improve circulating half-life and biological activity, and to minimize antibody response to neutralize the therapeutic value of the protein.
(iii) Clone is reasonably stable during manufacturing process—starting from working seed stage to the production stage.
(iv) Able to grow in medium with defined constituents without loss of cell yield and productivity of the final product.

Some of these important cellular properties are elaborated below.

9.3.1 Host cell engineering

Selection of host cells for the production of therapeutic is primarily governed by the nature of the product to be produced and the productivity of the cells. The biological and the regulatory criteria for selection of host cells have been discussed in Chapter 8. How the productivity and the quality of a recombinant protein are improved by introducing special properties to the cells has been discussed in this section. It is expected that high yielding strains used for large-scale production of therapeutic proteins are all modified forms of the normal continuous cell lines. The details of the modifications are not available in the literature due to commercial interests. Host cells can be modified with respect to the following properties:

1. Improvement of viability and growth-promoting activities: Host cells have been modified (Arden *et al.* 2004) by over expressing proto-oncogenes, cell cycle control genes (e.g., cyclins), anti-apoptotic genes (e.g., Bcl2, IL-12), and growth factor genes (e.g., insulin-like growth factor). Forced-expression of anti-apoptotic gene maintains viability of host cells for longer duration, even in the presence of adverse culture conditions (Simpson, 1999).

2. Improvement of protein glycosylation: Efficacy of antibodies is improved by enhancing their potency. Wild type CHO and NSO cells are used for the production of antibodies that are found to have limited glycosylation due to the absence of *N*-acetylglucosaminyltransferase-II. By over-expressing this enzyme in CHO and NSO cells, the antibodies become highly bisected and non-fucosylated form in the Fc region. Such modified antibodies are found to be 5 to 10-fold more potent in antibody-dependent cellular toxicity assay, as compared to that produced by the control cells (Umana *et al.*, 1999).

9.3.2 Stability of clone

One of the most important parameters for the success of large-scale cell culture is the stability of the clone. Stability of a clone means maintaining its genetic stability and constant productivity throughout the manufacturing process. For example, the specific antibody production rate by a clone may vary from 50 to 70 pg/cell.day, in laboratory experiments. It is crucial that the clone does not revert back to low yielding or wild type strain. The chance of reversion is much lower in case of recombinant clone than the clone obtained by natural selection process. Again, the rate of reversion and/or mutation is more apparent in hybridomas created by fusion of myeloma and spleen cells than a recombinant clone producing chimeric MAb. The frequency of mutations and alterations of hybridomas producing MAbs range between 10^{-2} to 10^{-7} per cell per generation, depending upon the type of mutation such as loss of heavy and light chain, domain deletion, loss of isotype, etc. (Neuberger and Rajewsky, 1981). Therefore, in hybridomas, the problem of clonal instability is a two-fold problem—decrease in the synthesis of desired product and the formation of altered antibodies. In recombinant cell line, the problem is associated with the reversion of high yielding clone to a low yielding clone. Recombinant clones grow slower than their wild counterparts, therefore, it is natural that wild type cells are accumulated after several rounds of doubling. The inherent tendency of immortalized cell lines towards genotypic and phenotypic heterogeneity warrant careful analysis of the clones before considering them as stable production clones. The presence of MTX in culture medium maintains phenotype stability of the clone with respect to the product yield, however, the drug promotes cytogenetic heterogeneity. A cytogenetically unstable cell line is not approved by the regulatory agencies as a host for manufacturing therapeutic products. Thus, it is recommended for the exclusion of selection drug in culture, once the promising clones are obtained (Wurm *et al.*, 1992; Kim and Lee, 1999). However, as a natural process, mutant clones may accumulate in large-scale culture of animal cells. This may affect the productivity of the culture, depending upon the frequency of mutation and the total number of generations that the culture had to grow. How the reverted cells accumulate in the culture has been explained in the following case study:

CASE STUDY: *clonal stability and productivity*

Assumption:

1. Average doubling time of the recombinant clone and the wild type cells is 24 and 18 h, respectively. Thus, the sp. growth rates (μ) of cloned and the wild type cells are 0.029 h^{-1} and 0.038 h^{-1}, respectively (calculated by using Equation 15.5).
2. The starting cell number is 1×10^7 from working cell bank.
3. Cells are grown in 4 stages (500 ml, 5 L, 50 L and 1000 L capacity bioreactor), and in each stage the maximum attainable cell density is 5×10^6/ml.

Therefore, the total number of cells grown in each stage are 5×10^8, 2.5×10^{10}, 2.5×10^{11}, and 5×10^{12} (considering negligible cell death), respectively, and the corresponding generation numbers are 5.6, 5.7, 3.3, and 4.3 (calculated by using Equation 15.6). Thus, the working cell stock needs to be multiplied for 19 generations to become 5×10^{12} cells (Figure 9.2).

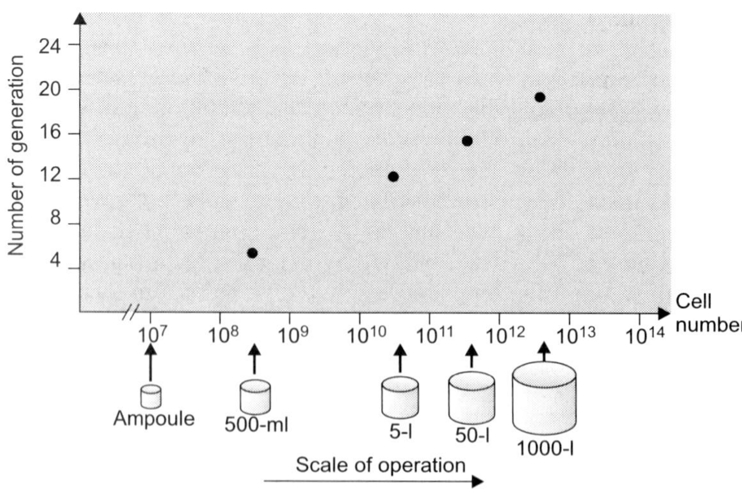

Figure 9.2 Effect of clone stability on increasing volume (scale) of operation.

Suppose, the frequency of reversion (to wild type), r, is 1/10,000 cell per generation and in general the culture is prone to reverse in third and fourth stages (where MTX is not added in the culture). The number of cumulative wild type cells will be approximately 2×10^{10} cells (calculated by using Equation 15.4), which is approximately 0.4% of the productive cells in the culture. Such an insignificant number of wild type cells will not affect the overall productivity, however, a 10-fold increase in reversion (1/1000 cell per generation) may affect the productivity of the culture. Higher the final culture volume, greater will be the generation numbers that cells have to divide, thus more and more wild type cells are accumulated in culture. The situation will be more critical, if selection pressure is not used in the second stage of the culture.

9.4 PRODUCTION OF MONOCLONAL ANTIBODIES (Mab)

The production of monoclonal antibodies involves *in vivo* (mouse ascites) or *in vitro* procedures or a combination of them. Although *in vitro* technique is followed for more than 90% of Mab production, it has been recognized that in some cases *in vivo* production is most beneficial. Other than animal-welfare issues, the major concern for the production of Mab by these two procedures is the economy of manufacturing. The advantages and disadvantages of mouse ascites and *in vitro* method for the production of Mab are discussed below.

Advantages of mouse ascites method:

(i) The mouse *ascites* method usually results in high concentrations of antibodies that often do not require further concentration steps, as required in antibodies produced *in vitro*.
(ii) Some hybridomas do not grow well in culture or are lost in culture.
(iii) *In vitro* method generally requires FCS for culture, the quality of which may vary from batch to batch ultimately influencing the product quality.
(iv) The change in glycosylation pattern in Mab produced *in vitro* as opposed to that produced in mouse ascites, makes the antibody not suitable for applications in human due to immunogenicity, reduced binding affinity, changes in biological functions, or accelerated clearance.
(v) *In vitro* produced Mab may be contaminated with dead hybridoma cellular constituents, which require expensive purification steps.
(vi) *In vitro* culture methods are generally more expensive than the ascites method for small-scale or medium-scale production of antibodies (Hendriksen and de Leeuw1998; Jackson *et al*., 1996; Peterson and Peavey, 1998).

Advantages of *in vitro* method:

(i) *In vitro* culture methods are usually the choice of the pharmaceutical industry for large-scale production because of economic considerations and ease of production.
(ii) *in vitro* method avoids or decreases the need for laboratory personnel experienced in animal handling.
(iii) *In vitro* method using semi-permeable membrane-based systems produce antibodies concentrated same as that obtained in the ascitic fluid, and are free of mouse ascitic fluid contaminants.
(iv) Ascites-based antibodies may contain xenoproteins, which require proper validation of the purification protocol.

(v) Ascites method involves continuous usage of mice, and requires daily supervision.

9.4.1 Comparative costs for the commercial production of Mab

Manufacturing cost is the major consideration for determining the method to be followed for commercial scale production of monoclonal antibodies. Ascites-based production of antibodies is economically viable in small scale manufacturing process. However, as the number and amount of each antibody in the company portfolio is increased, *in vitro* production method become economical. The fixed costs for *in vitro* production of antibodies is spread over a larger product volume, making cost per gram competitive as compared with the *in vivo* production process, where variable costs are comparatively higher (Figure 9.3). In small-scale production, *in vitro* production costs may be 0.5 to 6 times higher than the mouse ascites-based production (excluding the costs of animal house and technician) of Mab (Hendriksen and de Leeuw1998; Jackson *et al.*, 1996; Peterson and Peavey 1998). In large-scale manufacturing, *in vitro* production is economically competitive, as it reduces animal usage and decreases the presence of contaminating foreign antigens if serum-free media is used. In the diagnostic industry, the keen competitive factor is minimizing the cost of production, whereas the presence of foreign antigens is less important. As a result, ascites-based anitbodies are commonly used by the diagnostic industries. In therapeutic applications, the main concerns are manufacturing time, binding affinity of the Mab with the target antigen, and absence of foreign antigens. Thus, mouse ascites or *in vitro* method to be followed for the production of Mab is also governed by the end use. For small-scale production of antibodies, ascites-based method is often used because it is flexible than *in vitro* production and can be performed without optimizing *in vitro* culture condition.

The general guidelines for *in vitro* production of Mab using rat and mouse hybridomas are given below:

1. 10 mg -100 mg range: One-liter spinner flask or roller bottle culture.
2. 100 mg - 1 g range: Dialysis cartridge equipped with spinner flask.
3. 1 g -10 g: Hollow fiber bioreactor (commercial equivalent).
4. 10 g -10 kg: Air-lift bioreactor.

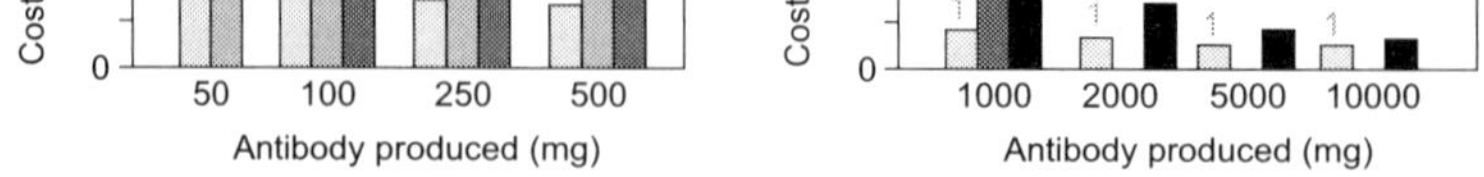

Figure 9.3 Comparative costs for production of antibodies by *in vivo* and *in vitro* methods. **A.** Manufacturing costs ($/mg MAb) at different scale of production in ascites (1), spinner flask (2) and mini hollow fiber bioreactor (3); **B.** Manufacturing costs ($/mg MAb) at different scale of production in ascites (1), mini hollow fiber (3) and midi hollow fiber (4) bioreactor.

9.4.2 Therapeutic antibody production systems

Production of therapeutic Mab involves large scale culture of cells either in homogeneous suspension culture or on a suitable immobilized system. Hybridoma cells are anchorage-independent; they can be trapped (immobilized) in a system through which cell free medium is collected. Chimeric or humanized Mab are produced by recombinant cell lines that could be either anchorage-independent or anchorage-dependent. For large-scale production, the trend is to use anchorage-independent cells (e.g., NSO) or adopt the anchorage-dependent cell lines (e.g., CHO, BHK) for suspension culture. The operational details of the two most popular configurations of bioreactors used for commercial production of Mab are explained below. Other configuration of bioreactors are also used for similar application.

9.4.2.1 Hollow-fiber bioreactor

Hollow-fiber bioreactor (heterogeneous system) has several advantages, which popularized the system for using in large-scale production of Mab by hybridomas or recombinant animal cells. Hollow-fiber bioreactor has following advantages:

1. High cell density (~ 10^8 cells/ml of extra-capillary space) culture is possible, thus the productivity is proportionally increased.
2. Both anchorage-independent and anchorage-dependent cells can be grown.
3. Consume minimum serum, efficient distribution of nutrients and removal of metabolic waste products are possible, which keep cells viable and productive for longer period.
4. Secreted product can be obtained in a concentrated form, because antibodies (molecular weight ~ 150 kDa) cannot pass through 10-50 kDa MW cut-off membranes.
5. Cells are immobilized on the fiber bundle, allowing cell-free harvest of product, which can be directly applied to the purification steps.
6. Secreted proteins can be harvested in purer form (60-70%), thus the purification process become easy.

Despite the above advantages of using hollow-fiber bioreactor for production of Mab or other therapeutic proteins, its application for large-scale manufacturing has been restricted due to the following drawbacks.

1. Poor scalability of the system.
2. Formation of metabolites and oxygen concentration gradients in the hollow-fiber bioreactor.
3. Process control becomes difficult due to heterogeneous nature of the bioreactor.

What is hollow-fiber bioreactor?

Hollow-fiber bioreactor is based on the principle of *ultrafiltration/microfiltration*, in which the permeability of a solute across the membrane is restricted on the basis of membrane pore size. Animal cells (e.g., hybridomas) are larger than

12-μm in diameter, and hence can be retained on the upstream side of the filters (0.1 - 0.2 μm pore size). However, owing to smaller size of medium constituents, e.g., BSA (65 kDa) or IgG (150 kDa), they are easily filtered out of the membrane as a result of pressure gradient between upstream and downstream compartments. Thus, many beneficial effects can be derived in such system: (a) cells are retained in the upstream, so their concentration gradually builts-up with the growth, (b) as the metabolic waste products are constantly removed, cellular inhibition is minimized, and (c) since cells are retained by the filter, a cell free downstream product is obtained.

A simple form of microfiltration device is shown in Figure 9.4A, where depending upon the pore size of the membrane (filter), upstream molecules are sheaved into two parts.

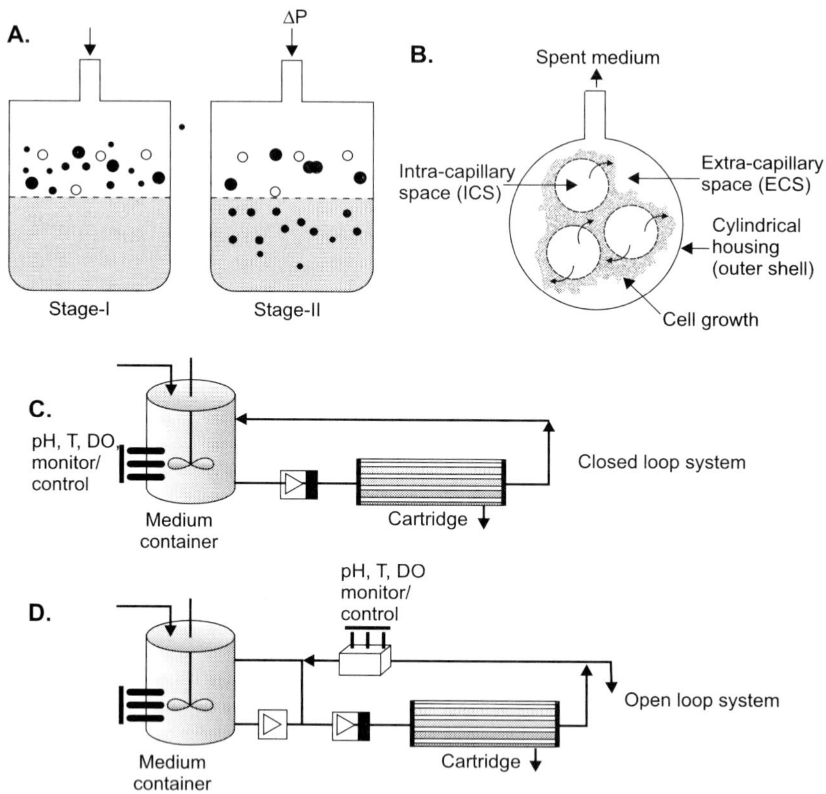

Figure 9.4 Basics of hollow-fiber bioreactor. **A.** Microfiltration device, in which different sizes of particles are retained in the up-stream side of the filtration device. The particles, those sizes are smaller than the porosity of the membrane filtered out due to pressure difference (ΔP) between up-stream and down-stream chambers; **B.** Cross-sectional view of hollow-fiber cartridge. Medium flows through ICS of the hollow-fiber, a part of that diffuses in the ECS. Cells, grow in ECS in attachment with the fiber surface, are nourished by the nutrients diffuse inside ECS. Products are accumulated in the ECS and finally removed with the spent medium; **C & D.** Close loop and open loop media supply systems.

Hollow-fibers are cylindrical-shaped filters of about 200 μm inner diameter. In hollow-fiber bioreactors, large numbers of these fibers are sealed into a cylindrical housing in such a way that any liquid entering from one end of the cartridge (end port) will go through the insides of the fibers, while access to the space outside the fiber is provided by side ports on the housing. Cells are grown and retained in the *extracapillary space* (ECS) of the fiber and the medium is flown through the lumen, *intracapillary space* (ICS) of the hollow-fiber (Figure 9.4B). Cells and the products are recovered from the ECS of the cartridge. Two most important determinants of the performance of hollow-fiber bioreactors are material used for making the fibers and the diffusibility/permeability of medium constituents through the membrane. In view of cell growth, the fiber materials should be non-toxic, act as a substratum for cell growth, and preferably hydrophilic in nature. Again, for efficient cell culture, high mass transfer across the membrane is desirable. Therefore, thin wall thickness hollow-fiber membranes are most popular, as the diffusive permeability is inversely proportional to the wall thickness. Physical characteristics of few commercially available, hollow-fiber bioreactors with comparable cartridges capacity are shown in Table 9.2. In some systems, *hemodialysis* cartridges have been successfully used for the growth of mouse hybridomas for antibody production.

Table 9.2 Physical characteristics of hollow-fiber cartridges

System	*Acusyst P (Endotronics Inc.)*	*Cell-pharm-I (CD Medical Inc.)*	*Vitafiber-III (Amicon)*
• ECS volume (ml)	100	75	250
• Hollow-fiber surface area (m^2)	1.3	1.4	1.0
• Fiber Material	Mixture of cellulose acetate and cellulose nitrate	Cellulose	Polysulphone or acrylic polymer
• Pore size (kDa)	7	10	10, 50, 100

Since hollow-fiber bioreactor is a heterogeneous culture system, control of temperature, pH, and dissolved oxygen concentration of the medium is not possible inside the cartridge (ECS). Again, due to heterogeneous nature of the culture, concentration gradients of dissolved oxygen and nutrients, gradient of temperature are easily established in the ECS, with the accumulation of toxic metabolic wastes. In order to minimize the fluctuation of culture environment inside the cartridge, the medium is continuously circulated through the lumen of the hollow-fiber. Two different modes of medium supply schemes usually followed in hollow-fiber bioreactors are close loop and open loop path. In the close loop path, media pH, temperature, and DO concentration are maintained at desired levels in a medium container. The conditioned medium is passed through the cartridge at a pre-determined flow rate, and the spent medium returns back to the container where it is further conditioned (Figure 9.4C). In the close loop path, the concentrations of nutrients are gradually depleted with the accumulation of metabolic wastes. These are taken care by batch-wise replacement of spent medium with the fresh medium. In the open loop path, medium is conditioned in a small manifold on the circulation loop. The conditioned medium is further circulated through the cartridge unless nutrients and metabolic waste concentrations are within their respective limits. The circulating medium is intermittently replenished with fresh medium (Figure 9.4D). In both systems, off-line analyses of nutrient and waste product concentrations are carrier out to replenish the medium.

Operation of hollow-fiber bioreactor for antibody production

All commercial scale hollow-fiber bioreactor operates on the same principle: allows cells to grow on ECS, while nutrients diffuse from the ICS to the ECS through the pores of the membrane. Monoclonal antibody produced by hybridoma is considered as a non-growth associated product, as the major fraction of antibodies is synthesized once the cells stop growing. Therefore, for the convenience of operation, the antibody production process is divided into two phases: cell proliferation and antibody production. The major advantage of splitting production phase from the cell proliferation phase is that serum-free medium can be used in the production phase, making Mab purification less cumbersome. Another distinct advantage in two-phase process is that trouble-free control of environmental parameters is possible, which may be difficult if the production phase is not separated from the growth phase.

Since cells proliferate in the ECS of the cartridge, the first step for starting hollow-fiber bioreactor is seeding the system with adequate number of hybridoma cells. In general, 50×10^6 to 100×10^6 cells are used to seed a 1 m^2 surface area (ECS volume ~ 100 ml) cartridge through the flow-path (shown by hidden line), as indicated in Figure 9.5. The cells grow on the fiber surface (ECS), where they are nourished with medium containing 5% FCS (or a suitable FCS replacement). In case, cells require serum for growth and the porosity of the fibers is more than 50 kDa, it is recommended to flow the growth medium through ECS. The growth medium ultimately enters into the lumen of the fibers and is collected in the holding tank. The medium may be circulated back to the cartridge after controlling pH and DO, and if needed it may be discharged from the system. The temperature of the culture may be maintained either by placing the system inside of an incubator or in a warm room. In a typical hollow-fiber run, cell density greater than 10^8/ml of ECS is achievable. The cross-section of a hollow-fiber cartridge holding lymphocytes in the ECS is depicted in Figure 9.6. In a typical large-scale hollow-fiber bioreactor (100 ml ECS), about 5×10^9 to 1×10^{10} cells may be achieved.

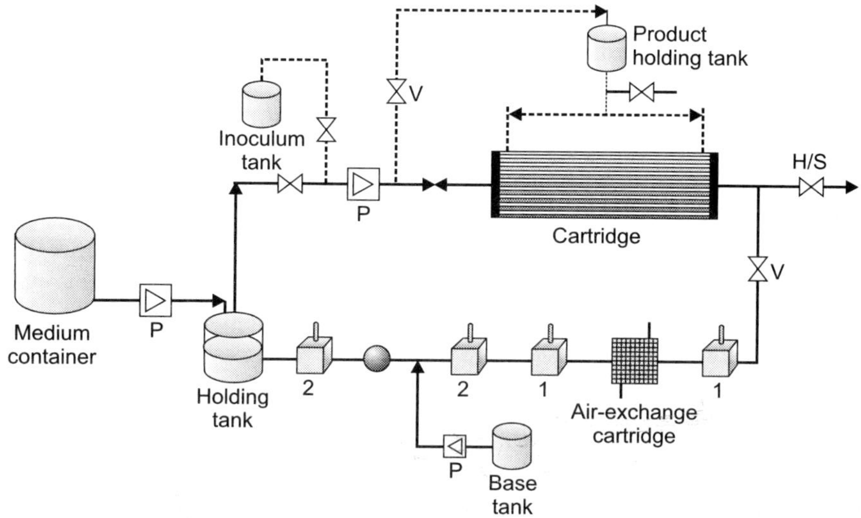

Figure 9.5 Flow-path of commercially available hollow-fiber bioreactor. Initially cells are pumped through (hidden lines) the ECS of the cartridge, cells start growing in the ECS. Once seeding is complete, medium is pumped (solid lines) through the lumen of the hollow fibers. The outlet medium is depleted with respect to oxygen and nutrients; pH of medium may also be changed. The spent medium is flow through air-exchange cartridge for oxygenation, pH is adjusted by pumping base. Finally, conditioned medium is collected in a holding tank before pumping into the cartridge. Product is harvested intermittently through the harvest/sampling port (H/S). Medium is supplemented by pumping fresh medium in the holding tank. P: pumps, V: valve, 1: DO electrode, 2: pH electrode.

In the production phase, serum-free or low serum medium is continuously pumped through the lumen of the cartridge, at a predetermined flow rate, to create a pressure gradient between ICS and ECS. The pressure difference is supposed to force the medium constituents equally along the length of the hollow-fiber membrane into the ECS and product holding tank. The flow rate of medium and the pressure at which it is pumped is governed by the design of the cartridge and the process requirement. One major problem in commercial scale hollow-fiber cartridges is the formation of oxygen and metabolite gradients due to *Starling effects*. These gradients occur in the ECS as a result of pressure difference from inlet to the outlet of the cartridge. This causes a fraction of cells in the ECS experienced with adverse environmental conditions. Cells are irreversibly damaged and the productivity is declined if these gradients persist for a longer time. In Acusyst-P (Endotronics Inc., Minnesota) model, a secondary medium flow through the ECS is established by controlling the flow through the lumen of the fibers past the cell mass (reducing pressure and flow rate of the medium). The medium in ECS and product holding tank returns to the lumen equally along the length of the cartridge. By this process the fluid dynamics in ECS is considerably improved, which causes an even distribution of nutrients and improves the removal of the waste products. Inward and outward flows of medium in ECS is carried out in cyclic fashion, maintaining the production run longer than the single path flow system. The production of Mab by hybridoma and hepatitis B surface antigen by L cells has been carried out continuously for 62 and 140 days, respectively.

Product harvest from ECS of a cartridge is carried out once the maximum growth of cells is established. For optimum productivity, the volume of the product to be harvested and the time of harvest are two most critical parameters. In hollow-fiber bioreactor, harvesting of product is performed in a batch or continuous mode by using a pump. The concentrations of antibodies usually vary from 0.5 to 5 mg/ml, depending upon the clone used.

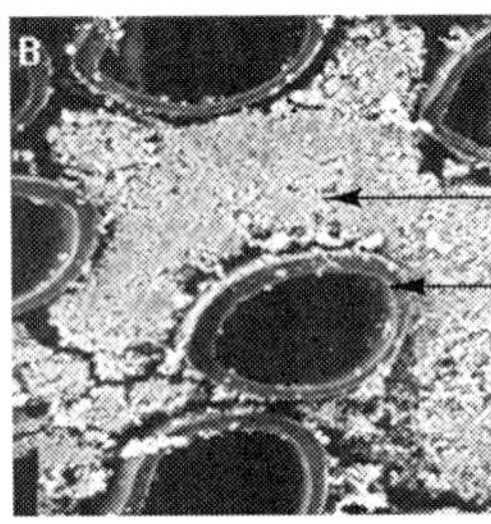

Figure 9.6 Cross-sectional view of hollow-fiber cartridge showing lymphocytes in the ECS (Reproduced by permission of Dr. JJS Cadwell, FiberCell Systems).

9.4.2.2 *Airlift bioreactor*

It is an oldest configuration of bioreactor after the stirred tank bioreactor. It is a homogeneous system, and thus has been used for large-scale production of many therapeutic proteins. It can be operated both in batch and in perfusion mode. In perfusion mode, cells are retained inside the bioreactor by means of a spin filter or *lamellar clarifier*. Airlift bioreactor has several advantages over other systems used for large-scale culture of animal cells. These are as follows:

1. The system has no moving part in direct contact with cells and medium, thus mechanical damage to the cells and microbial contamination are minimum.
2. It is a homogeneous system, thus easy control of environmental parameters is possible.
3. Airlift bioreactor is easily scalable to a very high capacity (10,000 L).

As the name suggests, air (or air-oxygen mixture) is introduced into the culture from a sparger tube at the base of a central *draught tube*. The sparged air causes a reduction in the bulk density of medium inside the draught tube, compared with that outside the tube. The lighter medium inside the draught tube moves upward and the empty space created is occupied by the heavier medium from outside the tube. Thus the medium keeps on circulating as long as air sparging is maintained (Figure 9.7). In addition to mixing of medium, sparged air also supplies oxygen in the culture medium. Oxygen transfer in the medium is regulated by increasing/decreasing airflow in unit volume of culture medium, by reducing the size of the air bubbles, and using air-oxygen-nitrogen mixture at inlet. The oxygen transfer in unit volume of air is gradually reduced as the bubbles move from bottom to the top of the draught tube. This is not only due to depletion of oxygen from air bubbles, but also due to an increase in their effective size (mass transfer area in unit volume of air reduces with increase in bubble size). The size of the air bubbles is increased due to (a) drop in *hydrostatic head* as bubbles move upward, and (b) *bubbles coalescence.*

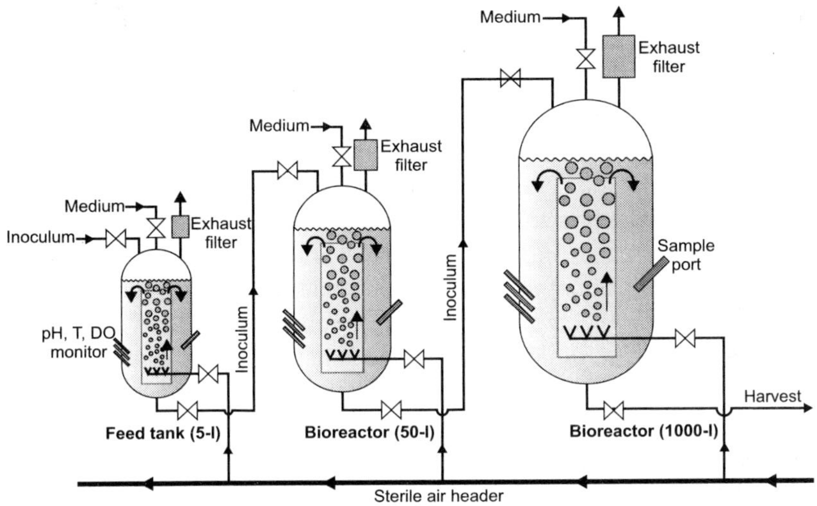

Figure 9.7 Three-stage batch air-lift bioreactor for culture of anchorage-independent cells. Cells are serially transferred from small to the large volume of the bioreactor. Transfer of cells/medium from one stage to another is carried out by positive air pressure inside the vessels. Direct sparging of air or air/nitrogen mixture carries out circulation of medium and oxygen supply in the culture. Each bioreactor is equipped with instrumentation to measure and control of temperature, pH, and dissolved oxygen level in the culture. The culture temperature is maintained by circulating chilled/warmed water through the external jacket (not shown in the figure).

Direct air sparging in serum containing medium may lead to the formation of foam, which may reduce the oxygen transfer through the headspace. Serum-free or low-serum containing medium solves the problem of foaming in direct sparged airlift bioreactor. Another problem associated with direct air sparging is the damage of cells at the region of bubble disengagement. At the liquid surface, air bubbles are rapidly broken and create oscillating force that may cause cellular damage. Damage may also occur in the trapped cells in draining liquid films in case of low protein containing medium. Cellular damage due to oscillating force of bubble bursting and film drainage can be reduced by stabilizing foam at medium-air interface (Handa-Corrigan *et al.*, 1989). Surface-active agent, Pluronic F-68, has been used in low serum containing medium to increase surface tension of medium, and to stabilize the foam (Handa-Corrigan *et al.*, 1989).

A typical flow diagram of batch airlift bioreactor is shown in Figure 9.7. In general, cell density of 1 to 2 × 10^7 cells/ml (per vial) is used to initiate the culture. Cells are grown in various stages to obtain sufficient numbers to seed a 1000 L capacity bioreactor. Initially, cells are expanded in tissue culture flask, followed by spinner bottle to produce enough cells to inoculate first stage air-lift bioreactor (5 L capacity) shown in Figure 9.7. Once the cells are ready, they are transferred into second stage bioreactor (50 L capacity). It is important to note that the purpose of first and second stage culture is to produce enough cells to seed 1000 L bioreactor. Hence, cells are cultured up to mid-exponential phase of growth in the presence of medium supplemented with FCS or equivalent substitutes. In batch culture, whether it is a hybridoma or engineered cell line, about 1.5 to 2.0 × 10^6 cells/ml can be achieved within 2-3 days of culture, depending upon cell type and medium used. In the third stage, cells are maintained first in the growth phase, which is followed by the production phase. Accordingly, the medium composition and the environmental parameters of these two phases of culture are separately maintained. In batch mode, the third stage of the culture lasts for 7-10 days in which maximum viable cell densities between 3.0 to 4.0 × 10^6/ml can be achieved. Finally, the culture is harvested to recover the product. The exact time of harvest is a trade-off between maximum achievable product concentration and extent of contamination by cellular components in the declining growth phase of the culture. Since the synthesis of antibody is maximized at the stationary phase, it is a normal practice to extend this phase to increase the yield of the products.

9.5 FLUIDIZED-BED BIOREACTOR

Therapeutic proteins are produced using two most popular bioreactors: *hollow-fiber* and *airlift*. Due to commercial interest, the manufacturing process of therapeutic proteins are not available in the literature, so no detail description regarding large-scale bioreactors has been documented. This sub-section deals with another bioreactor, potential for the production of recombinant therapeutic proteins. Fluidized/expanded bed bioreactor is one such kind, in which few operational shortcomings encountered in hollow-fiber and airlift bioreactors have been taken care of. In addition to that, the benefits of above two bioreactors are available in the fluidized-bed system. The specific advantages of fluidized-bed bioreactors are explained below:

1. Highly porous microcarriers can support enormous number of cells to grow and retain on the lattice of the microcarriers, thus high cell density culture is possible with an increase of product concentration. CHO and fibroblast cells have been grown up to a density of 3-4 × 10^8 cell/ml volume of microcarriers.
2. Since each microcarrier particle is continuously moved in the flow-path of medium, the concentration gradients of nutrients and metabolites between bulk of the medium and the microcarrier particles become least, as that prevailed in the heterogeneous hollow-fiber bioreactor.

3. Since cells are attached/entrapped on microcarriers ($\rho = 1.6 - 1.7$ g/cc), they cannot escape with the effluent stream, as in the case of continuous airlift bioreactor.
4. Due to porous nature of the microcarriers, cells entrapped inside the pores are protected from the shear effects on liquid-solid interface.
5. The system is easily scalable to the production level, and the culture stability is found to be greater than 50 generations.

However, the *fluidized-bed* bioreactor operation is not cost effective, unless:

1. the cells are anchorage dependent and remain strongly adhered on the surface of the carriers.
2. clones used in fluidized-bed bioreactor can be grown in very high cell densities, and they are highly stable in terms of the expression of target proteins for extended period in continuous perfusion mode.

The most important component of fluidized-bed bioreactor is porous microcarriers made of collagen or gelatin and weighted with a biocompatible material, e.g., titanium. The specifications of porous microcarrier particles have been described in Chapter 6. In order to keep in fluidized state, the microcarrier particles are exposed to high-velocity (70 to 100 cm/min) upward-flowing culture medium. However, owing to high density ($\rho \geq 1.6$ g/cc), the particles cannot escape the bioreactor and remain in suspension. The cultisphere G (Percel Biolytica, AB Sweden) and collagen microsphere (Verax Corporation, New Hampshire, USA) are two most popular porous microcarriers that have been successfully used for the cultivation of CHO, Vero, BHK, HEK293, MRC-5, hybridomas cells, etc. Porous microcarriers serve as 3D matrix, which induce the formation of suitable microenvironment for attachment, spreading, growth, and biosynthesis of products.

Scanning electron microscopy showed that cells might invade the open pores of the porous microcarriers and keep on growing as long as the cellular microenvironment is favorable (Vournakis and Runstadler, 1991). During the upward flow of medium, nutrients tend to diffuse through the pores of the microcarriers. It has been proposed that fluidization causes a pressure difference across the diameter of the particles, resulting in forced convective flow of nutrients from bulk of the medium to the interior of the particles through the inter-connecting pores. This improves the uniform supply of nutrients and oxygen throughout the particle and efficient removal of metabolic end products, leading to dense growth of cells. CHO cells have been grown on collagen microspheres at a density of 2.0×10^8 cells/ml of carrier volume with 80% viability (Vournakis and Runstadler, 1991).

The schematic diagram of fluidized-bed bioreactor with medium recirculation loop is shown in Figure 9.8. Initially, the sterile microcarrier particles are seeded with actively growing cells at densities of 0.5 to 1.0×10^6 cells/ml of carrier volume. Cells are allowed to attach to the microcarriers for a period of 3-6 h by intermittent circulation of medium at a low velocity. Once the cells are attached, the microcarriers are fluidized as a thick slurry inside the bioreactor. The recycle flow of medium surrounding the microcarrier particles provides nutrients to the cells by diffusion through the pores, whereas, the metabolic wastes and protein products diffuse out from the pores, into the bulk medium. The effluent medium is conditioned in the external loop for dissolved oxygen level, pH, and temperature, and circulated back to the bioreactor through its bottom. The loop consists of monitoring and controlling of dissolved oxygen, pH, and temperature of the effluent and inlet medium, base and medium feeding device, air exchange device to enrich oxygen level in the medium, heat exchanger to warm the inlet medium, and a circulating pump. In an intermittent perfusion culture, effluent medium is harvested at the exit of the bioreactor and fresh medium of similar volume is pumped in the circulation loop. The amount of effluent medium to be harvested depends upon the concentrations of residual nutrients and metabolic waste products in the effluent medium. In a continuous perfusion culture, effluent medium is continuously harvested with supplementation of equal volume of medium through the main feeding point of the bioreactor. In a fluidized bed bioreactor, the medium perfusion rate is often increased up to 5-reactor volumes per day to grow cells at a very high-density level.

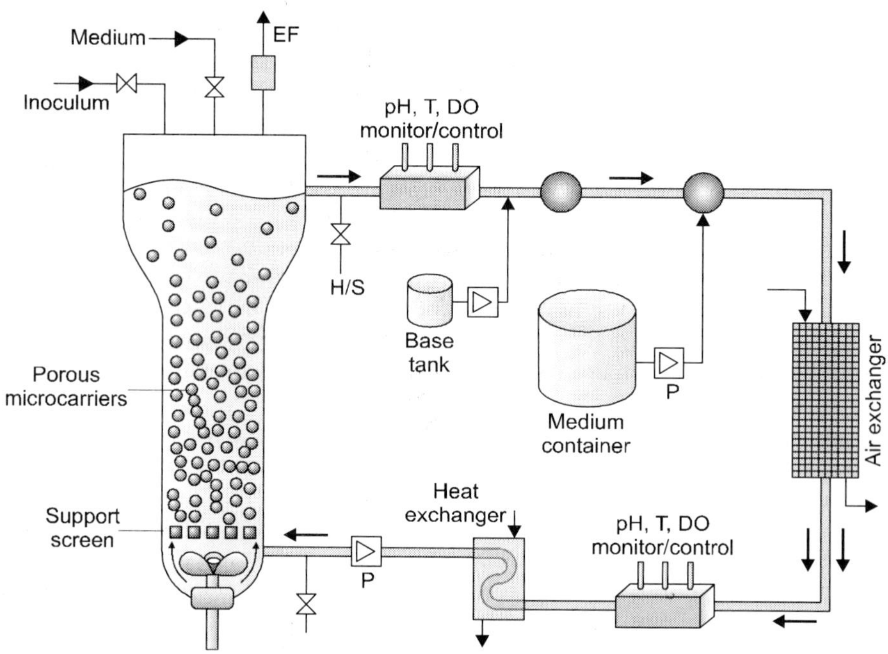

Figure 9.8 Schematic diagram of recirculating fluidized bed bioreactor. Cells attached on porous microcarrier particles are fluidized by upward flow of medium. Due to heavy density of the particles, they retained inside the bioreactor allowing cell-free spent medium in the exit. Spend medium either harvested (H) or sampled (S) at the outlet. The effluent medium is conditioned with respect to pH, temperature and dissolved oxygen concentration. Medium is oxygenated in a membrane-based oxygenator, where either air or air-oxygen mixture is flown through permeable membrane. The medium is warmed at desired temperature before entering into the bioreactor. Both measuring and controlling device at medium exit and inlet of the bioreactor are involved for fine control of pH, temperature, and dissolved oxygen concentration.

Therapeutic proteins produced in fluidized-bed bioreactor, using collagen based porous microcarriers, are found to be structurally and functionally comparable to the native protein, which may not be obtained in a batch culture process. This is mainly due to (a) the formation of tissue-like microenvironment on 3D matrix, (b) using low serum/serum free medium, and (c) lesser accumulation of the product, as once secreted from the cells is removed from the culture. Depending upon their physiological state, animal cells secrete many proteolytic enzymes. In serum-free medium, these proteolytic enzymes are fully active; they either degrade therapeutic proteins or reduce their function by modifying the structure. Cardiovascular therapeutic (t-PA) has been produced in CHO cells continuously for 6 months in a fluidized-bed bioreactor. In serum-free medium, the maximum product concentration (65-80 mg/l) was reported in the last 3 months of the culture (Vournakis and Runstadler, 1991).

9.6 STIRRED TANK BIOREACTOR

The most common form of bioreactor used for manufacturing various therapeutic products is stirred tank. In this sub-section, the stirred tank bioreactor has been described in special reference with the solid microcarriers based animal cell culture for production of viral vaccines. Viruses are obligate parasites. They require host cells as a substrate for their multiplication and subsequent maturation. In earlier time, viruses were cultivated on large numbers in Roux bottles or roller bottles. Although these culture systems can provide significant numbers of cells for the production of viral vaccines, there are a number of disadvantages to this approach of mass-scale culture of animal cells. These culture processes are labor intensive, prone to contamination, require large incubation room, and lead to the variation of product quality. The mass vaccination campaign against viral diseases provided the motivation for the development of large-scale anchorage-dependent animal cell culture process. Solid microcarriers-based unit culture process was one such development (van Wezel, 1967). Microcarrier particles offer large surface area in unit culture volume, thus more and more cells can be grown in relatively small volume of culture, which was not possible in roller bottle culture technique. Microcarriers have been used routinely on a large-scale by a number of pharmaceutical firms like Institute Merieux for the production of inactivated polio and rabies vaccines in Vero cells using 1000 L capacity bioreactor (Montagnon *et al.*, 1984).

Laboratory study showed that the inoculum size (cell-to-bead ratio), method of trypsinization, and the growth phase at which cells are harvested can dictate the performance of the culture. Following precautions are recommended for a successful cell culture using solid microcarriers as growth support:

1. Seed culture is harvested while cells are in exponential growth phase. Starved culture is never used as seed material.
2. Trypsin activity is neutralized and removed by washing with fresh serum containing serum.

3. Single cell suspension is used as seed culture, as cell clumps reduce the final growth of cells. Cell clumps may be removed by filtering the inoculum through pre-sterilized 200 mesh nylon or 316 stainless steel filters.
4. Pre-equilibration of microcarriers with respect to temperature and pH is recommended before seeding with cells. Pre-equilibration not only maintains the optimum temperature and pH of the culture, microcarriers surface is also conditioned by adsorbing cell adhesion glycoproteins from serum.

Serial propagation of cells on microcarrier beads

In a large-scale culture, the number of cells required to seed microcarrier beads is enormous. Therefore, the cells are grown in several stages with gradual increasing in numbers. The number of seed cells required for 1000 L charged volume bioreactor has been calculated in the following case study:

CASE STUDY: *Benefits of microcarrier culture*

Assumption:

1. 1000 L charged volume with 5 g/L cytodex-1 or equivalent microcarriers.
2. The approximate number of microcarrier beads will be $5 \times 6.5 \times 10^6 \times 10^3$.
3. Optimum cell-to-bead ratio for Vero cells is 5 (Mukhopadhyay, 1996).

Therefore, total number of cells required to inoculate the production batch (P2 stage) will be $= 5 \times 5 \times 6.5 \times 10^6 \times 10^3 = 1.62 \times 10^{11}$.

Number of roller bottles necessary for growing 1.62×10^{11} cells will be 1600 (each roller bottle supports the growth of approximately 10^8 cells). Such astounding number of roller bottles is unmanageable due to many technical reasons as mentioned earlier. Same number of cells can be easily cultivated in a small scale bioreactor (P1 stage). The capacity of P1 bioreactor is calculated as follows:

Vero cells grow at a density of 4×10^6/ml at about 80% confluent microcarrier beads (5 g/L). So, the volume of culture required for growing 1.62×10^{11} cells $= (1.62 \times 10^{11}) / (4 \times 10^6 \times 10^3) = 40.5$ L. On the safer side, 50 L charged volume bioreactor would be sufficient to grow the cells required for seeding a 1000 L culture. To inoculate 50 L culture, the number of cells required, as calculated above, will be 8.12×10^8, which can be easily grown in 8 to 10 roller bottles. It is clear from this analysis that a 50 L bioreactor is convenient to operate than handling 1600 roller bottles.

Harvesting of cells from microcarrier beads by trypsinization is a simple process for many cells. In laboratory practice, the beads are allowed to settle, culture medium is aspirated, and cells (eg. Vero) with beads are washed 2-3 times with sterile PBS. The microcarrier beads are re-suspended at 20-25% of the original volume in PBS containing 4 mM glucose, 0.2% trypsin and 0.02% EDTA. The cells are incubated at 37°C with intermittent agitation. If trypsinization is carried out as above, it takes about 10 minutes for Vero cells to detach from the beads. Examining beads under the microscope is recommended to monitor the progress in trypsinization process. This method of harvesting cells from microcarriers can be used for human diploid cells (MRC-5, FS-4), but it is difficult to remove viable cells from Cytodex-1 microcarriers. Once the trypsinization process is completed, the residual activity of trypsin is neutralized by trypsin inhibitor or by adding serum-containing medium. Finally, the microcarrier beads are separated from cell suspension by gravity settling or by filtration through a column packed with the glass beads (3-mm diameter).

In serial propagation, cells are cultured in P1 bioreactor till they are ready to seed bioreactor of P2 stage. Cells are trypsinized as above. Instead of separating microcarrier beads, entire suspension (cells and beads) is transferred to P2 stage containing fresh beads. In this case, along with new beads, the old microcarrier beads are further used to support the growth of cells. It is important to mention here that collagen/gelatin-coated microcarriers cannot be reused in P2 stage, as collagen/gelatin layer on the microcarrier beads are affected by the action of trypsin. One modified option of the trypsinization process is that instead of complete detachment of cells from beads, they are allowed to become round. The microcarriers are settle down and the trypsin is removed by aspiration. By this technique, the harmful effect of trypsin on cells can be minimized. An alternate to trypsinization method, sub-cultivation of cells, has been proposed by bead-to-bead transfer method (Crespi and Thilly, 1981). In exponentially growing culture (70-80% confluent culture) on microcarriers, cells are transferred to a fresh batch of equilibrated microcarrier beads. The mixture is left unagitated for some time. A major fraction of cells from old microcarrier beads is migrated to the fresh beads. The limitations of this technique are: (a) bead-to-bead transfer of cells occurs in presence of less calcium (<1.2 mM) containing medium, and (b) not applicable for all cell type (eg. human diploid fibroblasts).

High-density cell culture on microcarrier beads

High cell density culture on microcarriers has been found to be limited due to (a) depletion of oxygen and accumulation of bicarbonate, lactate, and ammonia in the cellular microenvironment, and (b) limitation of nutrient supply and synthesis of growth-stimulating endogenous metabolites in underfed cultures. These are the major causes for density-dependent inhibition of cell growth. Vero cells have been cultured at a density of 3×10^7 cells/ml in most optimized environment condition, where cells were assumed to grow in multilayer on microcarrier beads (Nahapetian *et al.*, 1986). Such a high-density growth was obtained at an elevated perfusion rate (8 volume/day) using 50% PBS-diluted L15 medium.

Production of viral vaccine

A schematic diagram for solid microcarrier-based cell culture process is shown in Figure 9.9. As mentioned earlier, the bioreactor at P2 stage (1000 L) is seeded with cells cultivated in P1 stage (50 L) at a cell-to-bead ratio of 5. Cells are cultured in perfusion mode by feeding sterile medium from feed tank. The growth of cells on microcarrier beads is enumerated by citric acid-crystal violet method (van Wezel, 1967). The culture at mid-exponential growth phase (70-80% confluent) is suitable for infection with virus. In case of Vero cells, it may take 6-7 days to attain this confluence. For infection, microcarrier beads are allowed to settle, the serum-containing medium is aspirated, and the cells are washed twice with serum-free medium. Infection is carried out at $1/10^{th}$ volume

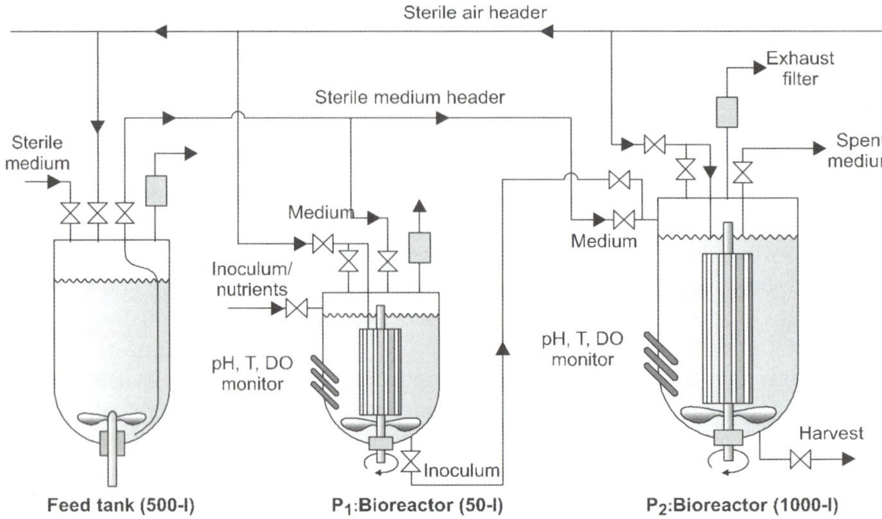

Figure 9.9 Schematic diagram of continuous perfusion stirred tank bioreactor. Example is shown for the manufacturing of viral vaccine in solid microcarrier culture of animal cells.

of the culture using viruses (e.g., rabies) that have multiplicity of infection (MOI) between 0.3 to 0.5. Cells are allowed to infect for 2-3 h, and the bioreactor volume is maintained to its initial volume by feeding 2% FCS containing medium. The production of rabies virus continues for 6-7 days. During this time, the viability of cells becomes absolutely low as lytic rabies virus causes cell death. Virus particles are accumulated in the culture medium. In batch process, at the end of production phase, the culture is harvested to recover virus particles. The production phase may be continued at low perfusion rate (0.5 batch volume/ day), where the virus yields have been reported to increase by 1.7-folds than the batch mode of production (Trabelsi *et al.*, 2006). The morphological features of Vero cells infected with rabies virus and the expression of rabies virus particles are shown in Figure 9.10.

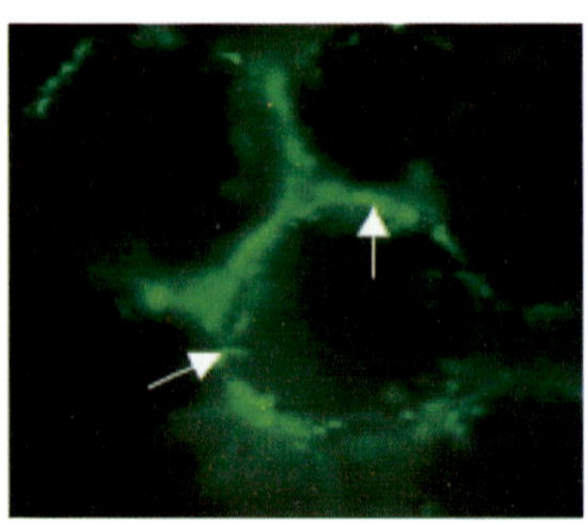

Figure 9.10 Infected Vero cells on microcarriers showing expression of rabies glycoprotein.

9.7 SUMMARY

1. Stability of the clone is paramount important for commercial manufacturing process.
2. In large-scale production of therapeutic proteins, system scalability is an important factor.
3. Different types of bioreactors have been developed for cultivation of animal cells: however, commonly used types in commercial applications (in decreasing order) are stirred tank, airlift, hollow-fiber, and fluidized bed.
4. The capacity of the bioreactor at each stage is decided by the number of cells required in subsequent stage or production of specific amount of biological.

ABBREVIATIONS

BSA: Bovine serum albumin
CHO: Chine hamster ovary
ECS: Extracapillary space
Epo: Erythropoietin
FCS: Fetal calf serum
IL-12: Interleukin-12
ICS: Intracapillary space
Mab: Monoclonal antibody
MOI: Multiplicity of infection
MTX: Methotrexate
PBS: Phosphate buffered saline.
t-PA: tissue plasminogen activator

REFERENCES

1. Arden N, Nivtchanyong T, Betenbaugh MJ (2004). *Bioprocessing* 3, 23.
2. Crespi CL, Thilly WG (1981). *Biotechnol. Bioeng.* 23, 983
3. Hendriksen C, W de Leeuw (1998). *Res Immunol* 149, 535.
4. Handa-Corrigan A, Emery AN, Spier RE (1989). *Enzyme. Micro. Technol.* 11, 230.
5. Jackson L, L Trudel, J Fox, N Lipman (1996). *J Immuno Methods* 189, 217.
6. Kim SJ, and Lee GM (1999). *Biotechnol. Bioeng.* 64, 741.
7. Monoclonal Antibody Production. A report of the committee on methods of producing monoclonal antibodies by Institute for Laboratory on Animal Research, National Research Council. Nation Academy Press Washington, DC, 1999.
8. Montagnon BJ, Vincent-Falquet JC, Fanget B (1984). Dev. Biol. Stand. 55, 37.
9. Neuberger MS, Rajewsky K (1981) *Proc. Natl. Acad. Sci. USA*. 78,1138.
10. Nahapetian AT, Thomas JN, Thilly WG (1986). *J Cell Sci.* 81, 65.
11. Peterson N, J. Peavey (1998). *Contemporary Topics Lab Anim Sci* 37, 61.
12. Pavlou AK, Reichert JM (2004a). *Nat. Biotechnol.* 22, 1513.

13. Pavlou A, Reichert J (2004b). *Nat. Rev.* 3, 243.
14. Simpson NH, Singh RP, Perani A, Goldenzon C, Al-Rubeai M (1999). *Bioeng.* 59, 90.
15. Trabelsi K, Rourou H, Loukil H, Majoul S, Kallel H (2006). *J Biotechnol.* 121, 261.
16. Umana P, Jean-Mairet J, Moudry R, Amstutz H, Bailey JE (1999). *Nat. Biotechnol.* 17, 176.
17. Vournakis JN, Runstadler PW (1991) In: *Animal cell bioreactors* (CS Ho, DIC Wang, eds.). Butterworth-Heinemann, Boston, p. 305.
18. van Wezel AL (1967) *Nature (London)* 216, 64.
19. Wurm FM, Pallavicini MG, Arathoon R (1992). *Dev. Biol. Stand.* 76, 69.
20. Wurm FM (2004) *Nat. Biotechnol.* 22, 1393.

SUGGESTION FOR FURTHER READING

1. Anderson BG and Gruenberg ML (1987) In: *Animal cell bioreactors* (CS Ho, DIC Wang, eds.). Butterworth-Heinemann, Boston, p. 175.
2. Lawrence S (2004) *Nat. Medicine*, 22, 1496.
3. Mukhopadhyay A (1996) Ph.D. thesis submitted to Indian Institute of Technology, Delhi, India.
4. Neuza M, Frazatti-Gallinaa, Regina M, *et al.* (2004) *Vaccine* 23, 511.
5. Paclou AK, Reichert JM (2004). *Nat. Biotechnol.* 22, 1513.
6. Reichert J, Pavlou A (2004) *Nat. Rev.* 3, 383.
7. Von Wedel RL (1987) In: DD Deaver ed. *Commercial production of monoclonal antibodies: A guide for scale-up.* Marcel.

10

Scale-Up Studies in Animal Cell Culture Bioreactors

The main problem for scale-up of animal cell culture bioreactor is related with adequate oxygen transfer. Animal cells are shear sensitive, they cannot be exposed to high shear force. Again, high-cell density cultures require more oxygen, and the solubility of oxygen is significantly dropped in serum containing medium. All these unfavorable factors made animal cell culture scale-up process become complicated. Animal cells grow slowly in normal batch culture and the final cell densities do not cross the required limit. Which means, with an increase in the demand of a product, the installation capacity of bioreactors is automatically escalated. Due to cost intensive cell culture process and high capital investment in large-scale applications, various investigators have put-forward the concept of *density scale-up*. The idea in density scale-up is to increase productivity of a small-scale bioreactor by proportional increase in cell density. Once the productivity touches the highest level using available technology, the next step is volume scale-up. Despite various advantages in density scale-up, suspension culture with low cell density is the production method for manufacturing vast majority of animal cell derived products. The primary reason of it is simple design of batch culture process. Oxygen transfer has been considered as one of the limitations in high-density culture processes. This chapter has been designed to discuss various constrains in animal cell culture processes, concept of density scale-up, and present state of scale-up techniques in different animal cell culture bioreactors.

10.1 DEFINITION OF SCALE-UP AND BIOLOGICAL CONSTRAINS

Scale-up is the study of problems associated with transferring data obtained in the laboratory and pilot-plant scale facilities to the production-scale. The objective of scale-up in animal cells derived processes is to maintain similar productivity of the small-scale bioreactor in the large-scale bioreactor, and the reverse in case of *scale-down*. The objective of scale-up and some relevant questions normally addressed to attain these objectives are shown in Figure 10.1. Often bioreactor scale-up is seen only as an engineering problem, and many empirical

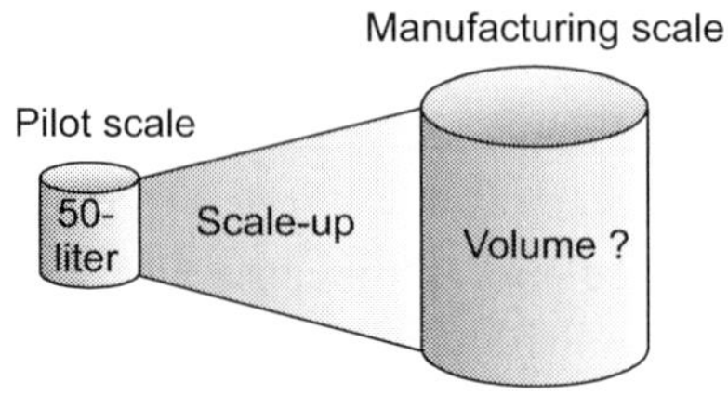

Assumptions: 1. Stable clone within reasonable generation numbers
2. Optimized manufacturing process available in pilot scale.

Objectives: 1. Maintain comparable productivity (mg/l/h).
2. To meet target capacity and quality of the product approved by the regulatory authority.

Questions: 1. Criteria to be followed to transfer pilot-scale experience?
2. How big will be the volume of the bioreactor?
3. Is single unit in (bioreactor) enough, or multiple units are to be required?
4. Adoptability with improved technology?

Figure 10.1 The objectives of animal cell culture bioreactor scale-up, and pertaining questions to attain the objectives. In order to undertake any scale-up problem, the answers of these questions should be available.

relations, used in Chemical Engineering design calculation are employed in designing bioreactors. It will be a serious fault if the *engineering principles* are not comprehended in view of *biological constraints*, and the results are not interpreted to validate the principles. The peculiarities in animal cell culture processes are due to: (a) involvement of shear sensitive organisms, thus sudden fluctuation of physical and biological process parameters are detrimental, (b) products are obtained through multi-variable and multi-step reactions; each step may have optimum reaction conditions, (c) every product is unique, produced by a unique clone, and (d) in case of therapeutic product stringent quality control is mandatory. Various characteristics of mammalian cells and specific requirements, which might change the yield of the product are shown in Figure 10.2.

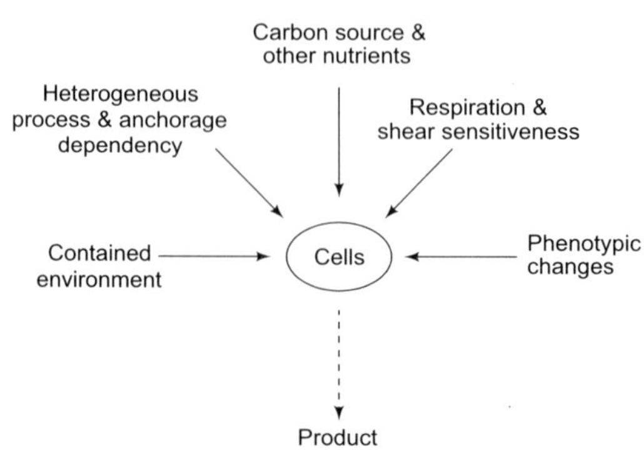

Figure 10.2 Characteristics of mammalian cells and their specific requirements may influence on the yield of the product.

10.1.1 Available scale-up criteria for fermentation processes

Two approaches are commonly followed to ensure similarity between the two different scales of bioreactors. These are:

(A) Geometric similarity of the physical boundaries. The shape of bioreactors is maintained same, and all linear dimensions of (H to D ratio, D_t to D_i ratio, etc.) the pilot-scale are related to the corresponding dimensions of the large-scale by a scale-up factor.

(B) Dynamic similarity of the flow fields. In this case the ratios of all forces acting on fluid particles in both the scales are maintained similar.

It is easy to obtain the *geometric similarity* in the two scales of bioreactor, but maintaining dynamic similarity between them is challenging. *Dynamic similarity* can be achieved, if the values of the non-dimensional parameters (e.g., Reynolds number, Froude number, power number, etc.) are same at geometrically similar locations of two different scales of bioreactors. In all practical purposes, gassed power input in unit volume (P_g/V), impeller tip speed ($\pi D_i N$), airflow rate in unit volume (Q/V), and few more scale-up criteria can be followed. It is important to find out the correlation of these dynamic/mixing properties with other physical parameter of the bioreactor (e.g., volumetric oxygen transfer coefficient), which would have direct relationship with the growth of cells and the productivity.

10.1.2 Limitation in animal cell culture processes

Animal cells differ many respects with microbial cells. The major structural difference between an animal and a microbial cell is that the former lacks the cell wall, and is therefore more sensitive to hydrodynamic forces (also known shear forces). Hydrodynamic cell death primarily occurs due to mixing of cultures (Chapter 6). Mixing of culture is essential: (a) to maintain homogeneous bioreactor environment for process control and cell growth, (b) for supply of nutrients, (c) to minimize local accumulation of by-products, and (d) for oxygen transfer. Hence, understanding the mechanisms by which hydrodynamic forces affect cell growth, death, and metabolism are important for design and scale-up of animal cell culture bioreactors. The effect of hydrodynamic forces on cells is mediated by fluid mechanical components that result in reduction of cell growth, alter cell cycle, cell death and even lowering product formation. It is essential to analyze the nature of fluid components in laboratory scale/pilot scale bioreactor, which is responsible to modulate above changes in cells, prior to the consideration for scale-up.

Another category of problem in cell culture is that two diverse classes of cells are used: (a) one that can grow in suspension, and (b) the other that cannot grow unless a solid surface is available to them (anchorage-dependent). Again, from the viewpoint of products, some are produced in the stationary phase (e.g., monoclonal antibodies), some are in the growing culture (e.g., viruses), and the others are affected structurally (e.g., glycosylation) due to unusual physiological state of the cells. Further, the problem is associated with extremely slow specific growth and product formation rates in animal cells, as compared to the microbial systems. Lastly, the animal cell culture is prone to microbial contamination, and thus commands stringent culture sterility than microbial-based processes.

Over the past three decades, much advancements have been registered in four major areas of cell culture process. These are: (i) mixing with low shear injury, (ii) aeration avoiding cellular damage, (iii) perfusion with maximum retention of cells, and (iv) large growth surface area in unit volume of medium for anchorage-dependent cells. These areas are linked each other, which require proper attention prior to a scale-up process is undertaken.

10.2 PROBLEMS IN CELL CULTURE PROCESS AND CORRESPONDING DEVELOPMENTS

It has been recognized for a long time that due to change in cell type and their process requirement, nature of product, and its kinetics of expression, no single bioreactor is suitable for all animal cell culture processes. Since suspension culture has several advantages over static culture, attempt has been made to remove the bottlenecks in static culture of animal cells by adopting a homogeneous system. The problems related to cell culture processes and the corresponding developments are depicted in Table 10.1. It is important to remember that different kinds of bioreactors have been proposed time to time to solve these problems.

Table 10.1 Problems in animal cell culture and the corresponding developments

Problems	*Developments*
1. Fragile	1. Immobilization of cells
2. Anchorage-dependency	2. Microcarrier beads
3. Slow growth rate and low level of product formation	3. Immobilization of cells and high cell density culture
4. Washing-off cells	4. Anchoring to solid surface or providing filtration device
5. Nutrients limitation and toxic effects of metabolites	5. Continuous perfusion
6. Process control and scale-up	6. Homogeneous culture
7. Microbial contamination	7. Unit culture process
8. Batch to batch variation of product quality	8. Unit culture process

10.3 DENSITY SCALE-UP VERSUS VOLUME SCALE-UP

Linear *volume scale-up* is most common in a typical fermentation industry, whereas in cell culture-based processes, volume scale-up is associated with escalation of investment cost, increase in contamination risk, decline in productivity, and rise in media cost. Thus, in cell culture, volumetric scale-up is not popular at least for anchorage dependent cells-based processes. Accordingly, time to time various specialized cell culture bioreactors have been developed to encourage high-cell density culture. These bioreactors are based on the following principles: (a) high perfusion rate without washout, (b) retention of cells by physical means, and (c) high cell number in unit surface area or volume of the culture. The relative capacities of these bioreactors in terms of ratio of cell culture surface area to bioreactor medium volume are shown in Figure 10.3. In earlier days, the process scale-up for cell culture derived product (e.g., viral vaccine) was on the basis of increasing the number of Roux/roller bottles. This approach has several

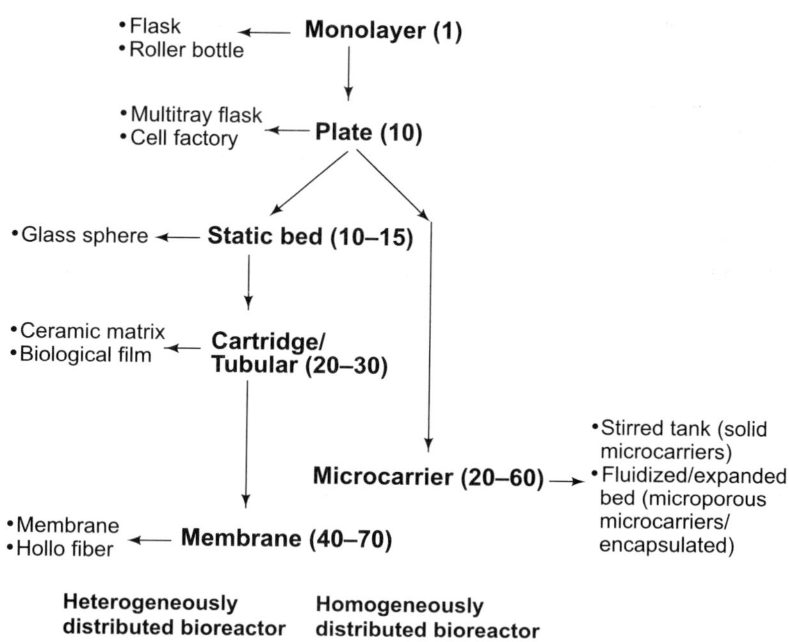

Figure 10.3 Density scale-up of different animal cell bioreactors. Density scale-up ratio, shown within the parenthesis, is based on surface area to medium volume. Analyses have been made for the culture systems that support the growth of anchorage-dependent cells. [Adapted from Griffiths, 1988]

disadvantages like product variability, intensive labor input, large incubation area, and microbial contamination. With the advent of microcarrier technology, it has become possible to replace small-scale multiple cell culture system by large-scale unit cell culture bioreactor. Furthermore, the idea of high-cell density culture simplified large-scale processes. The concept of high-cell density culture was developed on the basis of the principle that tissue (cluster of cells) functions better than isolated cells, and at high-density culture each cell behaves as a unit of the tissue. The merits and demerits of high-cell density culture processes are described in Table 10.2. Despite several advantages, high-cell density culture is not much popular in large-scale applications.

Table 10.2 Merits and demerits of high-cell density culture process

Merits	*Demerits*
• Small reactor volume, less investment • More concentrated product, less downstream processing cost • Reduced serum/growth factor requirement • Longer run time, lower labor cost • Saving of space	• Oxygen transfer difficult • In heterogeneously distributed bioreactor, process parameter control is not straightforward • Resterilization is not possible in heterogeneously distributed systems; disposable bioreactor is the only option (e.g., hollow-fiber, membrane bioreactor)

The advantage of high-cell density culture-based process over normal batch culture has been illustrated in the following case study:

CASE STUDY: Production of tissue plasminogen activator (t-PA)

Batch culture

Assumptions

1. Average yield: 20 mg/10^6 cells/day (Runstadler *et al.*, 1989)
2. Average cell density: 3×10^6 cells/ml
3. Batch run time: 4 days (production phase)

Suppose, the average treatment dose: 50 mg/patient (Builder and Grossbard, 1986), and number of heart-attacked victims: 500,000/annum.

The estimated annual requirement (purified form): $50 \times 500{,}000 = 25$ kg

Estimated annual requirement (crude form): 25/0.5 = 50 kg (assuming 50% purification efficiency)

Total culture volume to produce 50 kg t-PA = $50{,}000{,}000/(0.02 \times 3 \times 1000 \times 4) = 208333.33$ L = 210,000 L (say).

In case of a 10,000 L suspension culture bioreactor (charging volume 7,000 L), the number of batch run required to produce 50 kg t-PA will be 210,000/7,000 = 30. Therefore, the number of 10,000 L bioreactors required to produce 50 kg t-PA/annum will be 1.5 to 2.

High-cell density perfusion culture

Assumption

1. Average yield: 20 mg/10^6 cells/day (Runstadler *et al.*, 1989)
2. Average cell density: 3×10^7 cells/ml
3. Continuous run time (production phase): 80 days (perfusion)

Total culture volume to produce 50 kg t-PA: $50{,}000{,}000/(0.02 \times 30 \times 1000 \times 80) = 1041.7$ L = 1100 L (say).

In case of a 100 L fluidized/expanded bed culture bioreactor (80 L charge volume), the number of culture batches require to produce 50 kg t-PA will be 1100/80 = 14 (say). Due to extended run time, total number of run/bioreactor/annum will be 2.5 (approximately). Thus, 6 number of fluidized-bed bioreactors of 100 L size would be sufficient to produce 50 kg t-PA/annum.

Distinct advantages: Less capital investment on upstream and downstream processing equipments, less labor cost, less FCS and other consumables in high-cell density perfusion culture.

Distinct disadvantages: Porous microcarriers used as support for the growth of cells are expensive.

10.4 BIOREACTOR SCALE-UP

Two types of bioreactor are commonly used in animal cell culture—stirred tank and airlift bioreactor. Other two bioreactors, not much common in large-scale applications, are fluidized/expanded bed and hollow-fiber bioreactor. It may be remembered that the objective of scale-up is to translate laboratory data into unit production scale culture without additional cost inputs. The principles for scale-up followed in different bioreactors are shown below:

10.4.1 Airlift bioreactor

Like many other manufacturers of recombinant proteins and antibodies, Celltech Ltd., UK adopted perfusion airlift bioreactor for manufacturing therapeutic grade proteins from hybridomas and CHO cells (Broad *et al.*, 1989). The reasons for adopting airlift bioreactor are: (a) simple design and maintenance, (b) widely used in fermentation industries, (c) provide adequate oxygen transfer rates, (d) low shear effect than stirred bioreactors, (e) low capital investment, and (f) reduced risk of microbial contamination.

The aim of scale-up is to maintain key process variables of pilot scale experiments that have significant effect on cell growth and productivity constant in the production scale operation. One such process variable used in scaling up bioreactors is volumetric oxygen transfer coefficient (K_La), as described in the earlier Chapter. In a typical air-lift bioreactor (Figure 10.4A), K_La is found to be a function of many design/operating parameters as shown below:

$$K_La = f\,[U_A, U_L, D_D, D_R, H_D, D_{CL}, \rho, \mu, \sigma] \qquad 10.1$$

where, U_A: superficial air velocity in the riser based on empty cross-section

U_L: superficial liquid velocity in the riser based on empty cross-section

D_D/D_T: diameter of the downcomer/tank

D_R: diameter of the riser

H_D: height of the gas-liquid dispersion

D_{CL}: diffusion coefficient of oxygen in liquid phase

ρ: liquid density

μ: liquid viscosity

σ: surface tension at gas-liquid interphase

The ratio of cross-sectional area of *downcomer* and *riser* (A_D/A_R) is crucial in mixing and oxygen transfer in airlift bioreactor. The values of A_D/A_R between 0.59 to 0.75 give optimum mixing and oxygen transfer. In some scale-up applications, A_D/A_R has been maintained constant.

For an existing airlift bioreactor to culture animal cells,

$$K_La = f\,[U_A] \qquad 10.2$$

Thus, 'K_La' is a function of *superficial air velocity* in the riser. U_A is an important operating parameter that can be manipulated to maintain desired oxygen level in the

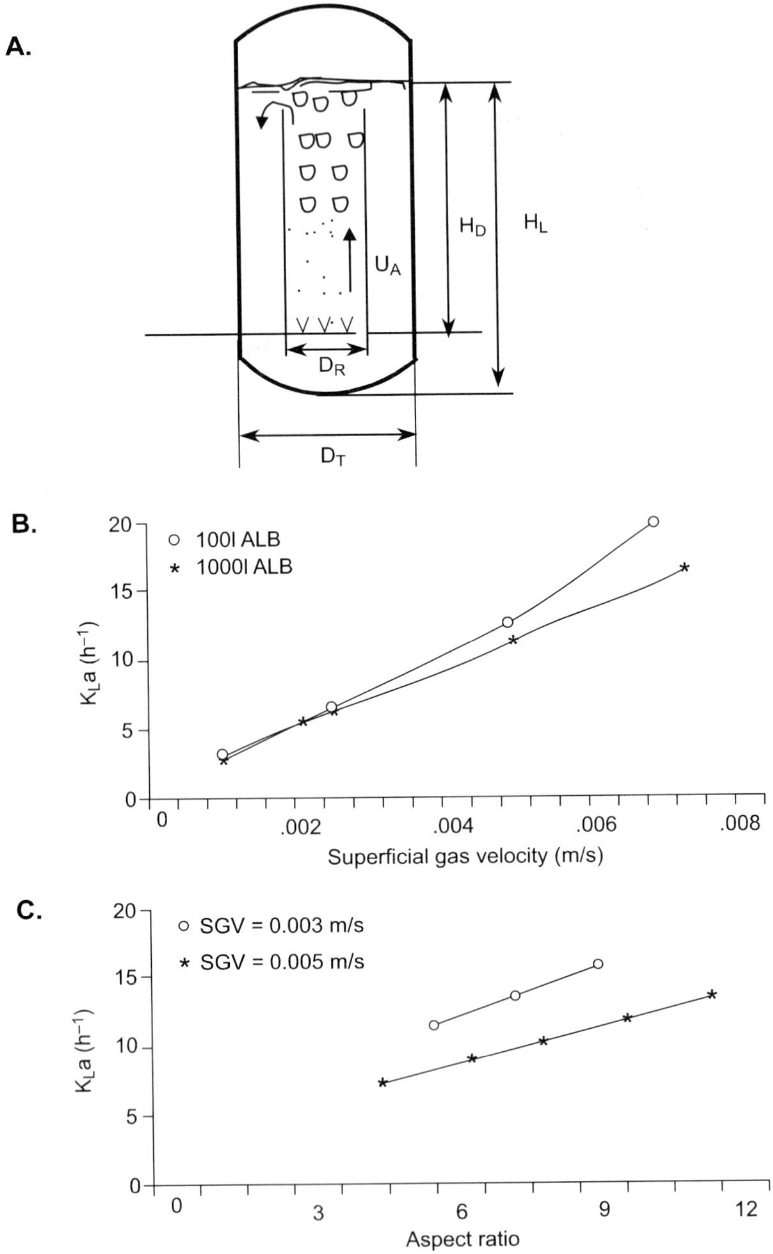

Figure 10.4 A. Air-lift bioreactor showing different critical geometric (dimension) parameters to be considered during scale-up studies; B. Effect of U_A on K_La at two different volumes of air-lift bioreactors. K_La is linearly increased with U_A, in both 100-l and 1000-l air-lift bioreactor (Adopted from Ref. 6); C. Effects of aspect ratio (H_L/D_T) on K_La at two different U_A. The overall K_La value depends on operating variable (U_A) and design parameters (H_L/D_T). [Adapted from Ref. 6]

liquid phase. It has been reported that K_La is linearly increased with an increase in U_A in both 100 L and 1000 L airlift bioreactor (Figure 10.4B), which has been successfully used as a *scale-up criteria* (Broad *et al.*, 1989). Again, the aspect ratio (H_L/D_T), K_La, and U_A are related. Figure 10.4C depicts how K_La increases with aspect ratio at two different superficial air velocities. These indicate that not only operating variable (U_A), but design parameter (H_L/D_T) also influences overall K_La value. Once the relationship between these design and operating variables are established, it will become convenient to maintain a desired level of K_La value by choosing a combination of aspect ratios and superficial air velocity. Manipulation between the operating and design parameters is primarily carried out to minimize physical damage to the cells. For example, the optimum K_La for a process is 10 h^{-1}. It can be attained by two ways: (a) lowering U_A (0.003 m/s) in a 12:1 aspect ratio bioreactor (for shear sensitive cells), and (b) increasing U_A (0.005 m/s) in a 6:1 aspect ratio bioreactor (Figure 10.4C).

The volumetric oxygen transfer coefficient (K_La) can be predicted from an empirical relation proposed by Bello *et al.* (1981) for air-lift bioreactor:

$$K_La = 5 \times 10^{-4} \, [1 + (A_D/A_R)]^{-1.2} \, [P_A/V_L]^{0.8} \qquad 10.3$$

where, A_D and A_R are cross-section area of descending and riser, respectively, P_A is power input by aeration, and V_L is volume of the medium. The ratio of A_D/A_R is found to be critical for mixing and oxygen transfer in airlift bioreactor. Air-lift bioreactors are normally tall; the height of a typical 1000 L vessel is about 5 m. *Hydrostatic pressure* increases dissolved oxygen concentration at the base than the upper levels of medium. Most importantly in a tall airlift bioreactors, cells are experienced a pressure cycle, highest at the bottom and lowest on the medium surface. The fluctuation of hydrostatic pressure may induce effects on physiology of animal cells, which should be taken into consideration while designing a tall airlift bioreactor. In order to study the effect of hydrostatic pressure on viability, NB1 cell (hybridoma) was exposed to 18 psig (equivlent to 11-m height for 10,000 L airlift bioreactor) for 60s followed by 60s in atmospheric pressure. No significant effect on cell viability was observed after 10 such pressure cycles (Birch *et al.*, 1987). Starting from 30 L capacity, Celltech designed (Rhodes *et al.*, 1991) progressively larger scale airlift bioreactors (100, 200, 1000, and 2000 L capacities).

The scale-up process is not limited to bioreactor. For a successful process scale-up, all interconnecting equipment should be considered. The peripheral equipment that need to be scaled-up are medium holding tanks, asceptic medium transfer, cell separator, etc. The process scale-up could be in different modes of operation, e.g., batch, fed-batch, and continuous perfusion. Batch process is simple in operation and design, whereas in perfusion process cells are separated from the effluent medium using various techniques mentioned in Chapter 6. The most simplified means of separating free cells from flowing medium is internal spin filter (suitable for stirred bioreactor) or an external cell settling device. One such system is *lamellar clarifier*, which is connected in the external loop of the bioreactor over-flow line. The flow diagram of airlift bioreactor with lamellar clarifier is shown in Figure 10.5. Lamellar clarifier has no moving part, and it works in normal medium circulation path. The retention efficiency of viable cells in this system was found to be 90% at 4-reactor volume perfusion/day; however, the system was tested only in 5.4 L airlift bioreactor (Stevens *et al.*, 1994).

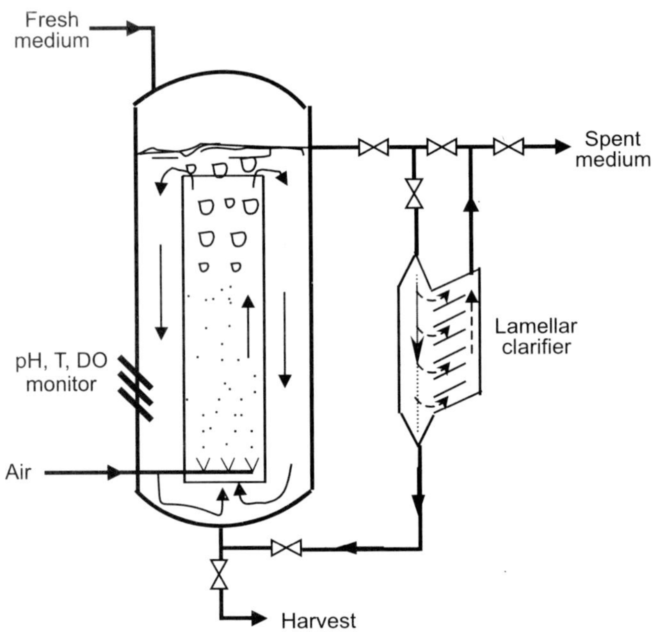

Figure 10.5 Schematic of air-lift bioreactor with counter operated lamellar cell clarifier. The medium is recycled back to the bioreactor through out-let (bottom) of lamellar clarifier. Cells present in the downward flow path of medium are settled-down on the plates, clear medium leaves the clarifier in the opposite direction, same as it entered inside the clarifier. The cells, which settle down on the plate, are gradually slide down and mixed with downward flow path of the medium and finally returned back to the bioreactor.

10.4.2 Stirred bioreactor

The experience of microbial fermentation helped investigators to design large scale stirred bioreactor for the culture of animal cells. Namalwa cell was grown in a 8000 L bioreactor for the production of interferon (Johnston *et al.* 1979). Many development that have been taken place during past two decades in designing stirred bioreactors, these are: (a) curved bottom for better mixing at low agitator speed, (b) using low shear impeller at minimum speed, (c) internal mirror finished (> 200 grains) stainless-steel tank without baffles, (d) outside water-jacket to control temperature, (e) top driven stirrer to avoid entanglement of cells between moving parts, in alternate very specially designed magnetic driven impeller, and (f) control of dissolved oxygen by injecting air or pure oxygen beneath the impeller just to maintain the oxygen demand of the cells.

Like other microbial bioreactors, oxygen transfer and mixing are the two most important operating variables in animal cell culture bioreactor. It is known that oxygen transfer has direct relationship with mixing, but due to shear sensitive nature of the cells it is recommended to consider these two operating variable not linked each other. For example, while designing impeller, care should be taken to achieve adequate mixing without disruption of air bubbles. The problem related to oxygen transfer through sparged air should be considered separately. A typical high mixing impeller with low shearing effect is shown in Figure 10.6A. Combination of such impeller with *hemispherical* bottom bioreactor is found to be ideal for greater mixing in low shear field system (Figure 10.6A).

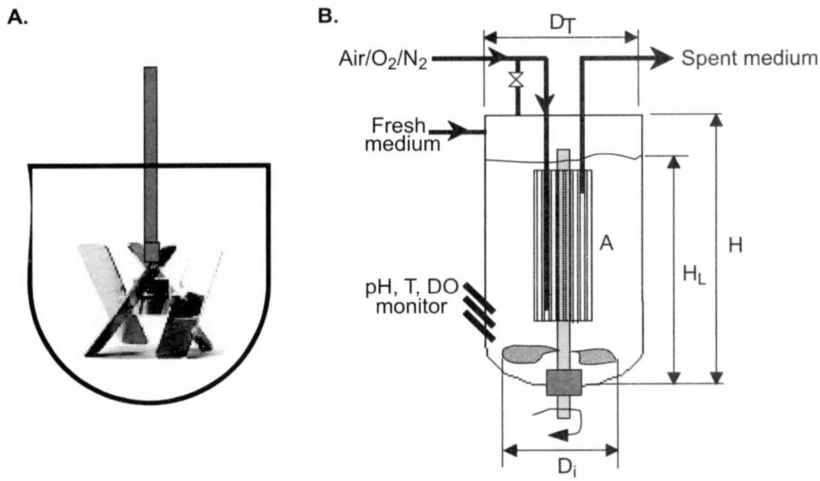

Figure 10.6 A. Hemispherical-bottomed animal cell culture bioreactor with flat-blade impeller. Efficient mixing of cell culture medium at low agitation speed (low-shear) has been obtained upto 450L stirred tank bioreactor (*Source*: SGi, France); **B.** Schematic of a stirred-tank bioreactor equipped with spin-filter. Different geometric (dimension) parameters influence the design of a bioreactor are shown (H: height of the bioreactor, H_L: height of liquid, D_T: Tank diameter, D_i: impeller diameter, A: area of spin filter). It is important to be remembered that a successful perfusion culture relies on spin-filter (its area, mesh size, and rotation speed) with respect to the density of microcarriers (in case of anchorage-dependent cell culture) and the perfusion rate.

Mixing time is an important parameter used to design cell culture bioreactor, which may be one of the criteria for scale-up. Many empirical relationships have been developed to determine mixing time of an agitated bioreactor, however, significant discrepancies have been reported when mixing times were calculated from these empirical relations and compared with that actually determined by the experiments. These differences are primarily due to the measuring method followed and the design of the mixing impeller used by the various investigators (Kawase and Moo-Young, 1990). The most realistic empirical relation to calculate *mixing time* using a disc turbine has been reported by Middleton (1979). It has been shown that bioreactors of different volumes and impeller sizes correlate well with the following relationship:

$$\tau \sim (V)^{0.3} (D_T/D_i)^3 N^{-1} \qquad 10.4$$

where, τ is mixing time, V, volume of the fluid in bioreactor, N, impeller speed, D_T and D_i are tank and impeller diameter, respectively.

It may be noted that the above relation establishes mixing time of bulk liquid of a bioreactor without spin filter. The mixing time calculated using Equation (10.4) is changed if the bioreactor is equipped with a spin filter. It may be recalled that spin filter is commonly used to separate cells in a continuous perfusion bioreactor. Mixing times have been studied in dimensionally similar 80, 250, and 1000 L bioreactors equipped with rotating *spin filters* (Jem et al. 1994). It was found that the actual mixing times in these bioreactors do not match with the values calculated using above empirical relation, and other available in the literature. The actual mixing time was found much higher than that calculated from these relationships. In the presence of spin filter, the mixing efficiency slows down as compared with the mixing obtained in the bulk of the stirred bioreactor. This is probably due to different mechanisms of mixing involved in the bulk liquid to that of inside and through the spin filter (25-50 μm pore size). The perfusion bioreactor improves overall mixing time by 20% as compared with the batch system. It appears that impeller used for bulk mixing is not enough to maintain homogeneity across the spin filter. Bulk mixing time in spin filter bioreactor has been significantly improved by providing buffles on outer perimeter of the filter and by increasing impeller diameter. However, before changing the design of the bioreactor in this regard, one has to consider the consequence of improved mixing efficiency on viability of cells. Slow mixing inside the filter is primarily due to low fluid exhange with the bulk. Improved mixing time can be obtained by increasing perfusion rate, pore size (~ 70 μm), air sparging inside the filter, and installing baffles inside the spin filter. Cells growing on microcarriers can be conveninently operated, as large pore size spin filters are used in these cases (typical size of a microcarrier beads > 150 μm), which prevents clogging of the filter and permits high perfusion rate. Spin filter bioreactor has been successfully used in a perfusion microcarrier culture of a recombinant CHO cells where maximum growth was achieved at a density of 8×10^7 cells/ml (Averginos *et al.*, 1990).

Mixing through the spin filter can be improved by the combination of (a) increasing perfusion rate, (b) providing buffles and/or impeller inside the spin filter, and, (c) sparging air inside the filter in addition to sparging underneath the impeller for the bulk medium mixing. Spin filter bioreactor gained acceptance in commercial scale vaccine manufacturing process using microcarrier culture technology, as scale-up process is straightforward. Killed polio vaccine has been produced in 1000 L microcarrier culture of Vero cells (Montagnon *et al.* 1984). The main advantage of microcarrier culture is that density scale-up is possible by increasing microcarrier loading in unit volume of the medium. Mixing could be a problem in culture containing high microcarrier loading (<10 g/L). Mixing problem in microcarrier culture can be minimized by using specially designed impeller blades, which allow proper bulk mixing without imparting much shear effect on microcarrier beads. A schematic diagram of a typical spin filter equipped stirred bioreactor is shown in Figure 10.6B.

10.4.3 Fluidized bed/expanded bed bioreactor

The fluidized bed bioreactor was developed on the principle of fluidized bed catalytic reactor for utilization of maximum catalytic sites to minimize mass transfer limitation, and to maintain homogeneous environment. The key features of fluidized bed bioreactor are as follows:

(a) Due to special nature of fluid dynamics around the individual matrix, intra-matrix transport of nutrients and removal of metabolic wastes and products are considerably improved.
(b) Due to porous nature of the matrices, cells are protected from fluid shear effect, at the same time mass transfer rates are improved.
(c) *Porous collagen*/gelatin matrices provide biomimetic support to the cells, thus high cell density growth is possible.
(d) Three-dimensional tissue like growth improves cell-cell contacts, thus specific productivity is enhanced as compared with that obtained in monolayer culture.

The most commonly used porous microcarriers have been derived either from gelatin or native bovine type I collagen. Gelatin-coated porous glass carriers (Siran™, Schott Glaswerke, Germany) are also used in culturing hybridomas in fluidized bed bioreactor (Rolef *et al.*, 1994). The specifications of Verax collagen microsphere are shown in Table 10.3. It may be seen that the specific gravity of porous microcarriers is much higher than the solid microcarriers.

Table 10.3 Specification of Verax microsphere used in fluidized bed bioreactor

Properties	*Specification*
Average diameter	500 μm
Interconnected pore size	20 to 40 μm
Specific gravity (wet)	1.6-1.7
Fluidized liquid velocity	70 cm/min
Internal surface area (estimated)	30 m^2/L
Number of microcarrier particles (25% solids)	$3.8 \times 10^6/L$

Source: Verax Microsphere™, Verax Corporation, Lebanon, New Hampshire, USA.

To avoid elutriation losses due to fluidization, high specific gravity porous microcarriers are used. Porous microcarriers are normally titanium imprignated to increase specific gravity higher than the medium. Porous collagen microspheres have been tested with many recombinant cells and hybridomas for making therapeutic proteins in fluidized bed bioreactors. A partial list of growth and viability of commercially useful clones/cell lines cultured on porous collagen microcarriers is shown in Table 10.4. It may be seen that despite enormous growth, the viability of cells remains very high in long-term studies. This was believed to be due to interconnecting pores, internal medium flow and nutrient/metabolite exchange between bulk medium and cells inside the pores of the microspheres. It is believed that fluidization results in change in medium flow dynamics and pressure difference across the microcarriers, which causes convective flow of nutrients through the porous microspheres.

Table 10.4 Cell density and viability of cell in collagen microsphere.

Cell type	*Product*	*Cell density (cells/ml)*	*Viability (%)*
Human/mouse hybridoma	IgG	8×10^7	68
CHO	Thrombolytic	18×10^7	79
CHO	Hormone	25×10^7	72
CHO	Interleukin	26×10^7	92
C127	Thrombolytic	9×10^7	85

[Taken from Ref. 16]

Process scale-up using fluidized-bed bioreactor has not been studied much. The design criteria of fluidized bed bioreactor is again related to the mixing and oxygen transfer efficiency (Fig. 10.7). Since, 25-40% volume of the fluidized bed is occupied by microcarrier particles, it is considered a process containing high density slurry in culture medium. As fluidization of solid phase (microcarriers) occur due to the action of vertical movement of medium, its superficial velocity (U_M) based on empty cross-section of the bed is critical. The uniform distribution of medium across the cross-section of the bed is of paramount important to avoid *channeling problem*. The supporting plate used to prevent microcarriers to enter in the impeller mixing zone also acts as distributor of medium across the diameter of the bed. Optinum design criteria of the distributor is another important issue for successful operation of fluidized-bed bioreactor. As microcarrier particles are fluidized, they may have tendency to escape with the effluent. The terminal velocity of each particle on the expanded section of the bed should be higher than the draging effect due to the flow of medium. In order to create a cell free zone on the upper section of bioreactor, its diameter is increased to reduce upward liquid velocity and hence the dragging force. The terminal velocity of the particles becomes higher than the upward dragging force, and as a result they settle near the inclined portion of the fluidized bed bioreactor. The diameter of the expanded section of fluidized bed bioreactor is related to the specific gravity of the particles and U_M.

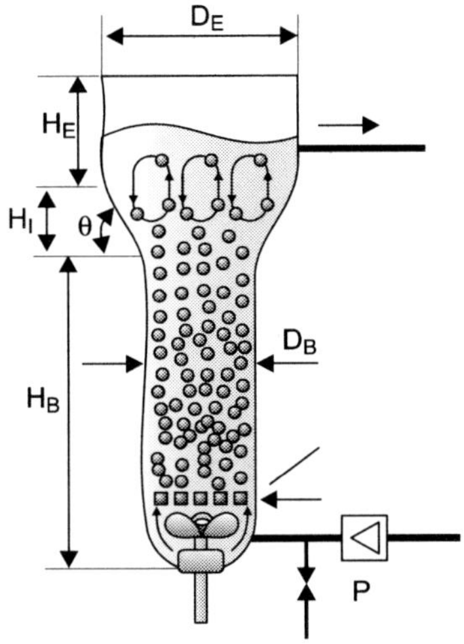

Figure 10.7 Fluidized-bed bioreactor showing critical geometric (dimension) parameters used to design the system (H_B: height of the bed, H_I: height of inclined portion of the column, H_E: height of the expanded portion of column, D_B: diameter of the bed, D_E: diameter of the expanded portion of column, è: angle of inclination). In the bottom of cell free zone, arrows indicating uniform distribution of medium through the supporting plate. Probable circulation path of each microcarrier bead is shown in the expanded section of the column. The gravitational force on the particles present on the upper section of the loop in more than the upward dragging force, therefore particles are settled down in the expanded section of the column. As soon as they entered in the narrow section, the upward dragging force of medium (relatively higher than the gravitational force) pull the particles in upward direction.

10.5 SUMMARY

1. In addition to physical scale-up criteria, biological criteria are to be considered as scale-up problems in animal cell based processes.
2. Instead of volume scale-up, density scale-up is an alternate way for increasing manufacturing capacity of biological.
3. The design and scale-up of animal cell culture bioreactor is still an art, where experience in cell culture processes is of most important.

ABBREVIATIONS

FCS: fetal calf serum
IgG: Immunoglobulin G
t-PA: tissue plasmitogen activator

REFERENCES

1. Aiba S, Humphrey AE and Millis NF (1973). *Biochemical Engineering*, 2nd edition, p 195. University of Tokyo Press, Japan.
2. Averginos GC, Drapeau D, Socolow JS, Mao J, Hsiao K and Broeze RJ (1990) *Bio/Technol.* 8, 54.
3. Bello RA, Robinson CW and Moo-Young M (1981). *Adv. Biotechnol.* 1, 547.
4. Birch JR, Lambert K, Thompson PW, Kenney AC, Wood LA (1987). In: *Large scale cell culture technology* (BK Lydersen, ed.), Hanser Publishers, Munich. p. 2.
5. Builder SE and Grossbard E (1986). *Transfusion Medicines: Recent Technological Advances*, p. 303, Alan R. Liss, New York.
6. Broad DF, Brown ME, Grant AP and Wood LA (1989). In: *Adv. In animal cell biology and technology for bioprocesses* (RE Spier, JB Griffiths, J Stephenne and PJ Crooy, eds.), p. 412, Buttherworth & Co. (Publishers) Ltd. UK.
7. Jem JK, Fateen S and Michaels J (1994) In: *Animal cell technology: Products or today prospects for tomorrow* (RE Spire, JB Griffiths and W Berthold, eds.) p. 392. Buttherworth-Heinemann, Oxford.
8. Johnston MD, Christofinis G, Bull GD, Fantes KH and Finter NB (1979) *Dev. Biol. Stand.* 42, 189.
9. Kawase Y and Moo-Young M (1990). *Chem. Eng. J.* 43, B19.
10. Middleton JC (1979) *Proc. 3rd Eur. Conf. on Mixing*. University of York, York, England, A2.
11. Montagnon B, Vincent-Falquet JC and Fanget B (1984) *Dev. Biol. Stand.* 55, 37.
12. Runstadler RW, Tang AS, Hayman EG, Ray NG, Sample J and DeLucis DE (1989). In: *Large-scale mammalian cell culture technology* (Lubiniecki AS, ed.), p. 139, Marcel Dekker, New York.
13. Rhodes M, Gardiner S and Broad DF (1991) *In: Animal cell bioreactors* (CS Ho and Daniel IC Wang, eds.) p. 253. Buttherworth-Heinemann, Boston.
14. Rolef G, Biselli M, Dunker R and Wandrey C (1994). In: *Animal cell technology: Products or today prospects for tomorrow* (RE Spire, JB Griffiths and W Berthold, eds.) p. 481. Buttherworth-Heinemann, Oxford.
15. Stevens J, Eickel D and Onken U (1994). In: *Animal cell technology: Products or today prospects for tomorrow* (RE Spire, JB Griffiths and W Berthold, eds.) p. 234. Buttherworth-Heinemann, Oxford.
16. Vournakis JN and Runstadler PW Jr. (1991) *In: Animal cell bioreactors* (CS Ho and Daniel IC Wang, eds.) p. 305.

SUGGESTION FOR FURTHER READING

1. Ho CS, Wang DIC (eds.) *Animal cell Bioreactors*, Buttherworth-Heinemann, Boston. 1991.
2. Lydersen BK (ed.) *Large scale cell culture technology*. Hanser Publishers, Munich. 1987.

Section IV
Advanced Cell Culture Technology

11

Embryonic and Adult Stem Cells

Stem cells have procreated more expectation at the same time hypes than any other areas in past two decades of scientific studies. Stem cells were discovered more than 6 decades ago from bone narrow cells. At present scientists around the world are striving on different kinds of stem cells for their potential to regenerate and repair diseased tissue for revolutionizing the medical sciences. Stem cell therapy is the novel form of medicine that will have a major impact on human health care in the next 5 to 10 years. Stem cells have been classified into two major groups— the *embryonic* and the *adult stem cells*. Embryonic stem cells are more promising for clinical applications than adult stem cells, because they have the ability to differentiate into any tissue of adult human being. However, their use in human may be limited due to safety and ethical issues. On the other hand, adult stem cells are beyond any controversies regarding safety and ethical dilemma. They are believed to be present in almost every tissue, and seem to survive for long time and in harsh conditions. From the viewpoint of developmental biology, adult stem cells were considered to be irreversibly committed to particular lineage. However, number of studies over the past several years have shown the possibility of adult stem cells to change in completely different lineages' a phenomenon known as *stem cell plasticity*. This Chapter has been written for the students inclined to the clinical aspects of the stem cell technology. Here, the main emphasis has been given on defining and classifying stem cells, identifying their origin, biology of stem cells, and their potential clinical applications.

11.1 BEGINNING OF LIFE

Fertilized egg (*zygote*) gives rise to a complete organism through various stages of development. The process by which a fertilized egg turns in an embryo and subsequently develop is termed as *embryogenesis*, which involve coordinated cell division, differentation, migration, and programmed cell death. In human, most of the information about embryonic development has been derived from the studies of embryogenesis in mouse. Still, many of the developmental aspects of mammals are not clearly understood.

Prior to fertilization, the *oocyte*, a haploid cell that contains half (23) of the chromosomes present in normal human somatic cells enlarges and divides by meiosis and matures in the ovarian follicle. The follicle releases the mature oocyte into the oviduct. The oocyte is covered with a protective layer of glycoproteins called *zona pellucida*. In order to fertilize, the sperm has to enter the oocyte cytoplasm and fuse with its *pronucleus*. Fusion of sperm and oocyte pronucleus maintains the number of chromosomes of normal diploid cells (23 pairs). After fertilization, zygote moves its way to the uterus for implantation; in case of mouse it takes 3-4 days, and for human 6-7 days. During this time the zygote divides into genetically identical two cells, and then to four cells, and so on. Mitotic divisions of cells are observed with brief or no G1 and G2 phases of cell cycle. At this stage no growth of cells occurs. In fact, cells become smaller with time due to simultaneous mitosis and *cytokinesis*. Also, at this stage the cells start differentiating into *trophoectoderm* and the *primitive endoderm*, and the whole embryo is termed as *blastocyst* (pre implant stage). Studies in mouse have shown that these early lineages give rise to *extraembryonic* tissues to protect embryo in the uterine environment. The development of the mouse blastocyst at embryonic day 3 (E3.0) involves the generation of polarized epithelium of the trophoectoderm, which pumps fluid to form the internal *blastocoel* (Rossant and Cross, 2002). Nonpolarized clump of cells called *inner cell mass* (ICM) is present on one side of the blastocoelic cavity. In case of human, equivalent developmental stage appears at days 5 to 6 of the fertilization (Fong *et al.*, 1998). The photomicrograph of a human blastocyst of a similar age is shown in Figure 11.1. At E4.5 in case of mouse, and day 7-8 of post-fertilization in humans (Figure 11.2), the blastocyst implants in the uterine wall.

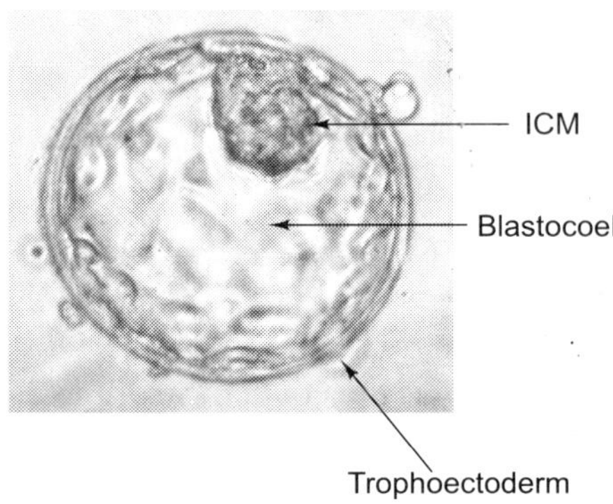

Figure 11.1 Human blastocyst showing inner cell mass (ICM) and outer trophoectoderm. (Reproduced by permission of Dr. Joe Conaghan, Director, ART Laboratories, Pacific Fertility Center, San Francisco, CA).

11.1.1 Development of blastocyst

Blastocyst develops as a result of early differentiation in the mammalian embryo. How does it occur from the genetically identical cells is not clearly understood. Studies in mouse showed that at 8-cell embryo stage, the compaction of cells is

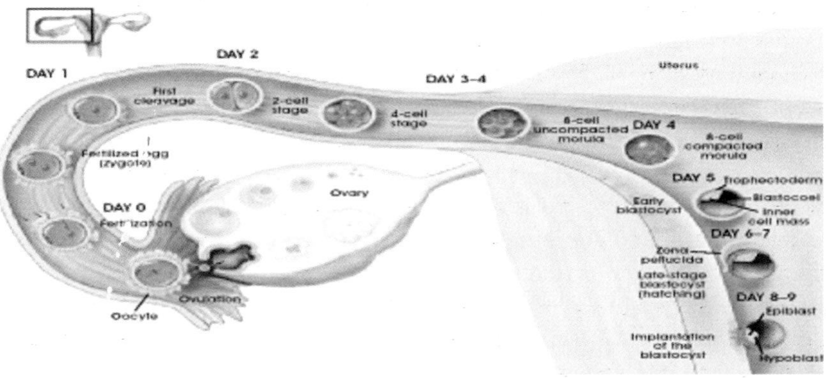

Figure 11.2 Development of pre-implanted blastocyst in humans. Fertilized egg (zygote) migrates towards uterus during different stages of development, and finally implanted on the uterus wall (*Source*: NIH Stem cells: Scientific progress and future research directions, 2002).

critical for blastocyst to develop (Pratt *et al.*, 1982). In embryo, the compacted cells adhere tightly by cell surface E-cadherin glycoprotein. Due to compaction, the blastomeres form closely adhered epithelial ball, which later divides *symmetrically* or *asymmetrically* depending on the direction of the plane of cleavage in relation to the outer polarized region of the blastomere (Figure 11.3). The asymmetric division generates two cell types, polarized outer cell (trophoectoderm) and an apolar inner cell, whereas the symmetric division generates two polarized outer cells (Figure 11.3). A combination of symmetric and asymmetric divisions result in the formation of blastocyst with required number of apolar inner cell mass (ICM) and polarized trophoectoderm cells (Fleming, 1987). In human, as in mouse, the 8-cell stage embryo undergoes the process of compaction.

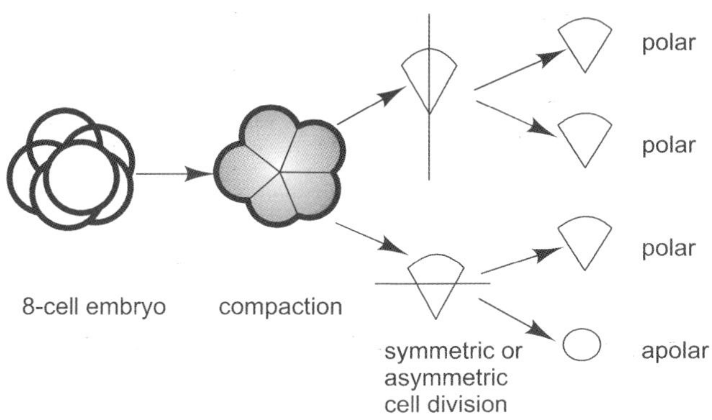

Figure 11.3 Generation of trophoectoderm and ICM by symmetric and asymmetric division of compact epithelial ball at 8-cell embryo stage in mouse. [Adapted from Ref. 45]

Formation of extraembryonic and embryonic tissues: The pre-implanted human embryo (blastocyst) at day 5 contains about 200 to 250 cells, of which 30 to 34 cells belong to ICM. Similarly, the pre-implanted mouse embryo of E3.0 contains about 200 cells consisting of 20 to 25 cells of the ICM (Hogan *et al.*, 1994). Embryonic and extraembryonic tissues are formed from various stages of blastocyst development (Figure 11.4). The extraembryonic tissues are made to provide nutrients to the embryo, to remove waste products from the embryo, and to provide protection to the embryo/fetus from any kind of shock. In the early blastocyst, trophoblast differentiates into syncytiotrophoblast, and is finally involved in the development of placenta. In human, at day 10 to 11 of post-fertilization, the *syncytiotrophoblast* is placed in direct contact with maternal blood vessels. The placenta is generated by fusion of the embryonic *chorion*, the maternal *deciduas*, and vascular tissue. The developing embryo is anchored to the uterine wall through placenta for bathing with maternal bloodstream. Later, the *umbilical cord* connects the embryo to the chorion portion of the placenta.

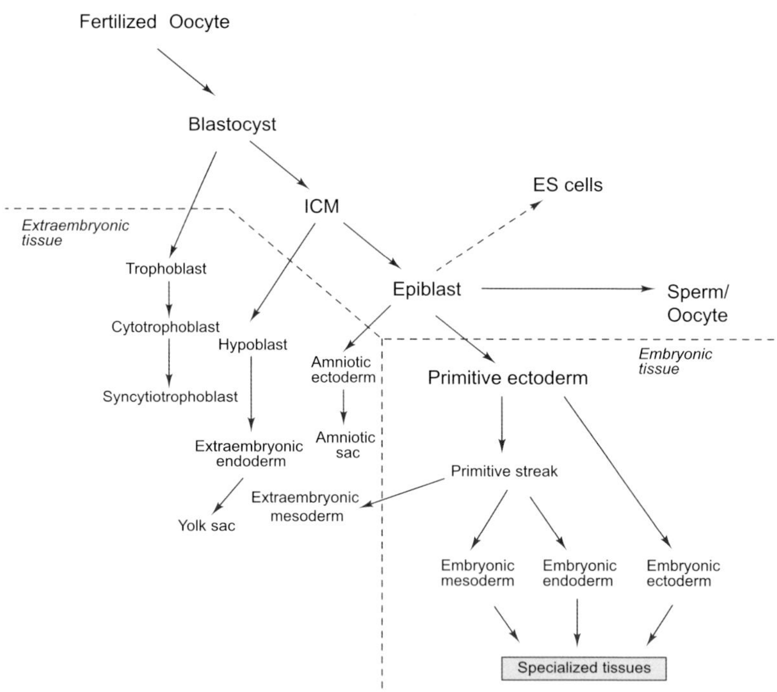

Figure 11.4 Development of human extraembryonic and embryonic tissues, and primordial germ cells from zygote.

The primitive ectoderm, developed on the surface of ICM of the blastocyst, gives rise to *hypoblast*. Extraembryonic endoderm cells, originated from hypoblast, are involved in the lining of primary and secondary *yolk sac*. The origin of the extraembryonic mesoderm cells is controversial, but it is believed that they arise from the caudal end of *epiblast* in the region of the presumptive primitive streak. The extraembryonic mesoderm lines the inner surface of the cytotrophoblast and the outer surface of the amniotic membrane. In humans, the amniotic sac in the developing embryo is composed of two different membranes—the inner membrane (layer) of amnioblast is derived from amniotic ectoderm and the outer thin membrane (layer), the chorion, is somatic mesoderm origin.

In humans, the process of *gastrulation* starts between day 14 and 16 of post-fertilization (about E6.5 in mouse). During gastrulation, the cells of epiblast begin to differentiate, and by the end of day 21 (E8.5 in mouse), three germ layers of the embryo— *endoderm, mesoderm*, and *ectoderm* are formed from the primitive ectoderm (Figure 11.5). Later, through various complex processes (e.g., cell-cell interaction, cell-matrix interaction, and activation/inactivation of specific genes), embryonic germ layers form different tissues and organs of mammals. Table 11.1 describes germ-layers specific cells and tissues in mammals.

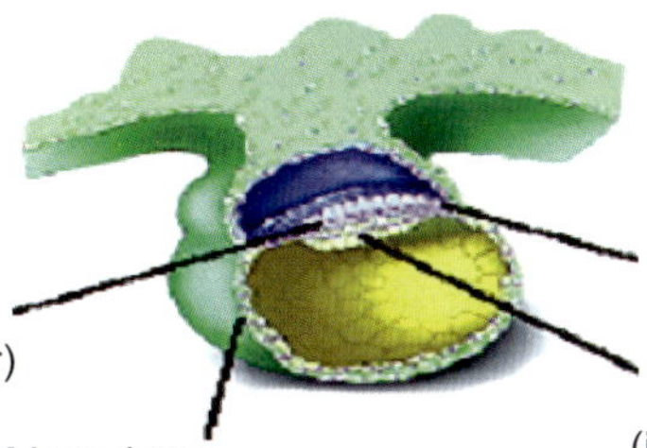

Figure 11.5 Formation of gastrula: the epiblast is specialized into three germ-layers specific tissues (ectoderm, mesoderm, and endoderm) and primordial germ cells.

Table 11.1 List of differentiated tissues from embryonic germ layers

Ectoderm (outer germ layer)	*Mesoderm (middle germ layer)*	*Endoderm (inner germ layer)*
1. Tissues of central and peripheral nervous system 2. Skin and epidermis 3. Cornea and lens of eye 4. Pituitary gland 5. Adrenal medulla 6. Pigment cells 7. Epithelia of mouth and nasal cavities	1. Bone marrow (blood and immune cells) 2. Skeletal, smooth, and cardiac muscle cells 3. Connective tissues (e.g., bone, cartilage, endothelial) 4. Urogenital system (e.g., kidney, ureters, gonads, reproductive duct) 5. Fat cells 6. Adrenal cortex 7. Vascular system	1. Epithelium of digestive tract 2. Epithelium of the respiratory tract 3. Liver and pancreas 4. Urinary bladder, vagina, and urethra 5. Thymus, thyroid and parathyroid glands 6. Epithelium of reproductive ducts and glands

Primordial germ (PG) cells are the precursors of *gametes* (sperm and oocyte) in adult mammals. Prior to gastrulation, these precursor cells get separated from the for epiblast and migrate to the extraembryonic mesoderm for committment to PG cells. Finally, PG cells migrate to their end destination, genital ridges, which gives rise to the gonads (testes in male and ovaries in females). In mice, the migration of PG cells to genital ridges is completed at about E11.5. The PG cells, which do not enter the genital ridges, can develop into germline tumors, later in life. In genital ridges, the PG cells come under various hormonal and cellular influences and finally differentiate into either oogonia or spermatogonia. In addition to these complex differentiations, PG cells retain the property of *pluripotency*, which requires for the gametes to form new progeny. Thus, PG cells are known as embryonic germ cells (EG cells). Germ cells are the only cells in the body to undergo meiotic cell divisions during their differentiation.

11.2 STEM CELLS AND CLASSIFICATION

A fertilized oocyte can give rise to a complete organism, made of more than 200 different cell types of specialized functions. In adult life, many tissues/cells die naturally, which are replenished to maintain tissue *homeostasis*. The remarkable ability of an embryo to form blastocyst and finally differentiates into adult (*somatic*) tissues, which self-maintain throughout life, is a direct consequence of stemness of the cells. The specialized cells having such properties are called stem cells. By definition, stem cells have both the capacity to self-renew, i.e., to divide for additional stem cells, and also able to differentiate along the tissue specific lineage. In mammals, the self-replication and the differentiation ability of these specialized cells are gradually compromised as they progress from embryonic to neonatal to adult life. On the basis of the ability of self-replication and differentiation into tissue specific lineages, stem cells can be classified into following four groups:

A. Totipotent stem cells: The most versatile form of stem cells, which are capable of producing all tissue types required to develop a complete mammal. Totipotent cells produce extraembryonic membranes and tissues, the embryo itself, and all post-embryonic tissues and organs that form a complete functional organism. Accordingly, fertilized egg (the zygote) is a *totipotent* stem cell. In first few hours of fertilization, the zygote undergoes few rounds of divisions to produce identical totipotent cells. Since these cells are still totipotent in nature, each of them can develop into an entire organism. In fact, identical twins are formed when two totipotent cells are separated and developed into two genetically identical embryos. All cells of the early embryo are totipotent until the 8-cell stage or so, before they begin to specialize and form blastocyst. In this regard, PG cells are also totipotent in nature. Interestingly, in plant, each cell behaves as totipotent, as a complete plant can be developed from any part of the tissue.

B. Pluripotent stem cells: These cells are like totipotent stem cells as they can give rise to all tissues of adult organism. However, like totipotent stem cells, they cannot develop into an entire organism as they cannot form all extraembryonic tissues. In humans, on the day 4 of development, the early embryo differentiates into blastocyst. The outer layer (trophoectoderm) of blastocyst become placenta, and the inner mass forms the tissues of the adult human body. These inner cells are relatively committed as compared to the totipotent stem cells, and are known as pluripotent stem cells. The pluripotent stem cells can divide and begin to specialize into germ layers during the gastrulation stage (Figure 11.6).

C. Multipotent stem cells: These are less potential and more differentiated stem cells. They can form limited range of cells within the tissue lineage. The differentiated progenies form the pluripotent stem cells are termed *multipotent* stem cells, such as, hematopoietic and neural stem cells. At this stage, they are multipotent stem cells as they can differentiate into several types of cells within the lineage. For example, multipotent hematopoietic stem cells can give rise to platelets, neutrophil, red blood cells, etc. (Figure 11.6). Similarly, neural stem cells give rise to neurons, oligodendrocytes and astrocytes.

D. Unipotent stem cells: These are an undifferentiated cells present in the differentiated tissue. *Unipotent* stem cells can self-renew and differentiate into specialized cells of the tissue, such as endothelial, muscle, epidermal stem cells, etc.

It should be remembered that totipotent and pluripotent stem cells belong to the embryonic life of the mammals, whereas multipotent and unipotent stem cells are found in the adult life (Figure 11.6). In general, multipotent and unipotent stem cells are termed as adult stem cells (ASCs) or somatic stem cells (SSCs). Totipotent stem cells have never been attempted to isolate, however, other stem cells of embryonic life (ES and EG) have been isolated and characterized. Similarly, many adult stem cells have been isolated and characterized. It has also been shown that ES, EG and adult stem cells can be grown in the laboratory. There are similarities and differences between these stem cells, which are depicted in Tables 11.2A and 11.2B.

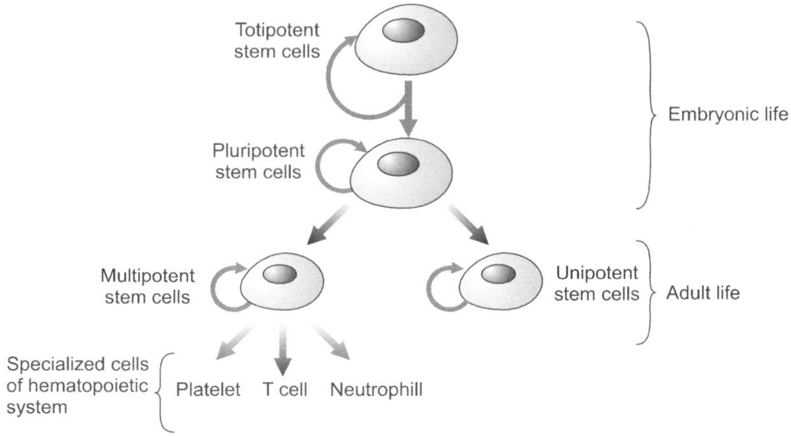

Figure 11.6 Classification of stem cells in embryonic and adult life of mammals. The totipotent stem cells self-replicate to the limited extend (up to 8-cell cells division), and finally differentiate into pluripotent stem cells. All adult tissues/cells are formed from the pluripotent stem cells (embryonic stem cells), which give rise to multipotent and unipotent stem cells.

Table 11.2A Similarities between ES, EG, and adult stem cells

Type of stem cell	*Ability to self-replicate*	*Ability to differentiate*	*Can be isolated*	*Homing ability in organ*
ES, EG, and adult	Yes	Yes	Yes	Yes

Table 11.2B Differences between ES, EG, and adult stem cells

Type of stem cell	*Source*	*Differentiation potential*	*Abundance*	*Ease of identification*	*Immune rejection*	*Replication potential*	*Formation of tumors*
ES	Embryo (ICM of blastocyst)	Pluripotent	Limited	Easier	High	Studied up to 300-population doubling	Undifferentiated cells form tumors
EG	Fetus (PG cells)	Pluripotent	Limited	Easier	High	Studied up to 80-population doubling	Do not form tumors
Adult	Adult life time (from specialize tissues)	Multi-/uni-potent	Limited to rare	Difficult, except in few tissues	None (in case of autologous)	Limited replication	Not known

11.3 PLURIPOTENT STEM CELLS: DERIVATION AND CHARACTERIZATION

Mammalian pluripotent stem cells are of three types: *embryonic* stem (ES) cells derived from pre-implanted embryos, *embryonic germ* (EG) cells derived from primordial germ cells of the post-implanted embryos, and *embryonic carcinoma* (EC) cells derived from the testicular carcinoma. The sources of ES, EG and EC cells in mouse are shown in Figure 11.7. In case of humans, these pluripotent stem cells can also be obtained from the respective sources as mentioned in Figure 11.7. Pluripotent stem cells have many common attributes, with some exceptions of uncertain significance. The comparative characteristic features of ES, EG, EC cells in human and mouse are shown in Tables 11.3A and 11.3B, respectively. Some of these markers are not unique for the pluripotent stem cells; nevertheless, they are used to separate pluripotent stem cells from the differentiated phenotypes. Initially, we shall discuss how EG and EC cells are derived in human. The derivation of mouse and human ES cells, their culture, and directed differentiation will be discussed later.

Shamblott *et al.* (2001) first isolated EG-like cells from PGCs, obtained from the gonadal ridge and mesentery of 5 to 9-week gestation human fetal tissue. The tissue was mechanically and enzymatically dissociated and cultured on a feeder layer of non-dividing mouse STO fibroblasts in FBS supplemented medium. The isolated EG cells require growth factors like LIF, SCF, and bFGF/FGF-2 in the culture (Matsui *et al.*, 1992).

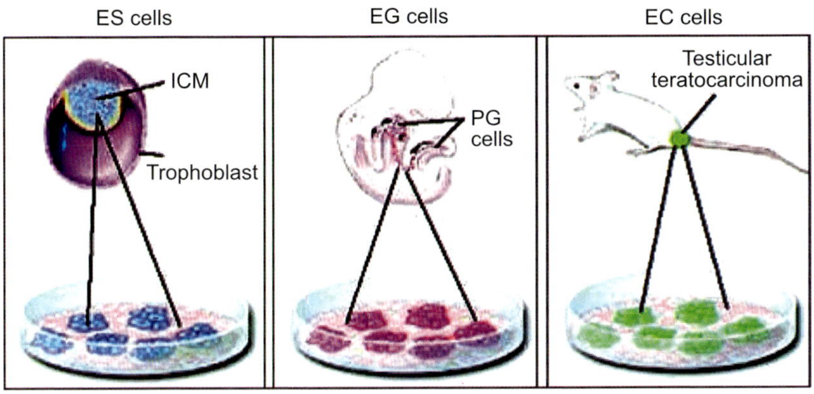

Figure 11.7 Source of mouse pluripotent embryonic stem, embryonic germ, and embryonic carcinoma cells. [*Source*: NIH Stem cells- Scientific progress and future research directions, 2002]

Table 11.3A Comparison among human ES, EG, and EC cells

Characteristics	*ES cells*	*EG cell*	*EC cells*
Derivation	ICM	PGC	PGC
Karyotype	Euploid	Euploid	Heteroploid
Chimera formation	Somatic and germ cells	Somatic and germ cells	Somatic cells
SSEA-1	–	+	–
SSEA-3	+	+	+
SSEA-4	+	+	+
TRA-1-60	+	+	+
TRA-1-80	+	+	+
AP	+	+	+
Telomerase activity	+	Unknown	+
Oct-4	+	Unknown	+
Teratoma formation	+	–	+
In vitro differentiation	+	+	+

SSEA-1/3/4: Stage specific embryonic antigen-1/3/4; AP: Alkaline phosphatase; TRA-1: Tumor rejection antigen-1; Oct4: POU-domain transcription factor.

Table 11.3B Comparison among mouse ES, EG, and EC cells

Characteristics	*ES cells*	*EG cells*	*EC cells*
Derivation	ICM	PGC	PGC
Karyotype	Euploid	Euploid	Heteroploid
Chimera formation	Somatic and germ cells	Somatic and germ cells	Somatic cells
SSEA-1	+	+	+
AP	+	+	+
Telomerase activity	+	-	+
Oct-4	+	+	+
Teratoma formation	+	+	+
In vitro differentiation	+	+	+

[*Source* (Table 11.3 A & B): NIH Stem cells- Scientific progress and future research directions (2002).]

Once EG cells are established, bFGF is no longer required in the growth medium. Mouse and human EG cells require same combination of growth factors, suggesting that some of the gene expressions regulating the development of the germ line have been conserved during mammalian evolution (Matsui *et al.*, 1992). Within 1 to 3 weeks of culture, the human PGCs form multilayered colonies of cells resembling mouse EG cells that express markers, as mentioned in Tables 11.3A. In the absence of growth factors, human PGC spontaneously forms *embryoid bodies*, which include derivatives of all three germ-layers specific cells, indicating that PGCs are pluripotent in nature. Despite the fact that human PGC-derived cells are pluripotent, unlike ES cells they form teratomas in mice (Shamblott *et al.*, 2001).

Germ cell tumors have been studied extensively in mice and humans. These tumors contain a wide range of tissues, naturally derived from three primary germ layers of blastocyst. Thus, EC cells are pluripotent in nature like EG or ES cells (Stevens, 1967). *Teratocarcinomas* are some form of tumors developed upon injection of human ES cells in immunocompromised mice. It is of particular interest because EC cells, which are similar to ES cells in many respects, can be isolated from them (Andrews, 1998). Like human ES cells, EC cells proliferate extensively in mouse feeder layers in the presence of FBS, and are not dependent to LIF. Unlike human ES cells, EC cells form embryoid bodies when they are grown in high density (Andrews, 1998). Cultured EC cells on mouse feeder layers express similar surface molecules and other characteristic features of pluripotent stem cells (Table 11.3A and 11.3B), but they are heteroploid. Inspite of the fact that EC cells are pluripotent, similar to the ES and EG cells, they have limited differentiation potential *in vitro* and *in vivo*.

11.3.1 ES cells

Mouse ES cells were first derived by two independent research groups from the inner cell mass of blastocysts (Evans and Kaufman, 1981; Martin, 1981). In 1998, James Thomas and his colleagues first derived human ES cells from donated blastocysts through *in vitro* fertilization (IVF). The properties of ES cells were established in subsequent *in vitro* and *in vivo* studies (Marshak *et al.*, 2001; Smith, 2001). The characteristic features of ES cells are as follows:

1. These are derived from the ICM or epiblast of blastocysts.
2. Exhibit stable diploid chromosomes.
3. Capable of self-renewal in culture without differentiation (so far, culture for more than 300 population doubling had been purused).
4. Pluripotent in nature, capable of differentiating into cell types of all three germ layers *in vitro* and *in vivo*.
5. These cells are integrated into all fetal tissues during development, if injected into blastocyst (yet to show in case of human ES cells).
6. These are clonogenic, i.e, a single ES cell produces a colony, which is genetically identical to that of original cells.
7. These cells express Oct-4 and telomerase activity; the decrease of Oct-4 expression is associated with differentiation of ES cells.
8. ES cells lack G1 checkpoint of the cell cycle.
9. The undifferentiated ES cells do not show X chromosome inactivation.

11.3.1.1 Mouse ES cell line

ES cells are derived from the inner cell mass of preimplanted blastocyst. Studies from various laboratories have confirmed that the genetic background of mouse may influence the efficiency of ES cell generation. For example, 129/SvJ strain can produce better quality of ES cells than that produced from C57Bl6/J mice, in terms of *in vitro* propagation and the efficiency of the chimeras formed (Brook and Gardner, 1997). Another important observation made was that the pregnant *diapause* mice tend to produce more ES cells than normal pregnant mice. Diapause is a naturally occurring delay in the process of blastocyst implantation, which causes an increase of epiblast cell number at the cost of delayed gastrulation of the embryo. The procedure for obtaining mouse ES cell lines is given below:

Step **1:** Mouse embryonic fibroblasts (MEF), derived from 12-14 days old embryos, are cultured in KO-DMEM containing 10% KO-FBS on a 100 mm plastic dish for 2-3 days ($1.5 - 2.0 \times 10^6$ cells/dish). At 80% confluence, the medium is replaced and cells are treated with mitomycin C (10 mg/ml of same medium) for 2-5 h. Cells are washed to remove mitomycin C and further incubated in the same medium. The day before harvesting the blastocysts to derive ES cells, MEF cells are trypsinized by treating with 0.25% trypsin/1 mM EDTA solution. The cells are then plated in 24- or 6-well plate at density of 1×10^5 and 5×10^5 cells per plate, respectively, in the same medium and incubated overnight. On the day of plating blastocysts, the medium is replaced with 0.5-2 ml of ES cell medium per well (depending upon the plate). Alternatively, the same tissue culture plate may be coated with 0.1% gelatin for 2 h in room temperature. Air-dried plates can be used to culture ES cells in place of feeder layer.

Step **2:** Embryos are recovered from 3.5 dpc (super ovulated or naturally mated) females by flushing the uterine horns with M2 medium. The blastocysts are rinsed for several times in M2 medium, and then one blastocyst is transferred to each well of 24-well plate containing MEF feeder layer in ES cell medium. In case of gelatin-coated plates, ES cell medium is supplemented with 1000 U/ml LIF. Transferring of blastocyst to feeder layer should be performed under dissecting stereomicroscope placed in a laminar flow hood. Blastocytsts are cultured, without any disterbance, at 37°C in 5% CO_2 incubator for first two days. During this period, the blastocysts should hatch from the zona pellucida, remaine adhered on plate, and grow spontaneously. Alternately, ICM is separated from the trophoectoderm by *immunosurgery* (complement-dependent antibody cytotoxicity). Mouse blastocysts with or without zona pellucida are treated with rabbit anti-mouse sera (1:100) for 30 min, followed by an exposure to guinea pig complement for 5 min. The antisera treatment results in killing of trophoblastic cells, leaving the cells of the ICM. The ICM can be easily separated from the remnants of trophoblastic cells, and cultured *in vitro* for targeted differentiation. Immunosurgery method can be adopted to obtain large number of pure inner cell mass in a short time (Solter and Knowles, 1975).

Step **3:** Day 3 onwards, the culture is examined daily to know the right stage at which the outgrowth ICM is ready for disaggregation (usually between 4 to 6-days after plating). The size of the ICM-outgrowth may vary from embryo to embryo, so there is no fixed time for first disaggregation. The outgrowth ICM, ready for the disaggregation, should be big enough but in undifferentiated form. On the day before first disaggregation, a 24-well plate with fresh feeder layer is kept ready and equilibrated with ES cell medium. The medium is aspirated from 24-well plate and cells are washed in 0.5 ml of sterile PBS. In a 60-mm plate, trypsin solution (4 drops of 30 ml each) is placed. Using a fine Pasteur pipette or micropipette, ICM clumps are gently removed from the trophoblast cells and placed on the drops of trypsin. ICM is incubated at 37°C for 3-7 min, till the clumps break into smaller pieces of 3-4 cells. Some clumps may require repeated pipetting. At this stage it is important to watch the cells in microscope to stop the treatment of trypsin. Thirty microliter of serum containing medium is added to each drop to inactivate the action of trypsin, and the cell suspension is transferred to fresh feeder layers.

Step **4:** The medium is changed 24 hours after splitting. The wells are observed in microscope to look for ES cell-like colonies as shown in Figure 11.8 (compact colonies with uniform edges and undistinguishable cells with multiple nuclei). Large ES cell colonies (about 400 mm diameter) are picked, disaggregated one by one, as described above, and again placed onto fresh feeder layer. Smaller undifferentiated colonies present in the wells can be dissociated next day. Colonies above 400 mm tend to flatten and differentiate, therefore should not be used for sub-culturing. The detached colonies are gently removed and transferred to fresh feeder layer and cultured as above in a 24-well plate. ES cell colonies are often formed at the perimeter of the wells.

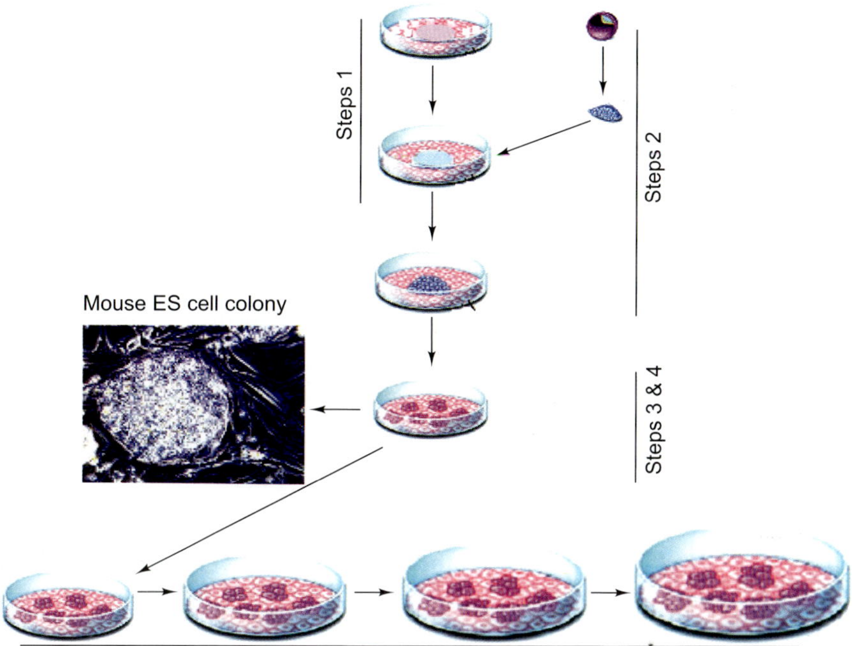

Figure 11.8 Derivation of mouse ES cell lines from blastocyst. ES cell lines are derived in various stages of passaging, starting from the isolation of ICM. Pluripotent ES cell colonies are tight with smooth shiny edged, and consisting of undistinguishable cells with multiple nuclei.

Step **5:** When ES cells reach near confluence, they are passaged in stages with increasing larger plates (12-well, 35 mm, 60 m m, 100 mm dish) containing feeder layer. The first subcultured cells in this step are considered at passage 1, and subsequent dissociation increasing passage number. Cells are fed everyday by replacing half of the medium, or by replacing complete media in alternate days. When enough cells are formed in 100 mm dish, they are cryopreserved for the future use. Once the cells are grown in 100 mm dish, the new cell line is usually established, which is splitted at 1:6 to 1:8 in subsequent expansion program. The overall efficiency for the establishment of ES cell line from permissive mouse strains is about 30%. The derivation of ES cells from blastocyst stage of the embryo is explained in Figure 11.8.

11.3.1.2 Human ES cell line

In vitro fertilization (IVF) experiment suggests that blastocysts having larger and distinct ICM (Figure 11.1) tend to yield ES cell line most efficiently than smaller size of ICM. Eighteen to twenty-four-hours after fertilization, the oocyte is considered to have the age of day 1. The zygote undergoes the first cleavage on day 2 to produce a 2-cell embryo. By the end of day 3, the embryo undergoes two more rounds of cleavages to produce a 8-cell embryo called morula. On day 4, the individual cells compact and form an epithelial ball. By day 5, the epithelial ball begins to divide to form ICM and the trophoectoderm, and finally the blastocyt is generated. For the derivation of human ES cells, day 5 blastocysts are used in the laboratory. The total number of cells in day 5 blastocysts varies from 200 to 250, out of which about 30-34 cells are as ICM (Bongso *et al.*, 1999). Human ES cell lines are derived as follows:

***Step* 1:** The ICM is separated from outer trophoectoderm of day 5 blastocysts. Two methods are generally followed to remove ICM from trophoectoderm layer: immunosurgery and microsurgery. In immunosurgery, blastocysts are treated with anti-human sera for 30 min, followed by 5 min treatment with guinea pig compliment. The trophoectoderm cells swell, detached, and are finally died in the culture.

***Step* 2:** The ICM is plated on g-irradiated MEF previously grown on a 60 mm dish. The purpose of g-irradiation is to maintain feeder layer in non-dividing state. Unlike mouse, human ES cells can be maintained undifferentiated phenotype without MEF. Thus, feeder layer free culture (LIF-supplemented medium) for human ES cells is not possible. The cells derived from the ICM are cultured for 10 to15 days in the presence of ES cell culture medium. The over-layered cells divide and form many clumps on the feeder layer during 2 weeks of culture (Figure 11.9A). In place of MEF, human ES cells have been propagated on mitotically inactive feeder layer of primary muscle and skin cells derived from normal human abortuses (Richards *et al.*, 2002).

***Step* 3:** Cells from the periphery of the clumps are isolated by mechanical dissociation or treatment with trypsin and replated on a new feeder layer.

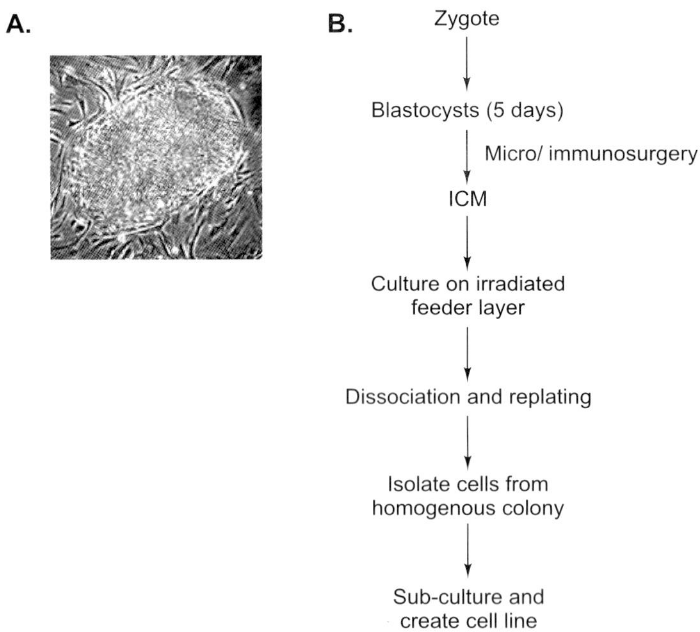

Figure 11.9 Derivation of human ES cell line. **A.** Human ES cell colony- 2 days after passaging (10X magnification) (*Source*: Thomson Lab. Medison, WI, USA); **B.** Sequential steps for derivation of human ES cells from blastocyst.

***Step* 4:** Colonies containing of homogenous cells are removed, expanded, and further passaged on the feeder layer to derive ES cell line.

Human ES cells, if not differentiated, express specific cell surface molecules and extend characteristic properties as shown in Table 11.3A. The basic steps for deriving human ES cells are shown in Figure 11.9B. Thomson and Odorico (2000) developed 14 blastocysts from frozen embryos, and finally derived 5 human ES cell lines: H1, H7, H9, H13, and H14. These cell lines have been continuously grown without differentiation for 5-6 months (Thomson and Odorico, 2000). H9 ES cell line has been grown for more than 300 population doublings. The techniques followed to establish human and mouse ES cell line and their culture are quite similar.

11.3.2 Maintenance of ES cells in culture

ES cells are pluripotent in nature and undergo unlimited replication without compromising with pluripotency. Unlimited replication by ES cells is understood, but

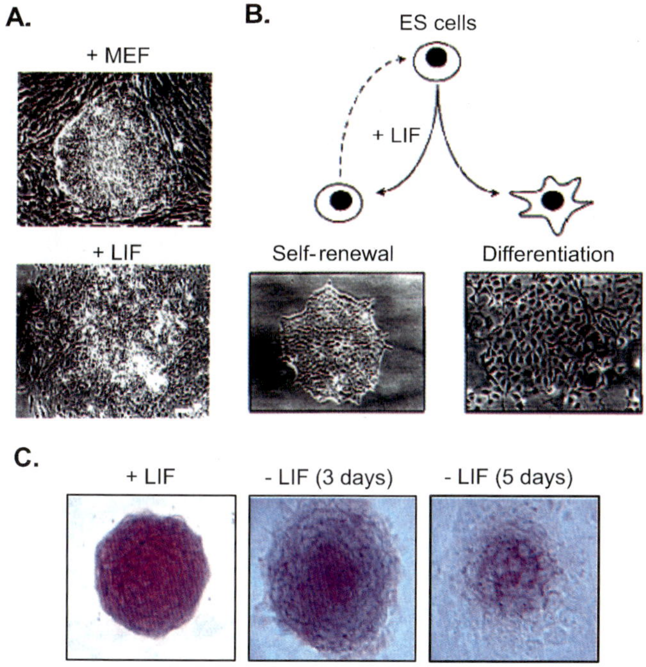

Figure 11.10 Maintenance of ES cells in culture. **A.** Colony of human ES cell growing on MEF (top), colony of human ES cell after 4-days of plating in absence of feeder layer, but in presence of LIF 2000 U/ml (bottom). The cells in the periphery undergo differentiation in absence of MEF [Taken from Ref. 48]; **B.** Effect of LIF in self-renewal and pluripotency of mouse ES cells. In absence of LIF, ES cells are differentiated into cells of different germ layers; **C.** Culture of mouse ES cells in the presence and absence of LIF, followed by staining for alkaline phosphatase. Undifferentiated ES colonies are structurally more compact, in which the expression of alkaline phosphatase become high. In absence of LIF, colony compactness is gradually lost, periphery cells undergo differentiation, and the colony express less amount of alkaline phosphatase.

what is the significance of the term "pluripotency"? The proof of the concept of ES cells has come from its pluripotent nature. Pluripotency of ES cells is determined by *in vitro* and *in vivo* experiments. *In vitro* experiments deal with targeted differentiation of ES cells into cells of three germ layers, whereas, the formation of teratocarcinoma and cellular integration are the two different tests that are followed to test the pluripotency of ES cells *in vivo*.

Studies in human and mouse cells have shown that MEF feeder layer is most crucial in expanding ES cells and maintaining their undifferentiated state. It implies that MEF culture supernatant may have certain growth factor(s), which prevent(s) differentiation of ES cells, and at the same time promote(s) self-replication. Scientists have used MEF-conditioned medium to culture ES cells, which turns-out quite effective in mouse ES cells, but not in the case of human ES cells. Later, it was found that leukemia inhibitory factor (LIF), produced by feeder cells, allows mouse ES cells to continuously proliferate without differentiation in the absence of feeder cells (Williams *et al.*, 1988; Rathjen *et al.*, 1990). However, in human ES cells, feeder layer has been found indispensable in maintaining undifferentiated phenotype in culture (Figure 11.10A). This indicates that besides LIF some other cytokines (membrane bound molecules) and/or interaction with MEF feeder layer are crucial for maintaining pluripotency in human ES cells. The essential role of LIF has been demonstrated in feeder free culture of mouse ES cells in the presence or absence of purified LIF. It was found that ES cells completely differentiate in absence of LIF, whereas, compact ES cellular morphology can be maintained in LIF supplemented medium (Figure 11.10B). It has been found that LIF provides extrinsic signal for self-renewal of mouse ES cells. LIF exerts its effect by heterodimerizing its receptors—the low-affinity LIF receptor (LIF-R), and common subunit gp130. The binding of LIF to LIF-R triggers activation of Jak-STAT3 pathway, which leads to nuclear localization of activated STAT3 transcription factor for regulating the expression of downstream gene products (Figure 11.11). Ligand-stimulated dimerization of cytoplasmic domain of gp130 leads to phosphorlyetion of several tyrosine residues present on it by associated Jak kinases. Four of these phosphorylated tyrosines have been identified as putative interaction sites with SH2 domains of STAT3 (Stahl *et al.*, 1995). It has been found that LIF may not be essential for maintaining pluripotency in mouse ES cells, but activation of STAT3 is crucial. An over expression of fusion constructs of the extracellular domain of G-CSF receptor with the transmembrane and cytoplasmic domain of gp130 in LIFR-deficient ES cells promotes proliferation in the presence of G-CSF rather than LIF (Niwa *et al.*, 1998). Here, the take home message could be that the activation of STAT3 is an essential event for the proliferation of mouse ES cells, using any ligand that can activate. The essential role of STAT3 in the development of mouse embryo and its involvement in maintaining pluripotency in ES cells has been shown in two separate experiments. Targeted disruption of *Stat3* gene in mice was found to be embryonically lethal. Again, ectopic over expression of DN-Stat3 in ES cells led to loss of pluripotency and induction of differentiation (Niwa *et al.*, 1998).

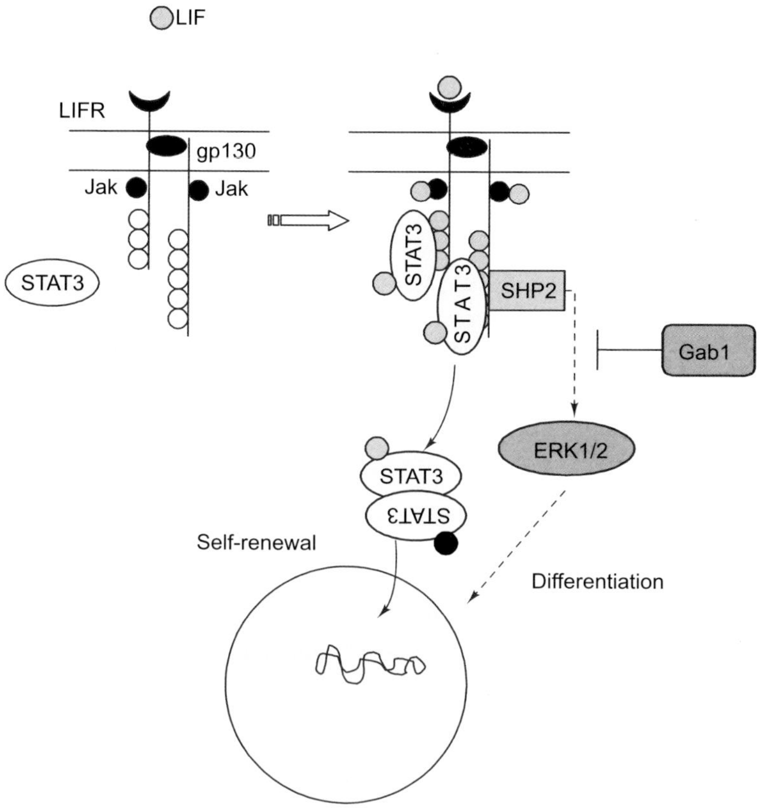

Figure 11.11 Signaling pathway for activation of STAT3 by LIF in mouse ES cells. Altered form of Gab1 prevents differentiation and thus promotes self-renewal of ES cells. (●): Jak kinase, (): Tyrosine, (○): Phosphorylated tyrosine and Jak kinase.

Activation of gp130 signaling pathway in ES cells also results in the phosphorylation of SHP-2, which leads to the activation of mitogen-activated protein kinase (MAPK) pathway. SHP-2 is a tyrosine phosphatase, which removes phosphate groups from the tyrosine residues in different cytoplasmic proteins. It also interacts with the intracellular domain of gp130 receptor. Although gp130 signal can activate both SHP-2 and ERK, such activation is more pronounced in downstream of receptor tyrosine kinases (RTKs). During embryogenesis, RTK signaling is crucial for differentiation and morphogenetic events for gastrulation. The activation of ERK is believed to counteract the proliferative effects of STAT3 activation on ES cells. Thus, in ES cells, blocking SHP-2 and ERK activation may block differentiation pathways promoted by MAP kinase signaling, which indirectly promotes self-renewal of cells. In mouse ES cells, altered form of Gab1 (adapter protein) is expressed in high level, which is believed to suppress the interactions of gp130 receptor with the ERK signaling pathway, thus inhibits differentiation pathway (Smith, 2001).

The hallmark of pluripotent ES cells is the expression of *Pou5f1* gene, which encodes the transcription factor, Oct-4. Oct-4 is constitutively expressed throughout oogenesis and preimplantation development. After that it becomes restricted to the ICM at blastocyst stage and is later expressed in the early epiblast (Palmieri *et al.*, 1994). Oct-4 is also expressed in PGCs of mice and mature germ cells (Scholer *et al.*, 1990). During culture, the expression of Oct-4 is downregulated with the differentiation of ES cells into embryoid bodies and other downstream tissues. However, downregulation of Oct-3/4 expression is not required for the differentiation of ES cells; it is the consequence of differentiation program. Inhibition of Oct-4 expression in mouse ES cells leads to differentiation into trophoectoderm. The Oct-4 mutant embryos initially resemble blastocysts, but do not form ICM, and in culture, only produce trophoblast cells (Nichols *et al.*, 1998). Artificial expression of Oct-4 leads to the differentiation of mouse ES cells into primitive endoderm and mesoderm. Thus, Oct-4 is considered as 'master regulator' of ES cell pluripotency (Nawa *et al.*, 1998). Although Oct-4 is constitutively expressed in mouse ES cells, the withdrawal of LIF leads to its differentiation. Thus, both intrinsic activity of Oct-3/4 and cytokine-induced STAT3 expression are prerequisite for maintaining pluripotency in mouse ES cells.

By expressed sequence tag (EST) libraries from ES cells, a new gene has been discovered that encodes divergent homeobox protein, known as Nanog. It is found to promote ES cell renewal, pluripotency, and epiblast formation (Mitsui *et al.*, 2003). Nanog is not only expressed in ES cells but also in EG and EC cells. Although, Nanog is considered one of the regulators for maintaining pluripotency in ES cells, an overexpression of Nanog in Oct-4 deficient cells does not prevent differentiation of ES cells. As Oct-4 is not required for Nanog expression, but Nanog cannot act in the absence of Oct-4. Thus, it is believed that combined signals from Oct-4 and Nanog lead to the renewal of ES cells and maintain pluripotency.

11.3.3 Targeted differentiation of ES cells into neurons and pancreatic islets

ES cells start differentiation when isolated from feeder layers and try to grown in suspension culture. Targeted differentiation is carried out in several stages. In the first stage, ES cells are induced to form embryoid bodies, which are then differentiated into specific cell types in the subsequent stages. The lineage specific differentiation is governed by the composition of the medium. ES cells form three-dimensional aggregates known as embryoid bodies (EBs) in a suitable culture condition. The procedure followed to induce EB differs from mouse to human. In mouse, undifferentiated ES cells are cultivated in hanging drops for 2-days and in suspension for additional 5-7 days. EBs are further cultured onto gelatin-coated tissue culture plates. The aggregates can be observed under light microscope. They are small and irregular objects upto 2 days, but by 5-7 days, the aggregates become larger and spherical with smoother outlines (Figure 11.12).

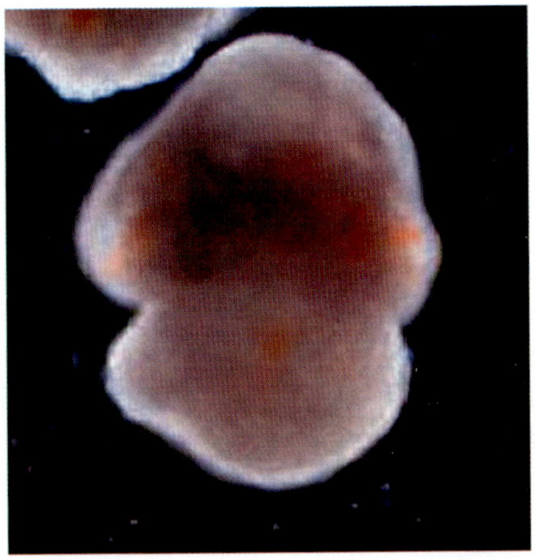

Figure 11.12 Mouse embryoid body towards differentiation into hematopoietic tissue. [*Source*: University of Texas Southwestern Medical School, TX]

The basic steps involved in inducing EBs from mouse ES cells are shown in Figure 11.13A. It is important to note that until the final stage of EBs is obtained, cells are cultured in ES cell culture medium containing KO-FCS and LIF. In mature EBs, cellular phenotypes of ectodermal, mesodermal and endodermal lineages are formed, which can be identified by examining gene expression patterns by quantitative RT-PCR, and examining the expression of protein by immunofluorescence.

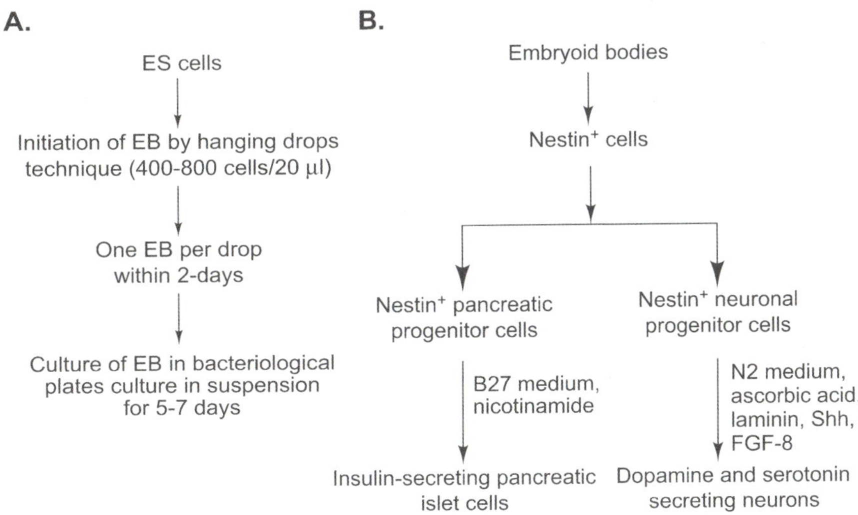

Figure 11.13 Targeted differentiation of mouse ES cells. **A.** Induction of mouse EBs from ES cells. **B.** Directed differentiation of EBs into insulin secreting pancreatic islet cells (left) and dopamine expressing neurons (right).

In the second stage, the medium is replaced with serum free medium containing insulin, transferrin, selenium and fibronectin, but not LIF. In the absence of serum, many cells die within 6 to 10-days, but the surviving cells express nestin—an intermediate filament marker protein for central nervous system (CNS) stem cells. In the third stage, the $Nestin^+$ cells are selected using immunomagnetic sorting or FACS, and cultured in the presence of N2 medium containing laminin and bFGF (for neuronal differentiation), and N2 medium containing bFGF and B27 medium supplement (pancreatic islet cells). The third stage is known as the phase of expansion, in which $Nestin^+$ neuronal and $Nestin^+$ pancreatic progenitor cells are expanded, depending upon the composition of the medium. The fourth stage is the specific differentiation stage of the respective precursor cells. For dopamine and serotonin secreting neurons, $Nestin^+$ neuronal progenitor cells are cultured for 6 to 15 days in N2 medium containing laminin, sonic hedgehog (Shh), FGF-8, and ascorbic acid (Figure 11.13B, right). To generate pancreatic islet cells, pancreatic progenitor cells are cultured in B27 medium containing nicotinamide (Figure 11.13B, left).

11.4 ADULT STEM CELLS

Adult stem cells are also known as *somatic* (Greek meaning of sôma is body) stem cells, because they not only available from adults, but also can be derived from children or umbilical cord. As mentioned earlier, multipotent and unipotent stem cells are known as adult stem cells. The adult stem cells have two basic properties: ability to self-renew without compromising with stemness, and differentiate into mature morphologically and functionally distinct tissue specific cells. Therefore, adult stem cells maintain cellular homeostasis throughout the lifetime of the mammals. Adult stem cells are generally rare, the highest frequency of adult stem cells is found to be in the bone marrow (1 in 10,000 to 15,000 nucleated cells). Most of the knowledge on adult stem cells has been derived from studies in mice. The adult stem cells may be present in all the organs and dispersed in tissues throughout the body, however, they have been isolated from limited organs/tissues. So far, adult stem cells have been identified in bone marrow, brain (e.g., hippocampus, ventricular zone), skeletal muscle, epithelia of skin, dental pulp, digestive track, cornea, liver, pancreas, blood vessels etc. Adult human stem cells, their location in the body, and primary direction of differentiation are shown in Table 11.4. On the basis of the location of adult stem cells and the mature cells/tissues obtained from them, it is concluded that the fate of each adult stem cells has been programmed in the postgastrulation stages. We shall now discuss hematopoietic and mesen chymal stem cells in detail.

Table 11.4 Adult human stem cells, their location, and primary direction of differentiation

Cell type	*Tissue-specific location*	*Cells or tissue produced*
• Hematopoietic stem cells	Bone marrow and peripheral blood	Bone marrow and all blood cells
• Mesenchymal stem cells	Bone marrow	Stroma cells, bone, cartilage, tendon, muscle, adipose tissue.
• Neural stem cells	Central nervous system (CNS)	Neurons, astrocytes, oligodendrocytes
• Oval cells	Canals of Hering (near bile ductules in liver)	Hepatocytes, bile ductular cells
• Pancreatic stem cells	Intra-islet	Beta cells
• Skeletal muscle stem cells or satellite stem cells	Muscle fibers	Skeletal muscle cells
• Keratinocyte (skin stem cells)	Basal layer of epidermis	Epidermis, hair follicles
• Epithelial stem cells of lung	Tracheal basal and mucus-secreting cells, bronchiolar Clara cells	Mucus and ciliated cells, type I and II pneumocytes
• Stem cells of intestinal epithelium	Epithelial cells located around the base of each crypt	Paneth's cells, brush-border enterocytes, mucus-secreting goblet cells, etc.

Source: Korbling M, Estrov Z. (2003). *N Eng J Med* 349, 570.

11.4.1 Hematopoietic and mesenchymal stem cells

Hematopoiesis, the regulated production of eight different blood cells is initiated from a small and district population of stem cells, known as *hematopoietic* stem cells. Hematopoiesis occurs mainly in the bone marrow of adult mammals. The bone marrow is a highly prolific tissue in the body; in an average, the daily output of blood cells by bone marrow is nearly 400 billion (Koller and Palsson, 1993). During embryonic development in humans, the site of hematopoiesis is believed to be in the yolk sac. In the second trimester, it is shifted to the liver and spleen, and later in the third trimester, hematopoiesis is predominated in the central and peripheral skeleton (William *et al.*, 1989). In adulthood, main sites of hematopoiesis are vertebral bodies, sternum, ribs, and pelvis (William *et al.*, 1989). Studies in mouse have confirmed that the origin of definitive hematopoiesis is *aorta/gonad/mesonephros* (AGM), originated from mesoderm layer of the blastocyst (Marella *et al.*, 2000). Definitive hematopoiesis refers to the adult hematopoiesis, which provides life-long support of blood cells. It has been proposed that a single primitive hematopoietic stem cell (HSC) is capable of more than 50 doublings, and produces about 10^{15} cells in 60 years of life span (Kay, 1965).

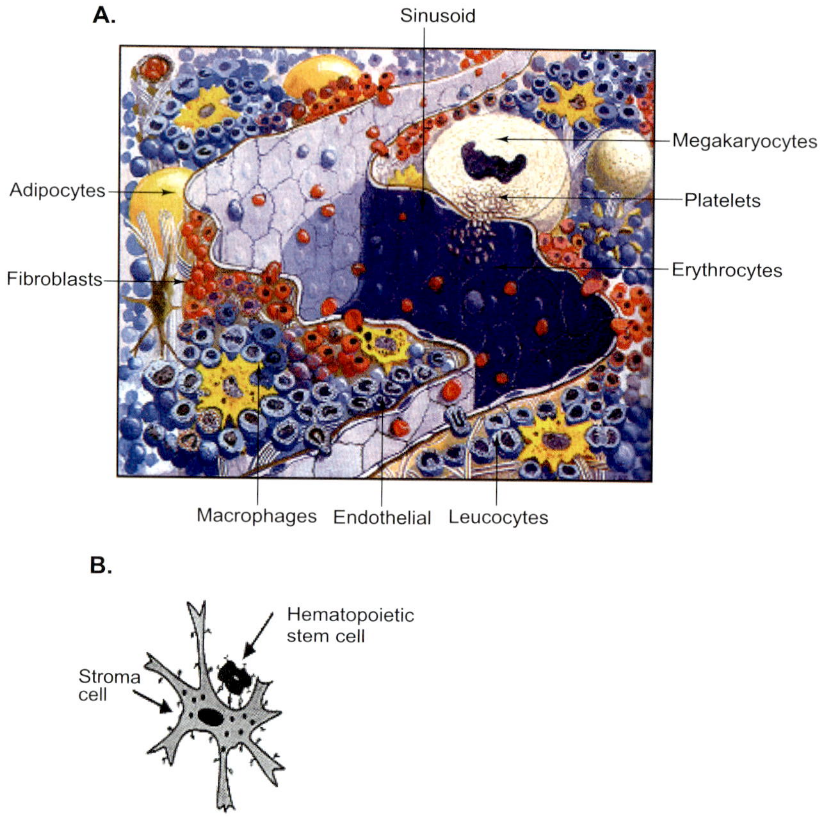

Figure 11.14 Bone marrow microenvironment. **A.** Microscopic organization of bone marrow tissue [*Source*: Gray's Anatomy, 1989]: **B.** Interactions between stroma and hematopoietic stem cells. Stroma cells produce growth factors and cognate ligands/receptors on which hematopoietic cells are implanted through the counter receptors/ligands.

In mammals, hematopoiesis occurs in the extravascular spaces between marrow sinuses. It has been shown that osteoblastic cells are present in the hematopoietic niche (Zhang *et al.*, 2003; Calvi *et al.*, 2003). The primitive stem cells reside on the osteoblastic cells of endosteum and trabecular bone areas. In bone marrow, osteoblastic and other stroma cells extend hematopoietic inductive microenvironment by synthesizing *extracellular matrix* (ECM) components and growth factors. Endothelial cells that line the marrow sinuses, adventitial reticular cells, macrophages, and adipocytes are the part of stroma cells (Figure 11.14A). The ECM components, produced by stroma cells, provide support and cohesiveness to the marrow structure. There is a growing body of evidence suggesting that ECM is involved in regulating hematopoiesis. For example, glucosaminoglycan side chains of heparan sulfate sequester and compartmentalize growth factors in local areas to present them to the HSCs (Roberts *et al.*, 1998). The stroma cells and the ECM form three-dimensional scaffold on which HSCs are implanted (Figure 11.14B).

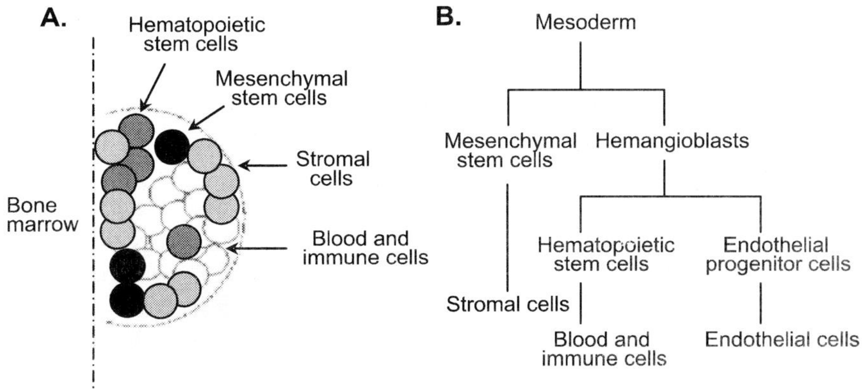

Figure 11.15 Heterogeneity and germ layer source of adult HSCs. **A.** Heterogeneity of bone marrow cells: bone marrow cells are composed of stromal cells, hematopoietic stem and progenitor cells, differentiated cells; **B.** Germ layer source of bone marrow cells. Both hematopoietic stem and endothelial progenitor cells are produced from a common progenitor, known as hemangioblasts.

Bone marrow is a highly heterogeneous population of both hematopoietic and non-hematopoietic cells, originated from common mesodermal precursor cells (Figure 11.15). Studies in mouse have shown a close developmental link between endothelial progenitor cells and HSCs in AGM. It has been found that during the development of definitive hematopoietic cells, clusters of cells appear on the ventral wall of AGM, and on vitelline / umbilical arteries. Initially, the cells lining the dorsal aorta express vascular endothelial growth factor-receptor 2 (VEGF-R2). When the clusters appear, these cells no longer express VEGF-R2, instead express CD45 and gain potential as *colony forming unit-spleen* (CFU-S). There is still a controversy surrounding endothelial progenitor and HSCs, regarding which appears first, but it is believed that hematopoietic cells are originated from hemangioblasts. The surface phenotypes of human and mouse HSCs are shown in Figure 11.16. In general, HSCs express few specific surface markers (e.g., CD34, CD133 for human, and CD34, Sca-1 for mouse), but do not express any lineage specific marker.

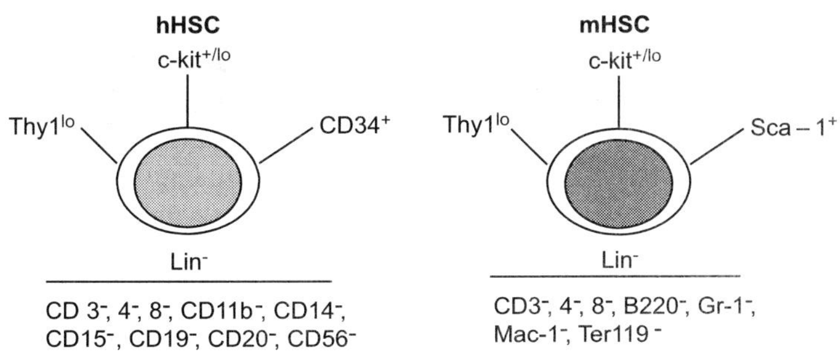

Figure 11.16 Surface markers of human and mouse HSCs. Hematopoietic stem cells express few specific surface markers, such as, CD34, CD133, Sca-1; however they do not express any lineage specific marker (Lin^-).

11.4.1.1 Hematopoietic stem cells

On the basis of extensive studies by various investigators, it has been consensually agreed that hematopoietic stem cells do not express any lineage marker (Lin$^-$). Thus, the phenotypes of human and mouse HSCs are Thy-1^{lo}c-kit$^{+/lo}$CD34$^+$Lin$^-$ and Thy-1^{lo}c-kit$^{+/lo}$Sca-1$^+$Lin$^-$, respectively. Study of Wu *et al.* (1968) lead scientists to believe that during hematopoiesis, a small fraction of primitive, self-renewing cells divide to generate progenies. Each of them are progressively restricted in developmental potential until mature effector cells are formed. On the basis of differentiation status, HSCs are divided into two broad stages: *long-term repopulating* (LTR) and *short-term repopulating* (STR) cells (Figure 11.17). The *multipotent/oligopotent progenitor* (MPP) cells are derived from the STR-HSC, and finally differentiated into *common lymphoid progenitor* (CLP) and *common myeloid progenitor* (CMP) cells. Hematopoietic stem cells are progressed along this differentiation tree (Figure 11.17). They gradually lose 'stemness' and become committed to produce effector cells. Transplantation experiments have shown that LTR-HSCs provide the support for functional cells in mice for more than one year.

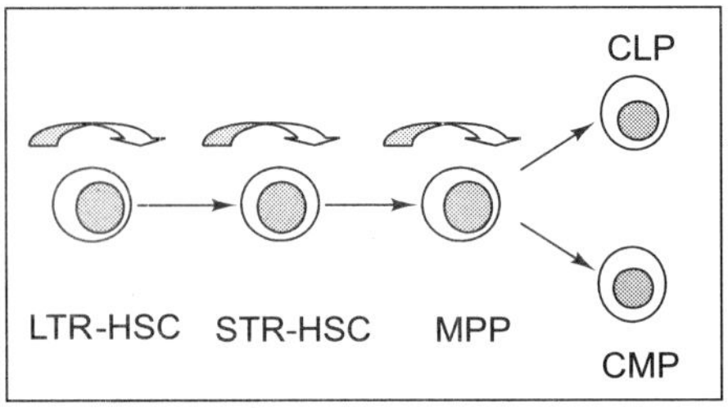

Figure 11.17 Pathway for differentiation of hematopoietic stem cells. Each differentiated myeloid and lymphoid cells are originated from primitive hematopoietic stem cells, through various stages of differentiation. LTR-HSC: long-term repopulating hematopoietic stem cell; STR-HSC: short-term repopulating hematopoietic stem cell; MPP: multipotent progenitor; CLP: common lymphoid progenitor; CMP: common myeloid progenitor.

In some literature, human primitive hematopoietic stem cells of bone marrow, peripheral blood or cord blood have been characterized as CD34$^+$CD38$^-$ (Payne and Crooks, 2002). These cells provide long-term multilineage engraftment in both non-obese diabetes/severe combined immunodeficiency (NOD/SCID) mice and in preimmune sheep xenograft models. Depending upon the age, human bone marrow contains about 2-3% CD34$^+$ cells, out of which CD38$^-$ cells comprise only 1%. Therefore, CD34$^+$CD38$^-$ LTR-HSC represent 0.02-0.03% of whole bone marrow cells. Recent studies have shown that a significant fraction of primitive human HSCs do not express CD34, i.e., they are CD34$^-$ (Goodell, 1999). Primitive HSCs are also found to express CD133. Primitive HSCs are also characterized by their ability to expel vital dyes like Hoechst 33342. In mouse that about 0.03-0.05% bone marrow cells are low side and low forward scatter in the dot plot (Figure 11.18). These cells express *mdr1* gene product P-glycoprotein pump, which prevents nuclear staining with the Hoechst dye and appear as *side population* cells (Goodell *et al.*, 1996). In human bone marrow, STR-HSCs are also known as early progenitors that acquire the surface phenotypes $CD34^{+}CD38^{+}$ or $CD34^{+}$c-kit^{hi}, but do not express lineage markers. The STR-HSCs provide support of hematopoiesis for 5 to 6 months (Harrison and Zhong, 1992).

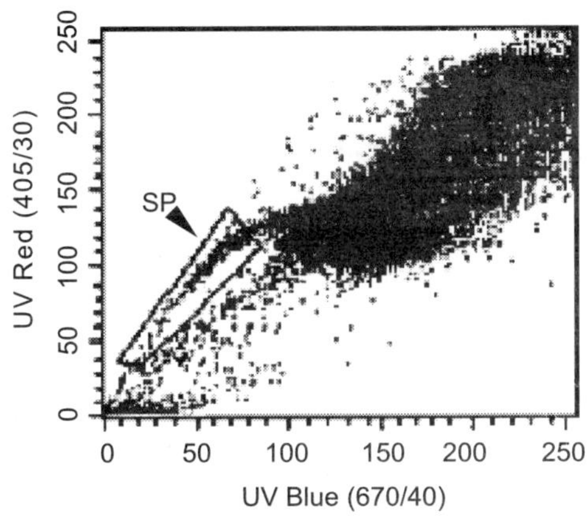

Figure 11.18 Side population (SP) cells of bone marrow tissue obtained from mouse. Hoechst 33342 nuclear dye stained BM cells are analyzed by flow-cytometer in UV-source using blue and red filters.

LTR- and STR-HSCs were better understood in phenotypic and functional analyses. It is important to recapitulate here that HSC is subset of cells of the crude bone marrow or in any other source (cord blood, fetal liver, peripheral blood), which extensively self-renew and provide all kinds of blood cells *in vivo* for reasonable period of time. Let us take an example of characterization of human bone marrow HSCs in terms of phenotypic and corresponding functional properties. Mononuclear bone marrow cells are treated with lineage depletion antibodies (see Figure 11.19), followed by separation of Lin^{+} cells by immunomagnetic sorting. The negatively selected cells are Lin^{-} cells (Figure 11.19). This heterogeneous population may contain many cell types including primitive stem cells. Let us assume that we want to analyze these cells in light of CD34 and CD38 phenotypes. Lin^{-} cells are labeled with CD34/PE and CD38/FITC, and sorted by using FACS. Four fractions (A, B, C, and D) of cells will be obtained, as shown in Figure 11.19. Each fraction of cells is tested using various *in vitro* and *in vivo* assays to determine STR and LTR potential. Purified fractions of cells are tested in NOD-SCID mice, which can form chimera of human cells for about 6 months. The $CD34^{+}CD38^{-}$ fraction of cells are LTR-HSC in nature; they will show highest potential for *long term culture-initiating cells*

(LTC-IC), *SCID repopulating cells* (SRC), and clonal B and myeloid cells. However, in case of STR-HSC, the LTC-IC and SRC repopulating ability will be declined with concomitant increase in CFU-S and colony forming cells. The results of each assay against four fractions of cells are shown in Figure 11.19.

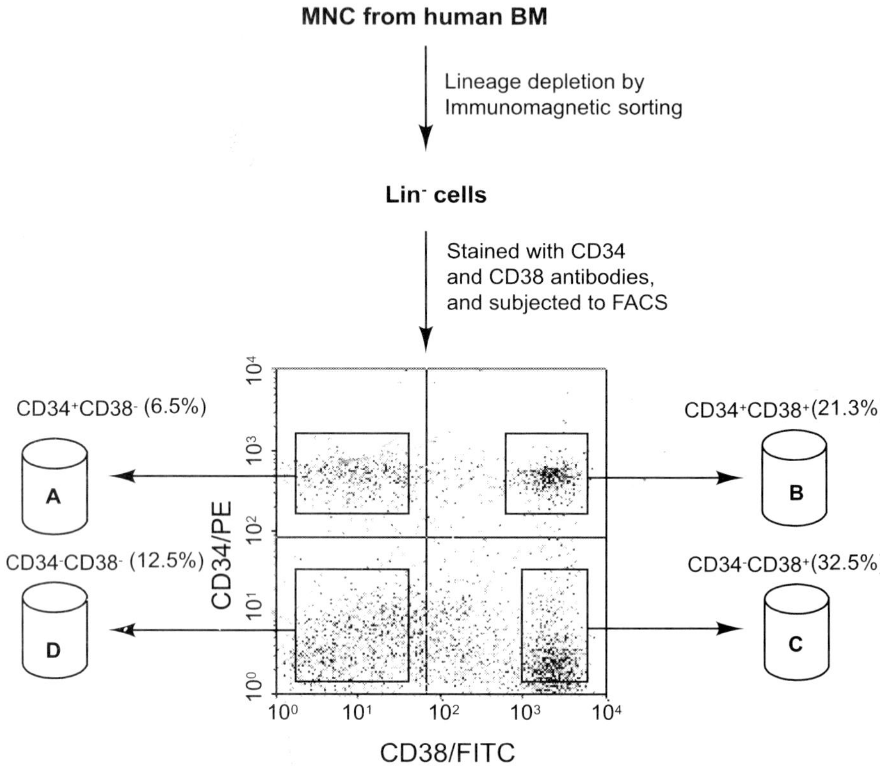

Cell fraction	Type of assays			
	SRC	*LTC-IC*	*CFU-S*	*Colony assay*
$CD34^+CD38^-$	+++	+++	+	+
$CD34^+CD38^+$	+	+	+++	+++
$CD34^-CD38^+$	-	-	-	++
$CD34^-CD38^-$	+	+	-	-

Figure 11.19 Functional characterization of different phenotypes of human bone marrow cells. SRC: SCID repopulating cells; LTC-IC: Long-term culture initiating cells; CFU-S: Colony forming unit-spleen. (+++): relatively high, (-): Nil.

11.4.1.2 Ex vivo culture of hematopoietic stem cells

HSCs are cultured in the presence or in the absence of stroma cells. For the first time *in vitro* culture in the presence of stroma cells was described by Dexter *et al.* (1973). Stroma cells synthesize ECM components and various cytokines to support the growth and maintenance of primitive hematopoietic stem cells. HSCs can also be cultured in the absence of stroma cells. Verfaillie *et al.* (1994) showed that different marrow stroma diffusible factors are responsible for maintaining of primitive stem cells. In contrast, Breems (1998) showed that the engrafting quality of mobilized peripheral blood $CD34^+$ cells is preserved when cultured in direct contact with stroma cells. From time to time, many investigators advocated in favor of using stroma cells as feeder layer to maintain HSC compartment. Brandt *et. al.* (1994) showed that $CD34^+CD38^-$ phenotype can be maintained among proliferating human HSCs by culturing on porcine endothelial cell line. It is important that the expanded cells are multilineage engraftable in human fetal bones implanted SCID mice (SCID-Hu). Despite these beneficial roles, some investigators attempted for partial substitution of stroma cells either by using stroma-conditioned medium or recombinant growth factors. In case of cord blood stem cells ($CD34^+CD38^-$), the LTC-IC has been found to expand about 60-folds in stroma free culture supplemented with fetal liver tyrosine kinase-3 ligand (Flt-3L), stem cell factor (SCF), interleukin-3 (IL-3), interleukin-6 (IL-6), and granulocyte-colony stimulating factor (G-CSF) (Zandstra *et al.,* 1997). The performance of *ex vivo* culture is influenced by the type and the quantity of growth factors used in the culture. Studies have shown that more than half-a-dozen growth factors are involved in hematopoiesis. Some are primarily responsible for self-renewal of the progenitor cells, some of them function as differentiating factors, and others act as inhibitors of hematopoiesis. On the basis of first two biological functions (i.e., self-renewal and differentiation), growth factors can be classified into three main groups:

A. ***Potentiating group*****:** Members of this group of growth factors potentiate the activity of other growth factors, e.g., SCF, Flt-3L, LIF, IL-6 and IL-11. SCF and Flt-3L prevent apoptosis and are primarily involved in the replication of primitive progenitor cells. Therefore, they are also known as stem cell growth factors (Leary *et al.*, 1992; McAdams *et al.*, 1995). The potentiating effects attributed to Flt-3L include an expansion of uncommitted progenitor cells ($CD34^+CD38^-$) in the presence of SCF and IL-3, and an increase of HPP-CFU and CFU-GM numbers when incubated with IL-3, IL-6 and Epo (McKenna *et al.*, 1995). SCF is found to act synergistically with G-CSF, GM-CSF, IL-3, and Epo to stimulate the growth of human progenitor cells (Papayannopoulou *et al.*, 1993).

B. ***Multilineage group*****:** Growth factors (e.g,. IL-3, GM-CSF) belonging to this group essentially act on colony forming unit-granulocyte-erythroidmonocyte-megakaryocyte (CFU-GEMM), colony forming unit-granulocyte monocyte (CFU-GM), burst forming unit-megakaryocyte (BFU-Mk), and burst forming unit- erythroid (BFU-E). IL-3 also prevents the apoptosis of committed progenitor cells (Brandt *et al.*, 1994).

C. ***Unilineage group***: Growth factors belonging to this group act on the precursors of mature cells (e.g., CFU-E, CFU-Mk, CFU-Eo). Some of the growth factors acting on unilineage group of cells are erythropoietin (Epo), thrombopoietin (Tpo), G-CSF, IL-5, etc.

11.4.1.3 Mesenchymal stem cells

The precursor of non-hematopoietic cells in bone marrow are known as *mesenchymal* stem cells (MSCs), originally derived from the mesoderm (Figure 11.15). Putatively, these are marrow-derived adhered fibroblastic cells that give rise to connective tissues like muscle, bone, cartilage, adipocyte, tendon/ligament, and marrow stromal cells. Plating studies have shown that MSCs are the rare population of bone marrow cells, representing 0.001% to 0.01% of the nucleated cells (about $1/10^{th}$ abundance than HSCs). MSCs can be easily isolated from bone marrow and are extremely proliferative. They can be cloned from bone marrow and expanded *in vitro* about 10^6-fold without compromising the ability to differentiate into different lineages (Pittenger *et al.*, 1999; Muraglia *et al.*, 2000). Besides the expansion potential, major advantages of MSCs are as follows

(i) Genetically stable
(ii) Avoid allo-rejection
(iii) Compatible in tissue engineering applications
(iv) Capable of repairing many vital tissues

MSCs are morphologically distinct being fibroblastic, expressing different surface markers, but do not express markers of the hematopoietic lineages (Figure 11.20). Adult bone marrow MSCs are believed to be quiescent.

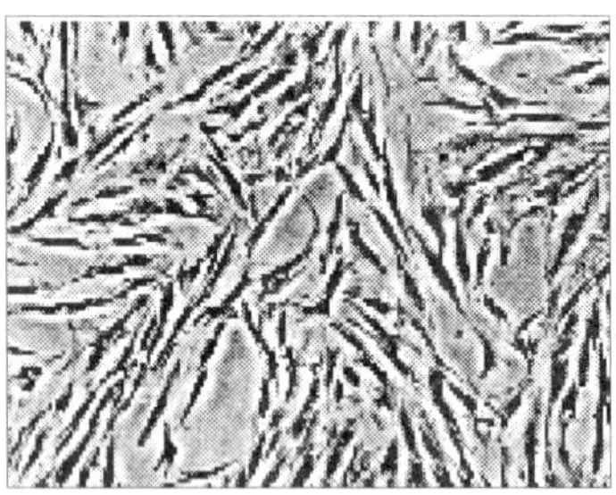

Figure 11.20 Morphological feature and surface markers, expressed/not expressed, in human mesenchymal stem cells. Mesenchymal stem cells (MSC) attain fibroblastic morphology in culture; they are homogeneous in size and shape. MSC express many cell surface markers, but interestingly do not express any hematopoietic makers.

11.5 PLASTICITY IN ADULT STEM CELLS

Over the past several years, a great deal of attention has been paid on the differentiation ability of adult (somatic) stem cells into other lineages. It has been suggested that the dogmas regarding the developmental potential of somatic tissue resident stem cells is needed to be revisited. Plasticity of adult stem cells is defined by their capacity to change into other type(s) within the lineage or outside the lineage against the fate, as explained in Figure 11.21. Pre-B cells are destined to differentiate into B cells with the help of Pax-5 transcription factor. Pax-5 is found to be critical for the maturation of B cells. Its depletion from pre-B cell (Pax-$5^{-/-}$) results in differentiation into other types of blood cells, depending on the specific culture conditions. For example, in the presence of IL-3, IL-6, SCF, and G-CSF, *Pax-5*$^{-/-}$Pre-B cells differentiate into neutrophils (Nutt *et al.*, 1999). Similarly, HSCs are shown to differentiate into neurons, skeletal muscle and other cells. Differentiation of one cell type, against its fate, into other type requires nuclear rearrangement/reprogamming within the cells, for which the competence of the responding cells (e.g., surface receptors, etc.) and microenvironmental cues are most important. For example, any bone marrow cells may not be competent for differentiation into neurons or hepatocytes. Similarly, the competent bone marrow cells do not undertake differentiation program against fate, unless they are exposed to inductive microenvironment.

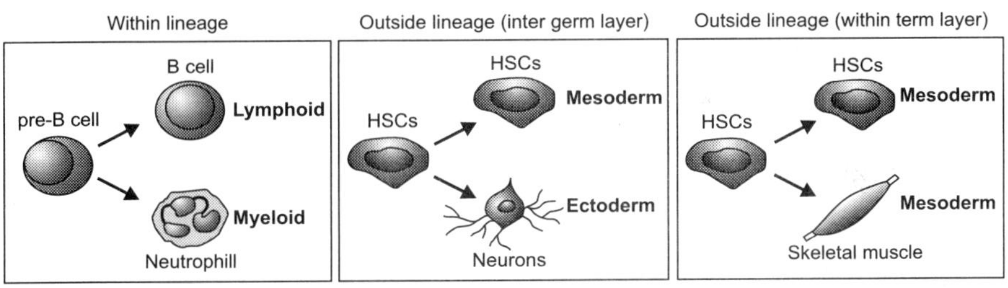

Figure 11.21 Plasticity in adult stem cells. Plasticity could be within and out side the lineage, against the predetermined program of differentiation.

At the nuclear level, it has been shown that when nuclei from mammary glands (Wilmut *et al.*, 1997), intestines (Gurdon and Uehlinger, 1966), egg-lining cumulus (Wakayama *et al.*, 1998), etc. are inserted into enucleated eggs of frogs, sheep (Wilmut *et al.*, 1997), and mice (Wakayama *et al.*, 1998), it results in partial or complete development of the respective animal. These experiments suggest that a cell or its nucleus is not ultimately committed, and their placement microenvironments would allow these cells or nuclei to change their apparent developmental fates. Despite these convincing studies, the capacity to change a somatic cell against its fate has been challenged by many investigators. Scientists have put forward five alternate mechanisms of fate changing by somatic cells (Figure 11.22).

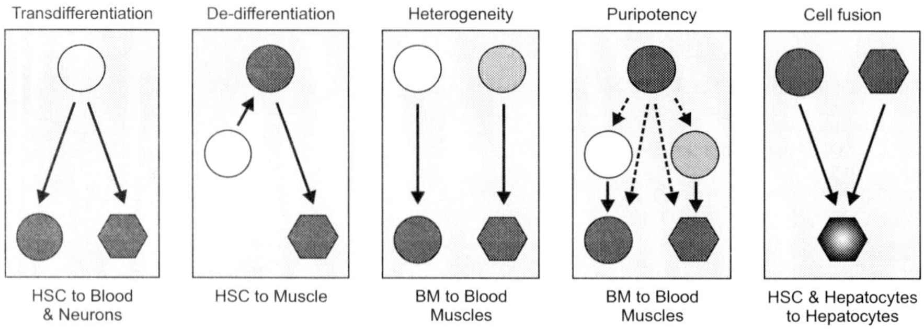

Figure 11.22 Alternate mechanisms for changing the fate of SSC. Fate change of SSC may occur due to one of the five mechanisms: transdifferentiation, dedifferentiation, heterogeneity, pluripotency, and cell fusion.

***Transdifferentiation*:** It is the most popular and debatable mechanism by which SSC of one germ layer changes the fate in favor of cells of other germ layer (Figure 11.22). For example, hematopoietic stem cells (mesodermal) are genetically programmed for delivering mature blood cells, but specific microenvironmental cues can influence them to differentiate into hepatocytes (endodermal) or neurons (ectodermal).

***Dedifferentiation*:** It is common in lower vertebrates but rare in mammals. By dedifferentiation process, SSCs reverted back to master cells from which they are originated. The master cells can differentiate into other cell types. Dedifferentiation process can be explained by a hypothetical example: mesodermal germ cells giving rise to hematopoietic and skeletal muscle cells. The conversion of HSCs into muscle cells *in vitro/in vivo* can be explained by concluding that HSCs first *dedifferentiate* into precursor mesodermal cells, which ultimately form skeletal muscle cells in the presence of specific microenvironmental cues.

***Heterogeneity*:** This is the most acceptable explanation for the plastic nature of SSC. One of the gold standards for plasticity of SSC is its clonal level purity. Normally, the purity of stem cells is determined by the presence and absence of specific surface markers, which do not affirm that all cells bearing particular surface markers are clonally identical. In other words, on the basis of surface markers, one cannot conclude that they are originated from single clone of cell. For example, highly purified murine bone marrow HSCs are phenotypically defined by Thy1.1^{lo}Lin^{-} c-kit^{lo}Sca-1^{+} or 'SP' cells. It is quite possible that this purified population of cells is *clonally heterogeneous*. Depending upon the site of application, these cells assume their fates, i.e., when injected in bone marrow, they behave as HSCs, but in the muscle or some other tissue, they emerge as tissue specific progenitor cells.

***Pluripotency*:** Plasticity of two cells originated from same germ layer can be explained by clonal level impurity. If *transdifferentiation* phenomenon does not exist in nature, how we explain the scientific evidence for differentiation of HSCs into neurons and vise-versa, and HSCs into hepatocytes, etc? It has been proposed that bone marrow is composed of *heterotropic* or *pleiotropic* stem cells. Inside the marrow, these cells are devoted to hematopoiesis, but outside the marrow, they behave differently depending on the local microenvironmental cues. Another hypothesis in this regard is that embryonic progenitors persist into adult life at a very low frequency. In normal situation they are naïve, but are activated in the presence of specific microenvironment.

***Fusion*:** The last mechanism proposed for the plasticity in SSCs is *cell fusion*. It has been shown that GFP expressing bone marrow cells is fused with ES cells in culture, and the resultant hybrid adopts some of the phenotypes typical for ES cells. Fusion of two cells means the formation of heterokaryons. By using sex mismatch donor and recipient mice, it has been shown that HSCs fuse with Purkinje neurons, cardiomyocytes, and hepatocytes *in vivo*, and finally acquire the cellular properties of the recipient cells.

Verfaillie and her group (2002) first showed pluripotency in adult SSC. They concluded that murine bone marrow contains a rare population of *multipotent adult progenitor cells* (MAPCs), which are extensively proliferative. *In vitro* and *in vivo* studies showed that clonally isolated MAPC contribute to the somatic tissues in mice. Almost similar results were obtained in case of human bone marrow derived multipotent stem cells (Douglas *et al*., 2005). Both mouse and human MAPCs showed some similarities with respective ES cells. The following criteria should be fulfilled to demonstrate transdifferentiation potential of SSCs (Table 11.5).

Table 11.5 Test for plasticity or transdifferentiation potential of SSCs

Status of the donor cell:

(i) clonally purified,
(ii) presence of donor cell specific marker (e.g., eGFP, LacZ), and
(iii) knowledge on donor cell specific gene expression and functional properties.

Recipient:

(i) target tissue specific gene expression and functional properties,
(ii) suitable diseased model.

Expected outcome:

(i) genetically marked cells loses donor specific gene expression and function,
(ii) donor cells acquire recipient tissue specific gene expression and function,
(iii) tissue level integration of donor and recipient cells, and
(iv) phenotypic correction in the recipient mice (k/o clone).

11.6 INDUCED PLURIPOTENT STEM (iPS) CELLS

Studies in the past have shown that in somatic nuclear transfer process or fusion between ES and adult cells lead to activation of several genes in the adult cells' nuclei. Later, series of studies have demonstrated that it is possible to change the fate of adult cells by introducing transcript factors. These suggest that the somatic cells, though terminally differentiated, gene expression remains dynamic and reversible, which is governed by epigenomic levels resulting in partial genomic reprogramming. Prof Yamanaka and colleagues worked on this approach to convert fibroblasts into a state of cells that is equivalent to ES cells. Initially, 24 different transcription factors, in different combinations, were used with an objective of converting them into pluripotent stem cells. In 2006, Prof Yamanaka first time reported the generation of a cell type by simply over-expressing four transcription factor genes encoding *Oct4*, *Sox2*, *Klf4*, and *c-Myc*. As the starting cells do not express these transcription factors and over-expressing lead to the development of pluripotency, the resultant cells are termed as iPS cells. Different tests are performed to qualify iPS as equivalent to ES cells are based on: (a) identical morphology and growth pattern in culture, (b) expression of ES-specific genes, (c) formation of teratoma upon transplantation in an immuno-compromised mice, and (d) stringent test of pluripotency.

The stages by which iPS cells are obtained and the function of four transcription factors have been described in Figure 11.23. Somatic cells have finite life span as they undergo senescence followed by apoptosis and cleared from the system. Furthermore, in somatic cells a major part of the chromosome is tightly packed. The first step of deriving iPS cells is to open the chromatin so that it can be accessible for genomic and epigenomic changes, followed by immortalizing the cells. The combining effect of *c-Myc* and *Klf4* in fibroblasts results in derivation of cells that are immortal and allow further genetic modifications. Despite, *Oct3/4* is considered as a pluripotent factor in ES cells; its over-expression in *c-Myc* and *Klf4* transduced fibroblasts could not generate pluripotent cells, comparable with ES cells. These cells are known as nullipotent ES-like cells as they were found to be developmentally blocked to form committed lineages. A complete pluripotent cells, known as iPS cells, has been generated once above cells are transduced with *Sox2*. The functions of individual transcription factors in generating iPS cells are described below:

c-Myc: It is a basic helix-loop-helix transcription factor, known as a *proto-oncogene*. It plays a significant role in many cancers by accelerating activation of cell cycle, involve in maintenance of pluripotency in ES cells, it also responsible for induction of pluripotency. There are 25,000 *c-Myc* binding sites in the genome. These bindings recruite histone acetylase complexes and thus considered that *c-Myc* could open-up chromatin structure. As *c-Myc* can promote progression from G1 to S phase, it counteracts the anti-proliferative effect of Klf4.

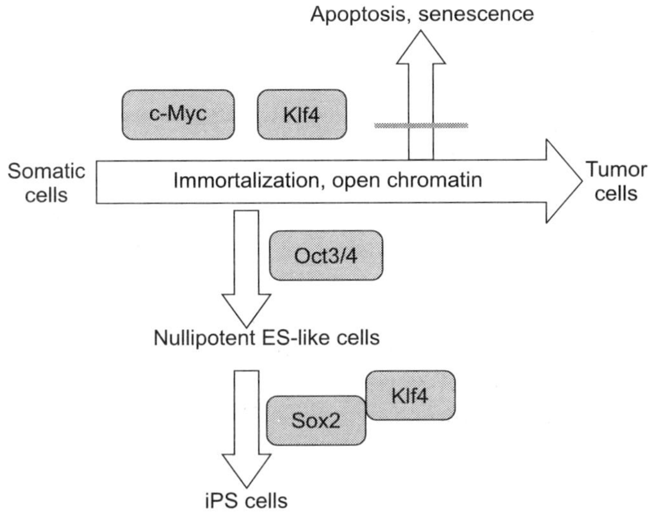

Figure 11.23: Generation of iPS cells from somatic fibroblasts by transfecting wth Yamanaka factors.

Klf4: It has been found that the over-expression of Kruppel-like zinc finger transcription factor inhibits differentiation of ES cells, co-operatively with *Oct3/4* and *Sox2*. It may act both as a tumor-suppressor by activating *p21* and promoting cell proliferation by down-regulating *p53*. By down-regulation of *p53*, *Klf4* may indirectly up-regulate *Nanog*, a key factor for maintenance of pluripotency. The overall roles of *Klf4* are considered to prevent early entry to senescence and cellular apoptosis.

Oct3/4: It is specially expressed in the pluripotent ES cells, thus it is considered to play a key role in the maintenance and self-renewal of pluripotent cells. It is also important in promoting differentiation, as loss of expression leads to differentiation of ES cells into mesoderm and endoderm lineages. It is know to regulate many genes in co-operation with *Sox2*.

Sox2: Both *Sox2* and *Oct3/4* are involved in the maintenance of pluripotency in ES cells. Its down-regulation in mouse ES cells leads to the differentiation into trophoectoderm and other lineages.

Later, both Prof. Yamanaka (2007) and Thomson (2009) demonstrate that by following similar approach, it is possible to generate iPS cells from human fibroblasts. Yamanaka's used *OCT3/4*, *SOX2*, *KLF4* and *c-MYC*, whereas Thomson's group replaced *KLF4* and *c-MYC* with *NANOG* and *LIN28*.

11.6.1 Mechanism for the formation of iPS cells

The efficiency of the derivation of iPS cells is extremely low, it may vary between 0.05 to 0.1%. This is largely depends in cellular level and the methodology adopted for such conversion. In this regards, individual programming (transcription) factors as well as those cellular factors support and antagonize epigenetic changes are crucial to understand. Two mutually exclusive models have been proposed by Yamanaka (2009), these are based on deterministic and stochastic principle. According to the deterministic model, the efficiency of iPS cells derivation is low due to low frequency of competent cells (stem or progenitor) present in any somatic tissue. In stochastic model, every cells of somatic tissue, whether stem or terminally differentiated, are equally potential to factor-mediated reprogramming. The overall derivation efficiency is low as each cell requires series of epigenetic modification to acquire pluripotency. The accepted working model of iPS cells generation is based on stochastic principle, as many terminally differentiated cells, such as immune cells (monocytes, T and B lymphocytes) have been used to generate iPS cells. It has been observed that in case of continuously proliferating cells, stochastic chances to become iPS cells is improved, even though the derivation requires months. In another study, the state of differentiation of cells has been found to govern iPS cell formation efficiency, as clonally plated HSCs and progenitors were shown much higher efficiency of conversion in much shorter time than mature monocytes and lymphocytes (Eminli et al., 2009). These diverse results on derivation of iPS cell lead to propose that terminally differentiated cells requires many stochastic events than stem cells in order to become a pluripotent cell, which requires longer time to gain characteristic properties. It has been found that many cells in fibroblasts are refractory to reprogramming as only few are able to down-regulate somatic transcription factors and activate pluripotency factors as they overcome genetic and epigenetic barriers that prevent induction of pluripotency.

11.6.2 Applications of iPS cells

Reprogramming of somatic cells into iPS has already reached to preclinical stages of evaluation against different diseases in which regenerative medicine is the only solution. The diseases that have been focused are pertaining to the dysfunction of a specific cell type for which an effective protocol for in vitro differentiation is available. Several disease models under investigations are in the areas of neurological disorder, such as amyotrophic lateral sclerosis (ALS), spinal muscular atrophy, Parkinson's disease (PD), familial dysautonomia (FD) and retinal degeneration. Sickle cell anemia and β-thalassemia are important among hematopoietic disorders. Metabolic disorder is another area in which iPS cells seems to have profound therapeutic potential. The mode of therapeutic applications of iPS cells has been described in the next chapter. Besides therapeutic applications, drug screening is one major area in which iPS cells will have role in drug discovery program.

11.6.3 Limitations in iPS technology

The major issue of deriving iPS cells is using retroviral or lentiviral vector to deliver the transgene. These viral vectors are preferentially integrated in the host genome and later may transform the cells. Thus the main focus in deriving iPS cells is to avoid transgenic sequences in host genome and insertional mutagenesis. This is critically important in case of clinical applications of iPS cell technology. Different ways by which iPS cells can be generate for therapeutic applications are: (a) using vectors that do not integrate into the host genome, and (b) easy removal of the integrated vectors from the genome. Alternate gene delivery systems, like episomal vectors, nonintegrating viral vectors, transient DNA transfection, etc. can overcome this problem. It has been realised that small molecules might have a role to activate the pluripotency program, if successful, integration-free iPS cells will derive from somatic cells. Another limitation is associated with the inherently low efficiency of iPS cells derivation. This can be overcome by incorporating a suitable selection method that can distinguish between simply transformed cells and reprogrammed clones. Activation of pluripotency genes, such as Nanog or Oct4 linked to drug selection cassettes has been found successfully for this purpose.

11.7 STEM CELL NICHE

Most of the organs replenish matured cells when they do not properly function or die. Thus, it is believed all many developing tissues harbour SSCs for life-term support that may retain some of the characteristics of early progenitors, including capacity to endless self-renew. Niches are subsets of tissue cells and extracellular matrices and growth factors that can indefinitely house one or more such stem cells and control their self-renewal and progeny production *in vivo*. The importance of the niche is perhaps best explained by the experiments shown in Figure 11.24. ES cells can be propagated indefinitely on MEF feeder layer without losing their pluripotency. The cultured ES cells do not behave as ES cells *in vivo,* as they cannot form germ layer specific cells without further stimulation. The same ES cells when injected in a nude mouse form teratomas and lost their unique properties. However, if injected inside of a mouse blastocyst, ES cells resume normal behavior and contribute to the generation of three germ layer specific cells of a healthy normal chimeric offspring (Figure 11.24). Inside blastocysts, ES cells behave normally as they are analogous to the native *niche*. Overall, these observations suggest that stem cells maintain their stemness by the combination of their intrinsic characteristics and the environmental cues offer by the niche.

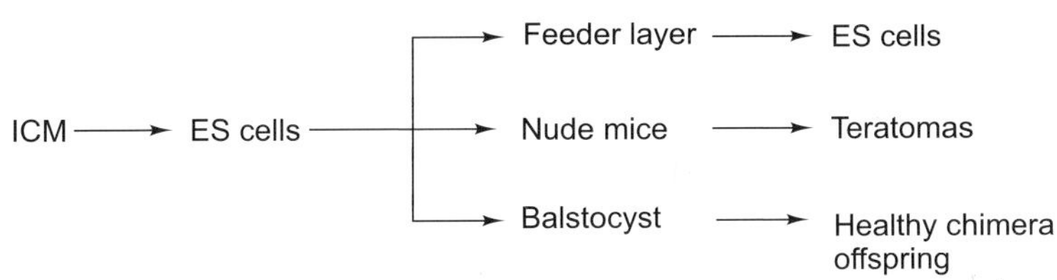

Figure 11.24 Blastocyst is the specific niche of ES cells, where they maintain functional properties. The unique properties of ES cells are lost, if they are not associated with blastocyst.

11.7.1 Identifying and classifying niches

Identifying niches and determining how they regulate stem cells is a challenging task. So far, few niches have been identified in somatic and germ line stem cells in mammals. Table 11.6 show the list of few stem cells and their location in corresponding niches in mammals. Identification of niche requires the removal of stem cells from the tissue and returning them back, which can be termed as *replacement assay* (Figure 11.25A).

It is generally believed that displacing endogenous stem cells is imperative for efficient acceptance of exogenous stem cells by the niche. In other words, niche is occupied by the endogenous stem cells. Unless some of them are removed, exogenous stem cells cannot find place in the niche to engraft. A simple example of reinvention of hematopoietic stem cells niche is described below. The green-colored cells are niche cells (supporting cells) and red-colored cells are HSCs (Figure 11.25A). In order to improve the grafting of exogenous HSCs (blue), the bone marrow is subjected to irradiation or treatment with chemotherapeutic agents (e.g., 5-flurouracil) to deplete host cells. The exogenous cells (blue) are

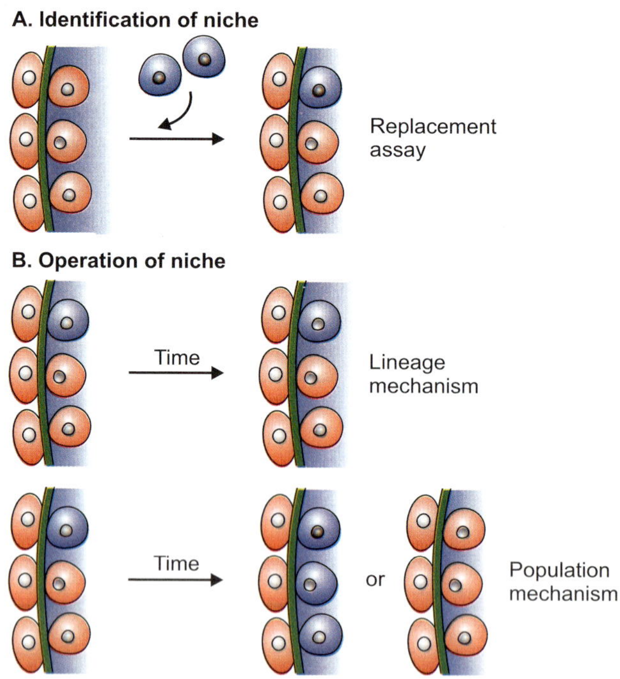

Figure 11.25 Identification of stem cells niche and the mode of its operation. [Adapted from Spardling A *et al.* (2001) Nature 414, 98]

transplanted into bone marrow through intravenous route. If one or more blue cells are engrafted in bone marrow tissue and subsequently function as stem cells, it is concluded that HSCs niche is present in the bone marrow. The transplanted blue cells may house other tissues of the body, but they will not function as HSCs due to lack of microenvironment (niche). This experiment can be conducted easily in mice model. Take few young mice (which do not express GFP or express allo-antigen Ly5.2) and irradiate them at 900 cGy to deplete the endogenous dividing bone marrow cells. To these mice, transplant *congenic* bone marrow cells either expressing GFP or allo-antigen Ly5.1. The presence of donor cells (GFP^+ or $Ly5.1^+$) in the recipient bone marrow, if any, can be detected by antibody based techniques (e.g., FACS, immuno-fluorescence) after a reasonable period of time.

Table 11.6 Stem cells and the respective niches in mammals

Type of stem cells and their matured cells	*Location of niche in the organ/tissue*	*References*
Hematopoietic: lymphoid and myeloid cells	Endosteum and trabecular bone area	Zahang *et al.* (2003); Calvi *et al.*, (2003)
Mesenchymal: muscle, bone, cartilage, and ligament	Bone marrow	Pittenger et al. (1999)
Skin epidermal: keratinocytes	Hair follicle in bulge below the sebaceous gland	Rochat *et al.* (1994); Oshima *et al.* (2001)
Intestinal epithelium: paneth cells, enteroendocrine cells, goblet cells, and adsorptive villus cells	Gut, crypts of intestinal villus	Bjerknes and Cheng (1999)
Spermatogonia (germ line cells)	Basement membrane of the seminiferous tubules of testis	de Rooij and Grootegoed (1998)
Neural: neurons, astrocytes, oligodendocytes	Hippocampus area of brain	Buttista *et al.* (2006)
Pancreatic progenitors: beta cells	Vascular basement membrane in pancreas	Nikolova *et al.*(2006)
Muscle satellite cells: skeletal muscle cells	Basal lamina surrounding each myofiber	Coolins *et al.* (2005)
Oval cells: hepatocytes and bile duct epithelial cells	Canals of Hering in liver	Alison *et al.* (1996)

Once the niche is identified, it is important to know how the niche operates. Niche regulates the fate of the stem cells through two mechanisms— lineage and population. In the *lineage mechanism*, the number of stem cells (*blue*) remains relatively unchanged over many divisions, as in every cycle one of the daughter cells leaves the niche (Figure 11.25B). Whereas, in *population mechanism,* both daughter cells may remain in the niche as stem cells or they leave the niche (Figure 11.25B). The operation of niche can be explained considering hematopoietic system. If the donor specific HSCs do not change in number, one can conclude that the lineage mechanism is predominanted. On the other hand, if donor specific HSCs number are increased over the time, or they are not detected in the niche, then population mechanism is believed to be operated.

11.7.2 Niche structure and function

Niches are subsets of tissue cells, extracellular substrates, and growth factors that house stem cells, and control self-renewal and progeny production. As shown in Figure 11.26A, the niche cells (*green*) underlying a basement membrane (basal lamina) hold the stem cells in position through adhesion molecules. The role of niche is to regulate symmetric and asymmetric divisions of cells. By regulating the balance between symmetric and asymmetric divisions of stem cells, niche maintains demand of the differentiated cells within the tissue. It has been mentioned earlier that by lineage mechanism, niche makes the stem cells divide asymmetrically to renew the stem cell pool and at the same time maintains the supply of differentiated progenies (Figure 11.25B). However, through symmetric division (population mechanism), niche restricts stem cells from differentiation and induces self-renewal in arithmetic progress (Figure 11.25B). It is believed that many *intrinsic* and *extrinsic factors* control the fate of daughter cells. For a daughter cell to be stem cell, it must retain self-renewal and differentiation inhibitor factors. On the other hand, if the daughter cell is destined for differentiation, it should have little stemness factors and inherit differentiation factors. The probable role of niche is orienting and defining the nature of stem cell divisions. Conceptually, the following sequence of events may take place before a epithelial stem cell divides:

(1) Generation of cell polarity
(2) Orientation of mitotic spindle
(3) Segregation of differentiation and stemness determinants equally or unequally between two daughter cells.

If the interaction between basal lamina and the neighboring stem cells establish the polarity as shown in Figure 11.26A (middle), it is expected that the fate determinants will equally split into two cells, resulting in symmetric division. In contrast, if the orientation of attachment of neighboring stem cells is such that the mitotic spindle is perpendicular to the major cell polarity axis (Figure 11.26A, bottom), asymmetric division of cells is expected. Figure 11.26B shows how the population of stem cells and differentiated progenies is expanded in symmetric and asymmetric divisions, respectively.

A.

B.

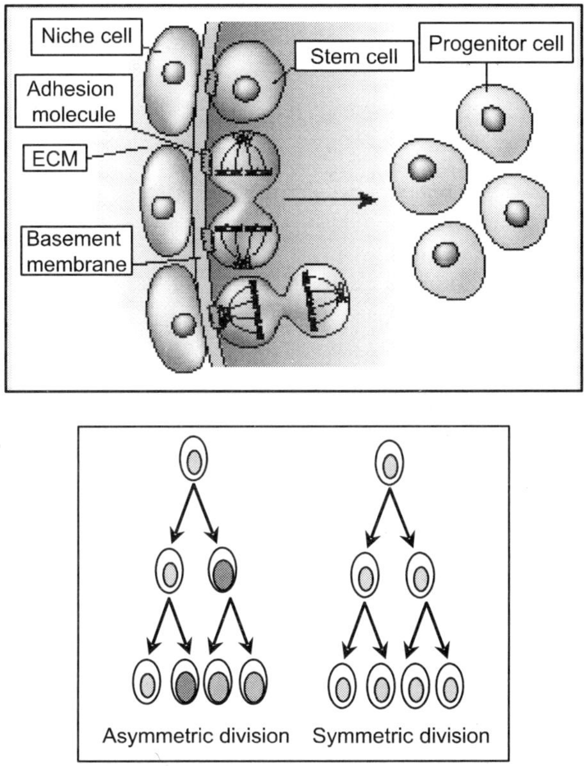

Figure 11.26 **A.** Composition of stem cell niche. Stem cell niches are composed of stem cell, niche cells (supporting cells), basement membrane, extracellular matrix, and the growth factors produced by the niche cells [Adopted from Spardling A et al. (2001) Nature 414, 98]. **B.** Asymmetric and symmetric division of stem cells. Yellow nuclei containing cells are the stem cells, which divide in each cycle and gives arise to two stem cells in symmetric division. In asymmetric division, the number of original stem cell remain constant, but their differentiated progenies (blue nuclei) are increased in each division.

11.8 SUMMARY

1. Till the stage of compaction, the bastomere cells maintain totipotency.
2. Extra-embryonic tissues are formed from trophoectoderm and epiblast.
3. During gastrulation, the ICM is specified into ectoderm, mesoderm, and endoderm tissues.
4. ES and EG cells are both pluripotent is nature, but EG cells do not form teratomas.
5. Both Oct-4 expression and STAT3 activation are imperative to maintain pluripotency in ES cells.
6. Aorta/gonad/mesonephros are the tissue of definitive hematopoiesis.
7. Primitive hematopoietic stem cells efflux vital dye, and provide life long protection.

8. MSCs are genetically stable, avoid allo-rejection, differentiated into all connective tissues in the body.
9. Stem cells cannot maintain stemness outside their niche.

ABBREVIATIONS

AGM	Aorta-gonad-mesonephros
AP	Alkaline phosphatase
ASC	Adult stem cell
BFU-E	Burst forming unit-erythroid
bFGF	Basic fibroblast growth factor
CFU-GM	Colony forming unit-granulocyte monocyte
CFU-GEMMk	Colony forming unit-granulocyte erythroid monocyte megakaryocyte
CFU-G	Colony forming unit-granulocyte
CFU-E	Colony forming unit-erythroid
CFU-Mk	Colony forming unit-megakaryocyte
CNS	Central nervous system
CFU-S	Colony forming unit-spleen
CD	Cluster of differentiation
CLP	Common lymphoid progenitor
CMP	Common myeloid progenitor
ECM	Extracellular matrix
EC	Embryonic carcinoma
ES	Embryonic stem
EG	Embryonic germ
EST	Expressed sequence tag
EB	Embryoid body
ERK	Extracellular signal-regulated kinase
FACS	Fluorescence activated cell sorter
Flt-3L	Fetal liver tyrosine kinase-3 ligand
Epo	Erythropoietin
FITC	Fluorescein isothiocyanate
GFP	Green fluorescent protein
G-CSF	Granulocyte colony stimulating factor
gp130	glycoprotein 130
HPP-CFU	Highly proliferating progenitor-colony forming unit
ICM	Inner cell mass
IL-3/6/11	Interleukin-3/6/11
IVF	*In vitro* fertilization
Jak	Janus kinase
KO-DMEM	Knock out- Dulbecco's modified Eagle medium
KO-FBS	Knock out- Fetal bovine serum
LIF	Leukemia inhibitory factor
LIF-R	Leukemia inhibitory factor-receptor
LTR	Long-term repopulating
LTC-IC	Long term culture-initiating cells

MEF	Mouse embryonic fibroblasts
MAPK	Mitogen activated protein kinase
MSC	Mesenchymal stem cells
MAPC	Multipotent adult progenitor cells
MPP	Multipotent progenitor
MDR1	Multi drug resistant-1
NOD	Non-obese diabetes
PBS	Phosphate buffered saline
PGC	Primordial germ cell
Pou5f1	POU-domain transcription factor Oct-4
Pax5	Paired box gene 5
PE	Phycoerythrocine
RT-PCR	Reverse transcriptase-polymerase chain reaction
SCID	Severe combined immunodeficiency
SSC	Somatic stem cells
SCF	Stem cell factor
SSEA	Stage specific embryonic antigen
Shh	Sonic hedgehog
SH2	Src homology 2
SHP2	SH2-Homology-Containing Protein-Tyrosine Phosphatase-2
Sca-1	Stem cell antigen-1
STR	Short-term repopulating
STAT3	Signal transducer and activator of transcript 3
SRC	SCID repopulating cells
SP	Side population
TRA	Tumor rejection antigen
Tpo	Thrombopoietin
VEGF-R2	Vascular endothelial growth factor-receptor 2

REFERENCES

1. Alison MR, Golding M, Sarraf CE, *et al.* (1996) *Gasteroentrology* 110, 1182.
2. Andrews PW (1998) *APMIS* 106, 158.
3. Battista D, Ferrari CC, Gaque FH, *et al.* (2006) *Eur. J Neurosci.* 23, 83.
4. Bjerknes M, Cheng H (1999) *Gastroentrology* 116, 7.
5. Brook FA, Gardner RL (1997) *Proc. Natl. Acad. Sci. USA* 94, 5709.
6. Bongso A (1999) *Handbook on blastocyst culture.* Sydney Press Indusprint, Singapore.
7. Breems DA, Blokland DAW, Siebel KE, *et al.* (1998) *Blood* 91, 111.
8. Brandt JE, Bhalla K, Hoffman R (1994) *Blood* 83, 1507.
9. Collins CA, Olsen I, Zammit PS, *et al.* (2005) *Cell* 122, 289.
10. Calvi LM, Adams GB, Welbreht KW, *et al.* (2003) *Nature* 425, 836.
11. Douglas WL, *et al.* (2005) *J Clin Invest.* 115, 327.
12. Dexter TM, Allen TD, Lajtha LG, *et al.* (1973) *J Cell Physiol.* 82, 461.
13. de Rooij DG, Grootegoed JA (1998) *Curr. Opin. Cell. Biol.* 10, 694.
14. Ebert AD, YuJ, Rose FFJr., *et al.* (2009). Nature 457, 277.
15. Eminli S, Frudi A, Stadtfeld M, *et al.* (2009) Nature Genet 41, 968.

16. Evans MJ, Kaufman MH (1981) *Nature* 292, 154.
17. Fong CY, Bongso A, Ng SC, *et al.* (1998) *Hum. Reprod.* 13, 2926.
18. Fleming TP (1987) *Dev. Biol.* 119, 520.
19. Gurdon JB, Uehlinger V (1966) *Nature* 210, 1240.
20. Goodell MA (1999). *Blood* 94, 2545.
21. Goodell MA, *et al.* (1996) *J Exp Med* 183, 1797.
22. Hogan B, Beddington R, Constantini F, Lacy E. (1994). In: *Manipulating the mouse embryo- a laboratory manual.* Cold Spring Harbor, New York.
23. Harrison DE, Zhong RK (1992) *Proc Natl Acad Sci USA.* 89, 10134.
24. Jiang Y, Jahagirdar BN, Reinhardt RL, *et al.* (2002) *Nature* 418, 41.
25. Koller MR, Palsson BO (1993) *Biotechnol Bioeng* 42, 909.
26. Kay HE (1965) *Lancet* 15, 418.
27. Leary AG, Zeng HQ, Clark SC, *et al.* (1992) *Proc Natl Acad Sci USA* 89, 4013.
28. Matsui Y, Zsebo K, Hogan BL (1992) *Cell* 70, 841.
29. Martin GR (1981) *Science* 209, 768.
30. Mitsui K, Tokuzawa Y, Itoh H, *et al.* (2003) *Cell* 113, 631.
31. McAdams TA, Sandstrom CE, Miller WM, *et al.* (1995) *Cytotechnol* 18, 133.
32. McKenna HJ, Vries PD, Brasel K, *et al.* (1995) *Blood* 86, 3413.
33. Marella *et al.* (2000) *EMBO J* 19, 2465.
34. Muraglia A, Cancedda R, Quarto R (2000) *J Cell Sci* 113, 1161.
35. Nutt SL, Heavey B, Rolink AG, *et al.* (1999) *Nature* 401, 556.
36. Nikilova G, Jabs N, Konstantinova I, *et al.* (2006) *Dev. Cell* 10, 397.
37. Niwa H, Burdon T, Chambers I, Smith A (1988) *Genes Dev.* 12, 2048.
38. Nichols J, Zevnik B, Anastassiadis K, *et al.* (1998) *Cell* 95, 379.
39. Niwa H, Miyazaki J, Smith AG (2000) *Nat. Genet* 24, 372.
40. Oshima H, Rochat A, Kedzia C, *et al.* (2001) *Cell* 104, 233.
41. Pratt HP, Ziomek CA, Reeve WJ, Johnson MH (1982) *J Embryo. Exp. Morphol.* 70, 113.
42. Palmieri SL, Peter W, Hess H, *et al.* (1994) *Dev. Biol.* 166, 259.
43. Payne KJ, Crooks GM (2002) *Immunol. Rev.* 187, 48.
44. Papayannopoulou T, Brice M, Blau CA (1993) *Blood* 81, 299.
45. Pittenger MF, Mackay AM, Beck SC, *et al.* (1999) *Science* 284, 143.
46. Rochat A, Kobayashi K, Barrandon Y (1994) *Cell* 76, 1063.
47. Rossant J, Cross JC (2002) In: *Mouse Development- Patterning, Morphogenesis, and Organogenesis* (J Rossant and PPL Tam, eds.) Academic Press, San Diego, p.155.
48. Rathjen PD, Toth S, Willis A, *et al.* (1990) *Cell* 62, 1105.
49. Richards M, Fong, C-Y, Chan W-K, *et al.* (2002) *Nature Biotechnol.* 20, 933.
50. Reubinoff BE, Pera MF, Fong C-Y, Trounson A, Bongso A (2000) *Nature Biotechnol.* 18, 399.
51. Roberts R, Gallagher JT, Spooncer E, *et al.* (1998) *Nature* 332, 976.
52. Shamblott MJ, Axeiman J, Littlefield JW, *et al.* (2001) *Proc. Natl. Acad. Sci, USA* 98, 113.
53. Stevens LC (1967) *J Natl Cancer Inst.* 38, 349.
54. Solter D, Knowles BB (1975) *Proc. Natl. Acad. Sci. USA* 72, 5099.
55. Stahl N, Farruggella TJ, Boulton TG *et al.* (1995) *Science* 267, 1349.
56. Smith AG (2001) *Annu. Rev. Cell. Dev. Biol.* 17, 2001.
57. Scholer HR, Ruppert S, Suzuki N, *et al.* (1990) *Nature* 344, 435.

58. Takoshastri K, Yamanaka S (2006) Cell 126, 663.
59. Takashashi K, Tanabe K, Ohmki M, *et al.* (2007) Cell 131, 861.
60. Thomas JA, Itskovitz-Elder J, Shapiro SS (1998) *Science* 282, 1145.
61. Thomson JA, Odorico JS (2000) *Trends. Biotechnol.* 18, 53.
62. Verfaillie CM, Catanzarro PM, Li W-N. (1994) *J Exp Med.* 179, 643.
63. Williams RL, Hilton DJ, Pease S, *et al.* (1988) *Nature* 336, 684.
64. William PL, Warwick R, Dyson M, Baunistner LH (1989) *Gray's Anatomy*, 37th eds. Churchill Livingstone, Edinburgh.
65. Wu A, Till J, Stiminovitch L, *et al.* (1968) *J Exp Med.* 127, 455.
66. Wagers AJ, Christensen JL, Weissman IL (2002) *Gene Ther.*9, 606.
67. Wilmut I, Schnieke AE, McWhir J, *et al.* (1997) *Nature* 385, 810.
68. Yamanaka S (2009) Nature 460, 49.
69. Wakayama T, Perry AC, Zuccotti M, *et al.* (1998) *Nature* 394, 369.
70. Zhang J, Niu C, Ye L, *et al.* (2003) *Nature* 425, 841.
71. Zandstra PW, Conneally E, Petzer AL, *et al.* (1997) Proc Natl Acad Sci USA. 94, 4698.

SUGGESTION FOR FURTHER READING

1. Donovan PJ, Gearhart J (2001) *Nature* 414, 92.
2. Embryonic stem cells- Methods and Protocols (K Turksen, ed.) Huamn Press, Totowa, New Jersey, 2002.
3. *Embryo-derived stem cell lines In Teratocarcinomaa and embryonic stem cells: a Practical Approach* Robertson EJ. (EJ Robertson, ed.), 1987, IRL Press, Oxford.
4. Fuchs E, Tumbar T, Guasch G (2004) *Cell* 116, 769.
5. Forbes SJ, Vig P, Poulsom R, *et al.* (2002) *Clinical Science* 103, 355.
6. Graf T. (2002) *Blood* 99, 3089.
7. *Manipulating the Mouse Embryo.* (B Hogan, R Beddington, F Costantini, E Lacy, eds), 2nd ed, 1994. Cold Spring Harbor Laboratory Press, New York.
8. *Manipulating the Mouse Embryo: A Laboratory Manual.* (A Nagy, M Gertsenstein, K Vintersten, R Behringer, eds), 3rd ed, 2003. Cold Spring Harbor Laboratory Press, New York.
9. Pittenger MF, Martin BJ (2004) *Circulation* 95, 9.
10. Raz R, Lee C-K, Cannizzaro LA, *et al.* (1999) *Proc. Natl. Aca. Sci. USA* 96, 2846.
11. Spardling A, Drummond-Barbosa D, Kai T (2001) *Nature* 414, 98.
12. Stadtfeld M, Hochedlinger K (2010). Genes Dev 24, 2239.
13. Weissman IL. (2000) *Cell* 100, 157.

12

Regenerative Medicine: Tissue Engineering

The adult human organ has limited regenerative potential, so cell loss due to chronic injury (also acute injury) cannot be efficiently replaced with organ specific parenchymal cells. This leads to the formation of noncontractile scar tissue results in progressive loss of function. The current medical practice is to replace the lost organ or a part of that. Whole-organ transplantation is limited by the inadequate supply of donor or cadaver organs, debilitating side effect of immune suppressant and acquired infection by donor organ. This leads to the development of cellular approach to repair and/or regenerate the organ, broadly known as *regenerative medicine*. Thus, regenerative medicine in general refers to the replacement/restoration of damaged body parts by transplanting cells/tissues isolated from the healthy organs, or inducing body's own tissue regeneration by providing suitable microenvironment. Currently, there is great excitement over the potential for repair of defective tissues/organs by regenerative medicine. The isolation of stem/progenitor cells, their expansion and differentiation into target tissues have futher improved the hope of regenerative medicine. Two approaches are followed in regenerative medicine: *tissue engineering* and *cell therapy*. This chapter deals for the development of basic concept on tissue engineering.

12.1 THE BIOLOGY OF AUTONOMOUS REGENERATION

Every organism regenerates, though the extent of regeneration ability varies among species and within same species from individual to individual (Goss, 1969). Certain amphibians (e.g., lizard and frog) can regenerate complex body structures like tails and limbs. Whereas, in case of adult mammals, the regeneration process is limited, which is termed as *maintenance regeneration*. In mammals, including humans, the maintenance regeneration is required to preserve the integrity of the tissues to perform normal functions of the tissues/organs.

As a whole, in normal individual, the regeneration can occur at three levels—molecular level, single cell level, and tissue level. The nature of regeneration is governed by the type of cells/tissues and the extent of injury. In molecular level regeneration, the original cells do not die but the turnover of protein synthesis increases to compensate its degradation in response to stress or load on them. For example, cardiomyocytes replace cellular proteins by freshly synthesizing proteins in response elevated blood pressure (Gevers, 1984). Single cell level regeneration is not common in vertebrates, except regeneration of axons in sensory and motor neurons after transection (Yannis, 2001). In regenerative medicine, tissue level regeneration is important, which is possible only when the following three criteria are satisfied. First, there should be intra-organ precursor or circulating progenitor cells that are competent for tissue specific differentiation. Second, the damaged tissue should have microenvironment cues (inductive microenvironment) that induce proliferation and differentiation of the above progenitor cells. Third, the damaged tissue should not produce any substances or allow fibrotic growth that suppresses normal regeneration process.

12.1.1 Mechanism of tissue level regeneration

In living organisms there are three known mechanisms by which regeneration of damaged tissues/organs takes place. These are *compensatory hyperplasia*, activation of adult tissue specific progenitor stem cells, and *dedifferentiation*. Dedifferentiation is the loss of phenotype and functional specialization of mature cells of one lineage by which they get converted into progenitor cells of the replacement tissue. Later, these progenitor cells proliferate and differentiate into mature replacement tissues/cells. As shown in Figure 12.1C, the mature cell 'C' dedifferentiates into progenitor cell 'A' before it is converted into cell 'B'. In mammals, including humans, dedifferentiation process is not common. However, in lower vertebrates, dedifferentiation process is belived to be one of the major

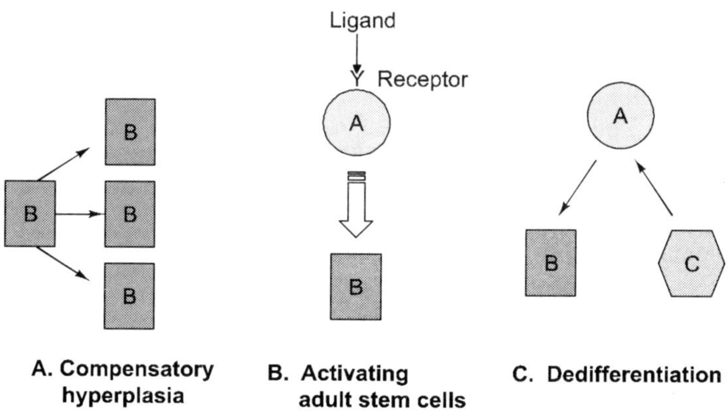

Figure 12.1 Mechanism of tissue level regeneration. **A.** Compensatory hyperplasia: adult tissue specific cells proliferate at the same time perform their normal physiological functions; **B.** Activating adult stem cells: circulating/mobilized stem cells of bone marrow origin differentiate into tissue specific adult stem cells with the help of local microenvironment cues; **C.** Dedifferentiation: differentiated cells reverted back to the progenitor cells, followed by redifferentiation into target tissue specific cells.

mechanisms by which the lost organ/tissue is regenerated. For example, lizards and fish regenerate tails and fins, respectively, by dedifferentiation process (Geraudie *et al.*, 1998). Detailed discussion on dedifferentiation is beyond the scope of this book, therefore, students are advised to consult reference section.

Compensatory hyperplasia is the most popular mechanism by which mature tissue specific cells proliferate to maintain homeostasis and at the same time perform normal functions (Figure 12.1A). The most common example is regeneration of liver after partial hepatectomy or acute damage following treatment with hepatotoxic chemicals. The existing hepatocytes proliferate to restore liver mass, while they continue normal functions like synthesis of serum proteins, drug metabolism, etc. (Trembly and Steer, 1998). It is also believed that cardiac muscle and blood vessels are restored from limited damage by a similar mechanism of compensatory hyperplasia (Bettencourt-Diaz *et al.*, 2003).

Regeneration by activating tissue specific adult progenitor stem cells is the main source of cells for tissue homeostasis and autonomous regeneration. It is believed that late in the *embryonic life*, tissue/organ specific adult stem cells (ASCs) of three germ layers reside in the respective tissue. These ASCs respond to various growth factors to proliferate and differentiate into mature cells, and are involved in regeneration throughout the lifetime of an individual (Weissman, 2000). Besides these resident ASCs, many progenitor cells circulate in the bloodstream that take part in tissue regeneration (Rafii and Lyden, 2003). ASCs isolated from different mammalian tissues are shown in Table 11.4. How adult stem cells are involved in regeneration process can be explained by considering of liver regeneration as an example. In liver, both oval cells (liver specific stem cells) and mature hepatocytes are involved in liver regeneration, depending upon the nature and the pathology of the damage. Oval cells are bipotential in nature; they multiply as well as differentiate into hepatocytes and billiary epithelial cells (Sell and Leffert, 1982). In normal liver regeneration model, the mature hepatocytes are involved to maintain tissue homeostasis by compensatory hyperplasia. Oval cells do not activate in autonomous liver regeneration unless the proliferation of mature hepatocytes is inhibited (Faust and Campbell, 2003). In oval cell activation model, rat is first treated with 2-acetylaminofluorene (2-AAF) to stop the proliferation of hepatocytes. Then liver is either *partially hepatectomized* or damaged with mepatotoxic chemicals. In this case, the oval cells present in the *canal of hering* undergo proliferation and differentiation into mature hepatocytes and biliary epithelial cells (Figure 12.1B).

12.1.2 Limitation of autonomous regeneration

In regeneration biology, maintenance regeneration is slow but a tightly regulated process. The main reason of maintenance regeneration is tissue homeostasis where natural cells/tissues loss occur, such as regeneration of skin following burn, regeneration of bone marrow incase of irradiation, etc. The failure of tissue regeneration after the damage is supposed to be due to mainly three reasons: (a) *lack of regeneration-competent cells*, (b) *absence of inductive microenvironment*, and (c) *synthesis of inhibitory factors*.

Following bone marrow ablation, acute injury of kidney and liver, the respective organ regenerates by compensatory hyperplasia and/or activating tissue specific adult stem cells, depending upon the tissue and the nature of injury. These regenerations occur mainly due to the presence of favourable microenvironmental cues. Sometimes, despite the presence of competent cells and microenvironmental cues, tissues do not adequatly regenerate. This could be due to the synthesis of inhibitory factors by the damaged tissues or inhibitory factors delivered to the site of regeneration through blood circulation. Therefore, it is important to identify the molecules produced in the damaged tissue that are inhibitory in nature. *Autonomous regeneration* may not occur if the damage is extensive or chronic in nature. If the extent of damage is beyond the capacity of tissue regeneration, instead of spontaneous natural regeneration, the damage is repaired by *fibrosis* or *scar* tissue formation (due to over growth of fibroblast cells). The fibrotic growth adversely affects autonomous regeneration of tissue. Similarly, in case of chronic injury, fibrosis prevents normal tissue regeneration process.

Currently, the terminal cases of organ failure are taken care by transplantation of donor's organ, which is not discussed here due to limitation of the scope of this book. Other promising treatment modalities to minimize such problems are cell and tissue therapy. In these cases, tissue specific donor/host cells or regenerative-competent donor/host cells are transplanted in the damaged site. It may be noted that in case of donor cells, proper matching (major histocompatibility complex molecules) with the recipient cells is a prerequisite to avoid *graft rejection*. The most successful application of cell therapy is the hematopoietic reconstitution in both malignant and nonmalignant hematological disorders, where bone marrow or mobilized peripheral blood or cord blood cells are transplanted to the patients. Besides hematopoietic reconstitution, cell or tissue therapy is yet to gain the status of regular treatment method due to many drawbacks. In some cases, cell/tissue therapy protocol is awaiting for the approval of the regulatory agencies for clinical trials and/or routine treatment.

12.2 CELL AND TISSUE THERAPY IN REGENERATIVE MEDICINE

The concept of regenerative medicine is decades old. The new additions in the recent past has been the sources of cells, improved culture techniques, and methods of delivering cells to the diseased tissue. Figure 12.2 explains different cell types and alternate routes for tissue regeneration. It is important to keep in mind that most of these techniques are at different stages of development. The first alternative is transplanting organ specific cells isolated from the patient or live or cadaver donor. The second to fourth alternative are based on cells derived from adult-tissue, ES or iPS cells. In case of ES/iPS cells, it is mandatory to expand the cells prior to differentiation into target cells. However, unrelated (different from the target organ) adult stem cells may be directly transplanted, provided that the cells are regenerative-competent (possess inherent plasticity). These adult stem cells may be grown and differentiate *in vitro* prior to transplant to the patients. *In vitro* cell culture systems used are either *two-dimensional*

(2-D) or *three-dimensional* (3-D) support, depending on the nature of the tissue. For example, bone and cartilage tissues are better regenerated if cells are cultured on a 3-D support system rather than 2-D system. Again, in 2-D support, cells are expanded/differentiated in suspension (e.g., bone marrow cells) or in adhered condition (e.g,. neuronal cells) in tissue culture flask. On the basis of culture support system used and mode of transplantation, regenerative medicine can be classified into two broad groups—cell therapy and tissue engineering. In cell therapy, single cell suspension is infused through the intravenous route or directly injected in the damaged site (e.g., endocardium or, pericardium region) of the infarcted heart muscle and brain in case of Parkinson's patients. In contrast, tissue engineering deals with *in vitro* culture of cells in the form of tissue on 3-D scafflolds, and subsequent implantation in the affected areas. This Chapter will cover tissue engineering aspect.

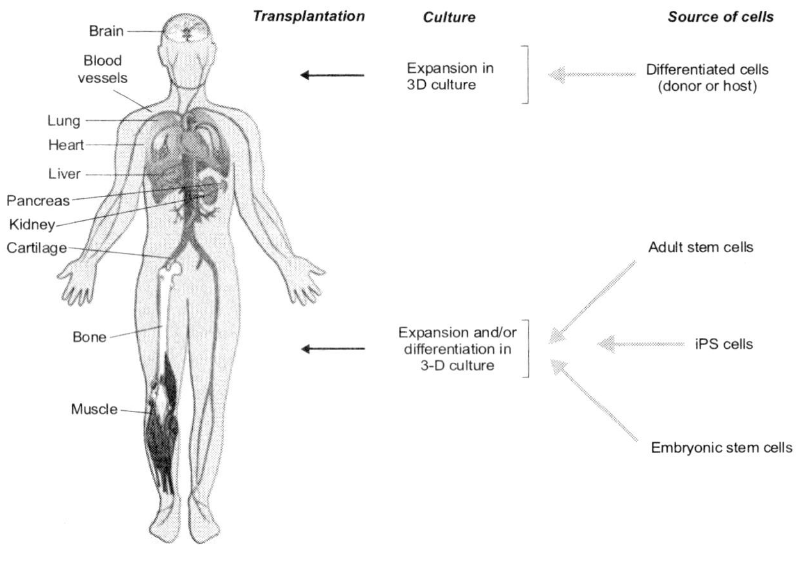

Figure 12.2 Potential alternate strategies for tissue engineering for the treatment of different orgam specific diseases. The sources of cells that can be transplanted are depicted.

12.3 TISSUE ENGINEERING

Tissue engineering brings together several specialized disciplines, including bioengineering, molecular and cellular biology, material science and clinical medicine, with the ultimate goal of producing tissue or organ substitutes. The spectrum of applications of tissue engineering is indeed wide from cartilage to living skin equivalent, artificial liver and pancreas to highly complex hematopoietic system. A successful *reconstitution* and *implantation* of tissue in human is a formidable challenge. Tissue engineering, or in other words tissue reconstitution, consists of the following four closely interlinked steps:

***Step* 1:** Biopsy is taken from the site of the damaged tissue, or cells are obtained from different sources. The desired cells are isolated and expanded in culture to produce a stock of resource material. The sources of cells may be from allogenic or unrelated adult tissue specific stem cells (Figure 12.2).

***Step* 2:** Cells are then seeded on a biomaterial frame called *scaffold.*

***Step* 3:** Once the scaffold has been seeded with cells, it is placed in a specially constructed bioreactor, which allows the cells to grow towards regeneration.

***Step* 4:** When the cells are stably grown within the scaffold, the cell-loaded scaffold is surgically implanted in the damaged site of the patient's body, where ideally they continue to grow and integrate with the surrounding host cells.

In tissue reconstitution process, all natural biological processes are maintained in their normal functions, only the extracellular environmental conditions are shifted. Thus, laboratory grown tissue is expected to be biologically and physically (mechanically) identical to the patient's own tissue. The interlinked steps in making engineered tissues are described in Figure 12.3. It will be easy to reconstract a tissue, if we understand the components of a tissue, that means different all types, biological states, extracellullar matrix components, and the relationship between each component.

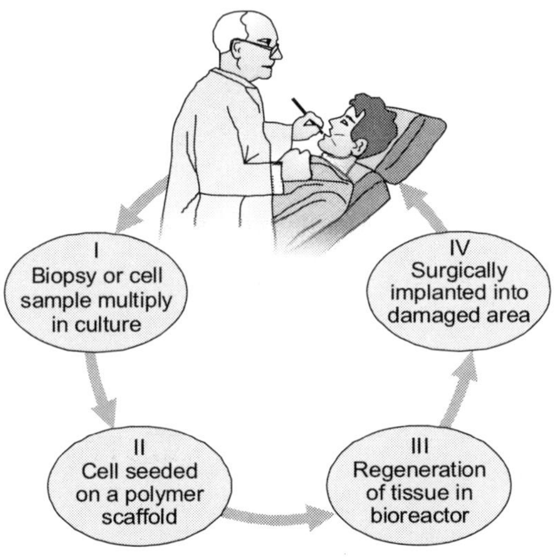

Figure 12.3 Four basic steps in tissue engineering: (I) isolation and expansion of cells, (II) seeding cells on biodegradable scaffold, (III) culture of tissue in a bioreactor, and (IV) implantation of laboratory grown tissue in the damaged site.

12.3.1 Tissue and its composition

The basic structure of a tissue, for example mature skin and its components are described in Figure 12.4. The primary function of mature skin is to provide a protective barrier to the environment. In addition, it forms appendages (hairs, nails, glands) that provide sensory and thermoregulatory functions. Structurally, skin can be envisaged as three distint layers: outer dermal layer (epidermis) in contact with middle *basement membrane*, and an inner dermis. Keratinocytes, the major epidermal cells, are derived from stem cells located in a basal epidermal layer. Keratin, produced by epidermal keratinocytes, protects epidermis from mechanical stresses, and in involved in the formation of stratum corneum. Stratum corneum cells (squames) anucleated flattened, filled internally with aggregated keratin matrix that originate from the basal layer. An impermeable cornified envelope that is cross-linked to the external lipid surrounds stratum corneum cells. The epidermis cells are adhered each other via adheren junctions and desmosomes, whereas, with the basement membrane, they are anchored via different integrins and other matrix-adhesion molecules.

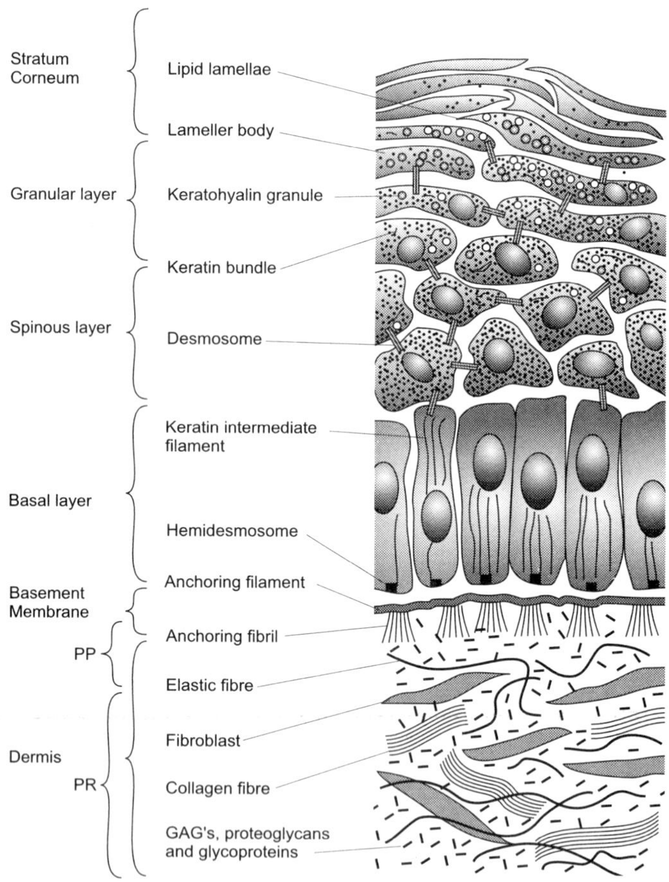

Figure 12.4 Sketch of adult skin. Skin has three distinct layers; dermin, basement membrane, and epidermin. The dermal layer is filled with different ECM and the major dermal cells, fibroblasts. Basement membrane divides dermis and epidermis layer, on which keratinocytes (epidermis cells) anchored with the help of different adhesion molecules.

The dermis provides mechanical support and nourishment to the entire structure of the skin and the parenchyma cells. The principal dermal cells are fibroblasts, responsible for the synthesis and secretion of dermal constituents. The extracellular space is confined between the dermal fibroblasts and its network of macromolecules. The extracellular matrix (ECM) such as collagen, elastin fibers, fibronectin, and an interfibrillar glycosaminoglycans/proteoglycan gel that impart massive water holding are confined between the space (Figure 12.4). Collagen (predominantly type I) provides tensile strength to the dermis layer. Deformation of collagen structure due to tensional forces is prevented by interconnecting bandes with elastin & fibrillin. The aqueous fraction of ECM allows diffusion of nutrients, hormones, and metabolites and creates *turgor* (rigidity due to abortion of water) pressure to withdraw considerable compressive forces. Molecular trafficking is regulated by the gel of different pore sizes, formed by assembling structurally divese proteoglycan chains. Proteoglycans also play a major role in cell signaling, binding different signaling molecules, etc.

12.3.2 Three-dimensional growth of tissues

Skin is a relatively simple two-dimensional (2-D) tissue with multilayer celluar arrangement, whereas most of the tissues and organs in our body are complicated. For example, cartilage, bone, muscle, etc. cannot be cultured *in vitro* like skin. Organs are more difficult to grow due to their structural complexity and specific biological functions. Most importantly, in organ, a structured network of blood vessels provide oxygen and nutrients for nourishing the cells. Tissue type growth of cartilage, bone, etc. *in vitro* is possible only when biomaterials are shaped into three-dimensional (3-D) scaffolds to provide mechanical support and guide for cell growth as new tissue. The scaffolds are expected to be highly porous to allow seeding of cells at high densities, and once implanted they should allow the formation of large numbers of blood vessels for supply of nutrients to the entrapped cells and the removal of waste products. The basic difference in tissue architect grown on 2-D and 3-D supports is shown in Figure 12.5. It should be remembered that 3-D growth is not possible unless cells are allowed to grow on a scaffold. There are several advantages of growing cells on scaffold for tissue engineering applications, which are as follows:

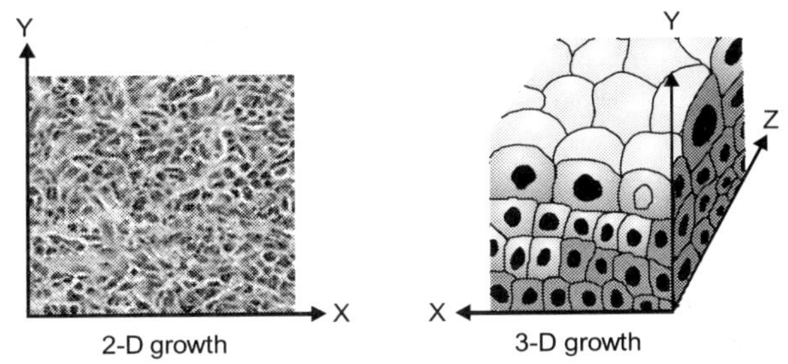

Figure 12.5 Nature of tissue growth on 2-D (left) and 3-D (right) support.

1. It provides large surface area for attachment of protein, and the growth of cells.
2. The 3-D scaffold allows for a spatial distribution of cell and other important ECMs that are analogous to that found in the counterpart tissue.
3. The increase of local volume for cell growth in the 3-D system may allow to establish microenvironments.
4. The 3-D scaffold provides desired cell-to-cell interactions, which are important for the maintenance, growth and differentiation of cells.

12.3.2.1 Biomaterials for 3-D growth of tissues

Biocompatible materials are used to fabricate scaffolds for growth of cells in the form of tissue like structure. Two different classes of biomaterials are commonly used in tissue engineering applications: *nonbiodegradable* polymers – materials that do not degrade by any biological actions, and *biodegradable* polymers – materials that are degraded at the site of application by enzymatic action or by *autohydrolysis*. For selection of biomaterials in manufacturing scaffolds, it is important to consider following properties in terms of material, engineering, biological and medical. Material and engineering properties provide information on surface properties, degradation rate, porosity, mechanical strength, and machinability/moldability of the polymers; whereas, biological and medical properties provide information on immune response, biocompatibility, toxic effect, support for growth of specific cell types, etc. The ideal materials for tissue engineering applications should not impart any adverse effect on the growth of cells. It should be non-immunogenic but imparts inflamatory respones for its own degradation, biodegradable (depending up on the applications), hydrophilic in nature, and having a desired mechanical strength. It should be easily processed into scaffolds of desired shape, size, and porosity. Every individual tissue reconstruct is unique interms of applications, therefore all biomaterials may not be suitable for the same purpose. A list of biomaterials for potential applications in tissue engineering, and their chemical structures are shown in Figure 12.6. Besides these, other biocompatible materials have been developed for specific applications, of which hydroxyapatite, tantalum-coated carbon material, hydrogels are important.

***Design of 3-D scaffold*:** The majority of scaffolds designed in tissue-engineering applications provide temporary support to the seeded cells to foster the growth. It is expected that during the initial growth phase, the scaffold degrades, and at the same time the growing tissue produces own ECM to maintain mechanical integrity. Balancing scaffold degradation with the growth of integrated tissue is crucial in support of bearing mechanical load, for example incase of bone and cartilage. It is important to study the degradation kinetics of the scaffold in implant-simulated conditions *in vitro*. The degradation of scaffold will depend on local pH and the release of enzymes from the growing tissue and macrophages migrated due to inflamation. Thus, uniform distribution of cells inside the scaffold is desirable for the steady degradation.

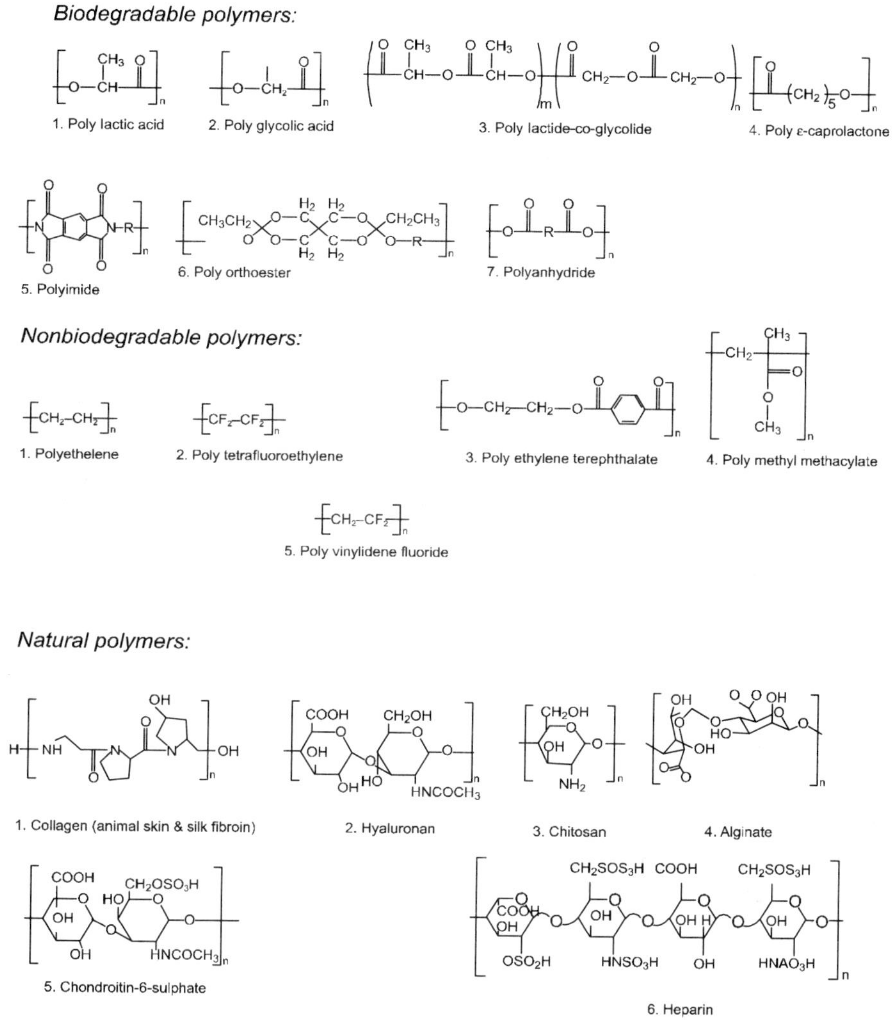

Figure 12.6 Potential biocompatible materials (biodegradable, nonbiodegradable, and natural polymers) for tissue engineering applications, and their chemical structures.

The ability to control the pore size, its distribution, and interconnection within a scaffold are key factors for the development of a scaffold. For designing scaffolds, few engineering or physical parameters are to be considered, these are pore size and shape, distribution of pore sizes, degree of interconnectivity of pores, surface area, mechanical properties of the scaffold, and the surface texture of the pore walls. Time to time, several alternate methods have been developed to create highly porous scaffolds from biodegradable polymers. The principles of these processes are described below:

Unwoven meshes (fiber bonding): This is the oldest method to manufacture scaffolds for tissue engineering applications. Polyglycolic acid (PGA) fibers are immersed in a poly L-lactic acid (PLLA) solution. The composite of PGA fibers embedded in PLLA is obtained after evaporation of the solvent, which is then heated to melt PLLA first and then PGA. Thus, fibers at the cross-point are joined to obtain highly porous structure. PLLA is removed by leaching with methylene chloride to obtain 81% porous structure with average pore size of 500 μm. Although the fiber bonding method gives highly porous interconnected scaffold, the residual solvent may be toxic to the cells.

Solvent casting/particulate leaching: This method also provides highly porous scaffold for tissue engineering applications. In this case, polymer (e.g., PLLA, PGA) is dissolved in chloroform or methylene chloride and mixed with NaCl. The solvent is allowed to evaporate to obtain polymer-salt composite. Later, the salt is leached out in water, leaving highly porous scaffolds. The scaffold's porocity can be controlled by increasing or decreasing the amount of salt in making the composite. In alternate, porous scaffold can be manufactured from a mixture of polymer (polylactic glycolic acid, PLGA) pieces and salt by compression molding into cylindrical form at just above the glass transition tempertrature of the polymer. The cylinder is then cut into discs of certain thickness, and later the salt is leached out with water to obtain highly porous scaffold.

Gas foaming: This technique includes both gas foaming and particulate leaching technology. In this case, ammonium bicarbonate is added to a solution of polymer in methylene chloride or chloroform. The resultant mixture is highly viscous, so it can be shaped according to the requirements. Later, the solvent is evaporated and the composite is immersed in warm water for concurrent gas evolution and salt leaching. High porosity (> 90%) with pore sizes ranging from 200-500 μm can be obtained by this technique.

Phase separation and emulsification: This technique is based on the concept of phase separation rather than using water-soluble salts in making the composite. Emulsion of water and solution of the polymer (PLGA in methylene chloride) is made prior to cast in a mold, which is then quenched in liquid nitrogen. The scaffold is freeze-dried at -60°C for removal of the dispersed water and polymer solvents. By this technique, about 95% porosity with much smaller pore sized (13-35 μm) scaffold can be achieved.

Porous matrix from collagen: Among the natural polymers, collagen is the most commonly used in manufacturing scaffolds for tissue engineering applications. There are several possible sources of type I collagen (e.g., dermis, tendons, bone, amnion, blood vessels, placenta of mammals, silk fibroin); all are useful for making scaffolds. The method for the extraction of collagen from dermis is as follows: collagen is initially extracted by acid treatment, and is then treated with pepsin to remove antigenic determinants situated in the *telopeptide* regions of the molecule. Collagen-based biomaterials are used in many forms: solution gel, sponge, hollow fiber tube, powdered sponge, rigid sponges, etc., depending on the nature of applications. The method for fabricating scaffold made of collagen and glycosaminoglycan (GAG) is shown in the flow diagram (Figure 12.7). The size

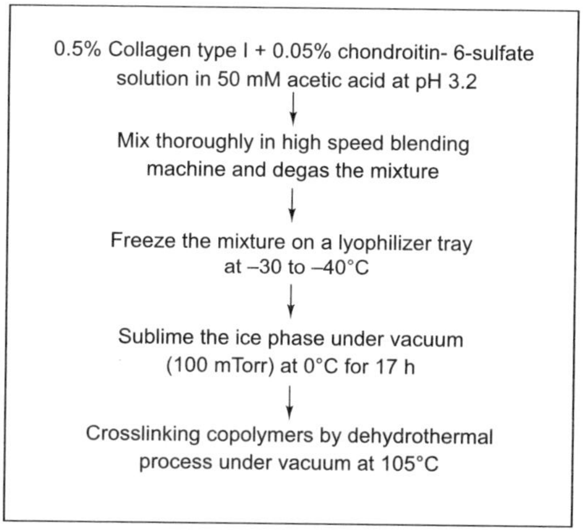

Figure 12.7 Fabrication of porous scaffold made of collagen and chondroitin sulphate copolymer. [Adapted from Ref. 33]

of the pores and their interconnection depends on the rates at which ice crystals are formed and sublimed. The scaffold is cross-linked for stabilization of its structure. In place of *dehydrothermal* process, glutaraldehyde solution (bifunctional reagent) can be used. However, care should be taken for the complete removal of glutaraldehyde or neutralizing its activity, as glutaraldehyde is highly toxic to the cells. Figure 12.8 shows some of the porous matrices have been developed for tissue engineering applications. Except tantalum-coated carbon matrix, all are biodegradable.

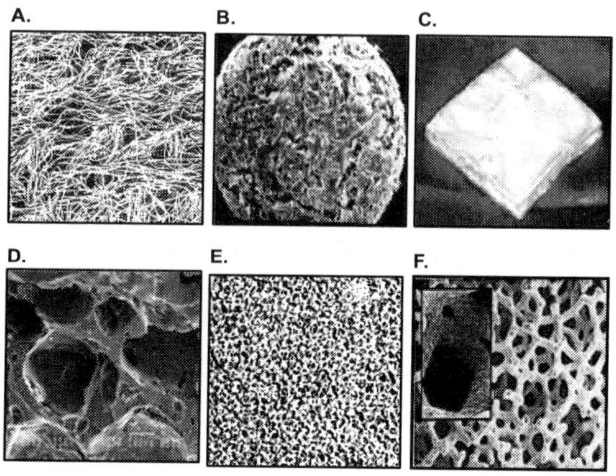

Figure 12.8 Different scaffolds for tissue engineering applications. **A.** Non-woven porous matrix made of PLA [*Source*: Langer RS, Vacanti JP (1999) Scientific American, April, p. 87]; **B.** Porous collagen composite scaffold made from bovine hide [*Source*: BD Biosicences, USA]; **C.** Collagen sponge scaffold [*Source*: Int. J Artif. Organs, 2001: 24, 50]; **D.** Poly(lactic-co-glycolic acid) porous scaffold [*Source*: Prof. W. Bonfield, Department of Material Sciences and Metallurgy, Cambridge Center for Medical Materials]; **E & F.** Porous tantalum-coated carbon matrix at different magnifications [*Source*: Cytomatics Corporation, USA].

12.3.2.2 Extracellular matrix (ECM)

Extracellular matrix components are natural support materials for the growth and differentiation of cells in the form of tissue. The main drawback of culturing cells on a conventional tissue culture flask is that they grow on synthetic materials rather than on their natural biological support. Growing of cells outside the body requires condition that mimic as closely as possible to the *ex vivo* environment. Therefore, it is essential to culture cells for tissue reconstruction on a natural extracellular framework. ECM holds cells and tissues together and provides a highly organized support in which cells can migrate and interact with each other. The matrix plays an active and complex role in regulating the behavior of cells that are in contact with it, thus influencing their shape, migration, proliferation and metabolic functions (Kratochwil, 1972; Folkman and Moscona, 1978; Hay, 1981). In contrast, cells growing on synthetic biomaterials lose many of their natural differentiated properties due to lack of interaction with ECM, provided that the cells do not produce their own ECM. To understand the role of ECM in maintaining natural differentiated phenotype, we will examine the culture of epithelial cells in the presence or absence of collagen layer on a plastic plate (Ingber *et al.*, 1987; Michalopoulos and Pitot, 1975; Reid, 1990). Epithelial cells maintain more differentiated phenotype in contacts with collagen than on plastic surface alone (Figure 12.9). Later, it was observed that a far more complex preparation of ECM, the Matrigel^R (composed of collagen IV, laminin, entactin, and heparan sulfate proteoglycan) exhibits better response in maintaining differentiated phenotype (Li *et al.*, 1987). Not only that, the extent of differentiation can be further improved if the Matrigel is allowed to retract from the solid surface, as compared with its attached form on the plastic surface (Li *et al.*, 1987). These series of study suggest that not only ECM components but mechanical force also play major regulatory roles in deciding between proliferation and differentiation of epithelial cells. It is believed that mechanical interactions between cell surface receptors and immobilized ECM molecules and corresponding change in cell shape are most critical determinants for switching between growth and differentiation (Ingber, 1993). It has been found that cultured endothelial cells secrete large quantities of ECMs on the plastic surface. These materials are comparable in organization and chemical composition to naturally occurring basement membranes on which cells migrate, proliferate, and differentiate *in vivo*. When these cultured endothelial cells are removed, the underlying ECMs components remain firmly attached on the plastic surface. If any other cells are placed on this ECM-coated plate, they attach rapidly and exhibit high plating and cloning efficiencies than plastic surface plate alone, as control. These cells can proliferate rapidly up to high saturation density, exhibit low serum and growth factors dependent, express differentiating functions, and show a longer life span. This study further demonstrates that ECM components are essential for equivalent *in vivo* activities of cells/tissues, whether produced by the cells itself, or supplied from outside.

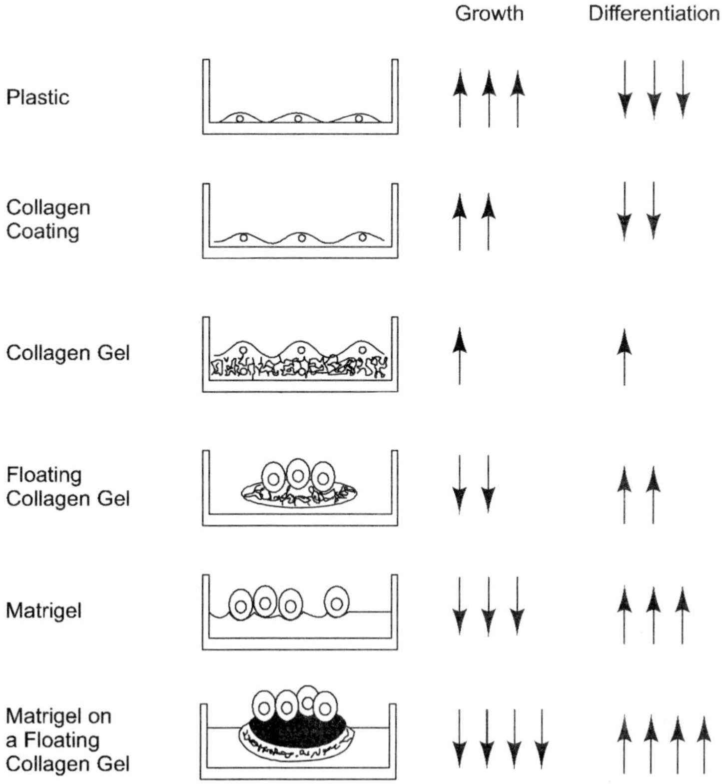

Figure 12.9 Effect of ECM components and malleability of substratum on growth and differentiation of epithelial cells. In presence of collagen-coated plate, the growth of cells is reduced as compared to the plastic plate. The malleable nature of adhered collagen gel and floating collagen gel further enhance differentiation process over the growth. Matrigel on floating collagen further improves differentiation than the gel form of marigel. [Adapted from Ref. 20]

The composition of different ECMs secreted by the cells may vary from one type to other. Table 12.1 describes the composition of ECMs and their respective functions in maintaining specific functional roles in cartilage, basement membranes, and endothelial cells. In general, ECMs can be divided into three major groups, which are as follows.

Table 12.1 ECM components present in cartilage, basement membrane, and endothelial cells.

Biological materials/cells	*Composition and function*	*References*
Articular Cartilage	About 1% chondrocytes of total volume of cartilage consists 60-80% water (by wt.). *Collagen-II, XI, IX* (60% of dry wt., of which 90% is type *II*): Provide mechanical integrity in cartilage. *Proteoglycans* (25-35% dry wt.): Interact with water, provide viscoelastic and compressive properties.	Buckwalter and Mankin (1997); Mayne *et al.* (1995); Netter Frank (1987).
Basement membranes: An amorphous sheet-like structure present between a cell layer and thick collagenous stroma (dermal-epidermal junction of human skin), between two layers of cells (renal glomerular and lens cells)	*Collagen IV and VII:* Meshwork scaffolding, containing multiple cell binding sites *Collagen VII:* Anchoring of dermal epithelial cells for underlying stroma. *Laminin:* Interact with collagen IV and entactin, and heparan sulfate proteoglycans; cell binding to ECM; involve in development, differentiation, nerve generation, and angiogenesis. *Entactin:* Binds to collagen IV, laminin, and fibronectin.	Husdon et al. (1995); Von der Mark and Goodman (1993); Timpl and Aumailley (1993).
Endothelial cells	*Collagen VIII:* Function unknown *Thrombospondin (TSP):* Attachment of keratinocytes; regulating proliferation of fibroblasts, smooth muscle, and endothelial cells; aggregating platelet; provide binding sites for fibrinogen, laminin, thrombin, etc. *SPARC/Osteonectin:* Binds collagen II, V, TSP, and cells. (also located in bone and basement membrane).	Sage and Bornstein (1995); Von der Mark and Goodman (1993); Heingard and Oldberg (1993).

A. Collagens: It is most abundant in animal kingdom, representings about one-third of all protein in tissues. So far, 18 structurally different collagens have been identified; except few, most of them are available in minute quantities in the cells. Students are suggested to consult the references for detail study on structures and functions of different collagens.

B. Proteoglycans: Proteoglycans are available on the surface of the cells as well as in the ECMs. Cell surface proteoglycans have transmembrane and cytoplasmic domains. Proteoglycans consist of a protein core and one or more negatively charged glycosaminoglycan (GAG) side chains covalently attached to the core protein (Heinegard and Oldberg, 1993). GAGs are long unbranched polysaccharides containing a repeating disaccharide unit, contain either of the two modified sugars: N-acetylgalactosamine (GalNAc) or N-acetylglucosamine (GlcNAc) and an uronic acid such as glucuronate or iduronate. The general structure of GAG linkage to protein in proteoglycans is shown in Figure 12.10. Proteoglycans are involved in maintaining the transparency of the corneas, the tensile strength of skin and tendon, the viscoelasticity of blood vessels, and the compressive properties of the cartilage. When proteoglycans contain many GAG molecules, they become highly polar, allowing strong interaction with water molecules. In ECMs, proteoglycans may contain three different sulfated GAGs: (1) chondroitin/dermatan sulfate, a polymer consisting alternate galactosamine and glucuronic/iduronic acid units (present in bone, cartilage, heart valves, and neural tissues); (2) heparin/heparan sulfate, a polymer consisting of alternating glucosamine and glucuronic/iduronic acid units (present in basement membrane, components of cell surfaces, and in liver and Schwann cells); (3) keratan sulfate, a polymer consisting alternate glucosamine and galactose units (present in cartilage, bone, and cornea).

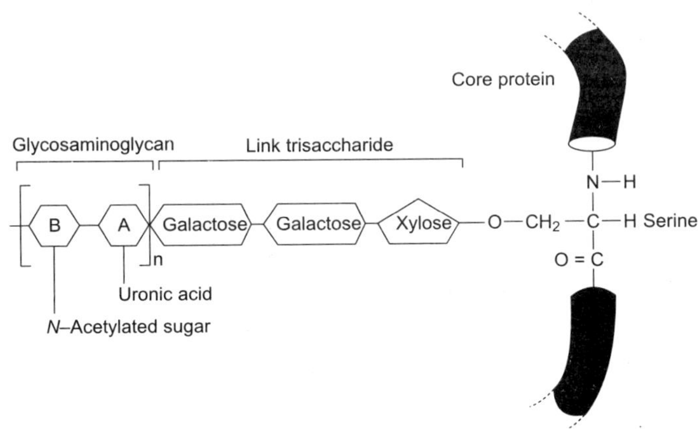

Figure 12.10 Structure of GAG linkage to protein in proteoglycans.

C. Structural glycoproteins: The members of this class are heterogeneous in size, structure, and tissue distribution (von der Mark and Goodman, 1993). There are eight structural glycoproteins found in ECM:

(1) Fibronectin—contains two subunits of molecular mass 220 kD (produced by hepatocytes, fibroblasts, epithelial and endothelial cells)
(2) Laminin—composed of α, β, and γ chains of combined molecular mass 850 kD (found in basement membranes)
(3) Thrombospondin—contains three identical chains of 150 kD molecular mass (produced by platelet, endothelial, glial, smooth muscle and fibroblasts)
(4) Entactin—a 150 kD glycoprotein (found in basement membranes)
(5) Tenascin—a hexameric glycoprotein of molecular mass 1240 kD (present in skin, bone, cartilage, tendon)
(6) Osteonectin—a 32 kD protein (found in basement membrane, produced by endothelial, fibroblasts, osteoblasts and smooth muscle)
(7) Vitronectin—a 75 kD protein (present in liver and platelets)
(8) Elastin—the elastic fiber which confers flexibility and distensibility of tissues (found in thoracic aorta and skin).

12.3.2.3 Biological roles of ECM components

Some peregrine signals in mammalian cells are modulated by ECM components via *cell adhesion molecules* (CAM). CAMs are cell surface molecules that involve cell-to-cell and cell-to-ECM interactions, thereby mediated the process of adhesion, migration, differentiation and intracellular signaling (Gumbiner, 1996). Four classes of cell-adhesion molecules are found to interact between two cells, and between cells and the surrounding ECM. These are integrins, cadherins, members of the immunoglobulin, and selectin families. The integrin molecules mediate interactions of cells with the ECM components. Following the activation of integrin receptor with a ligand (ECM component), the cytoplasmic domain of it recruits cytoskeleton and other cytoplasmic proteins, which anchor newly formed complexes to the actin cytoskeleton. This ultimately leads to the formation of special additive structures known as focal adhesions.

Integrins control the cell survival of endothelial and epithelial cells. Detachment of these cells from ECMs results in apoptotic death, even in the presence of serum. Apoptosis induced by cell detachment is known as anoikis, the Greek meaning of homelessness (Frisch and Francis, 1994). Many cell signaling molecules (e.g., Akt/PKB kinase, FAK, Ras/Erk) that regulate cell survival through different receptors are found to be activated by adhesion of cells through integrin receptors. Detachment of cells from ECMs shows inactivation of these signaling molecules and in the activation of pro-apoptotic proteins (e.g., Bax, caspases) or death ligands (e.g., Fas, TRAIL), and inhibition of anti-apoptotic protein (e.g., Bcl-2) expression (Zhang *et al.*, 1995; Aoudjit and Vuori, 2001; Rytomaa *et al.*, 1999).

Studies in animal cells showed that cell cycle progression requires attachment of adhered cells to ECM. Synthesis of cyclin A, cycline D, activation of Cdks, as well as the degradation of p21 and p27 takes place when integrin molecules are engaged with the counter ECM components (Schwartz and Assoian, 2001; Huang and Ingber, 1999). Different ECM components produced by animal cells, and their counter integrin receptors are shown in Table 12.2.

Table 12.2: Major mammalian ECM components and their counter receptor

ECM ligand	*Integrin subunit*
Collagen, laminin	β1 α1
Collagen, laminin, tenascin	β1 α2
Collagen, laminin, entactin, fibronectin	β1 α3
Fibronectin, VCAM-1	β1 α4
Fibronectin	β1 α5
Laminin	β1 α6
Basal lamina	β1 α8
Fibronectin	β1 αv
ICAM-1, 2, 3	β2 αL
Vitronectin, fibrinogen, fibronectin, TSP	β3 αIIb
Vitronectin, fibrinogen, fibronectin, TSP, vWF, osteopondin, collagen	β3 αv
Laminin	β4 α6
Vitronectin, osteopondin	β5 αv
Fibronectin, tenascin	β6 αv
E-cadherin	β7 αE
Vitronectin	β8 αv

12.3.3 Mechanical Force and Tissue Remodeling

How mechanical force influences cell fate determination has been the subject for many years. It is now known that bone remodelling is a dynamic process involving osteoblast and osteoclast cells. These two cells respond to changes in mechanical force and remodel bone formation to minimize local stress. The influence of mechanical environment on tissue growth and remodeling can be further explained by another example of *vasculogenesis*. Vascular endothelial cells (ECs) are exposed to pulsating mechanical environment caused by blood flow. In veins or arteries, ECs experience two kinds of forces—tangential (shear stress), due to blood flow, and normal stress, due to pressure. *In vitro* study showed that non-reversing pulsating flow dramatically inhibits the proliferation of EC compared to the static culture or culture under steady state flow (Levesque *et al.*, 1990). The average wall shear stress in arterial vessels is maintained at 15-20 dynes/cm^2 by an autoregulatory mechanism (Kamiya and Togawa, 1980). Endothelial cells autoregulate wall shear stress by changing blood vessel size through the recruitment of smooth muscle cells in arterial wall. In the presence of shear stress, ECs express more platelet derived growth factor (PDGF), a mitogen for smooth muscle cells (Zerwes and Risau, 1987).

Mechanical stimuli also play a critical role in determining muscle mass and its phenotype in both skeletal and cardiac muscles *in vivo*. Chick skeletal muscle cells become hypertrophy instead of dividing if the elastic substratum, on which the cells are cultured, is stretched. Similar observation was noted in case of neonatal cardiac myocytes in culture (Komuro *et al.*, 1991). Stretch causes the activation of several kinases in cardiac myocytes, leading to the transactivation of *c-fos* protooncogene. Articular cartilage is another tissue which undergo changes in its structure and composition in the presence of diverse mechanism stimuli (Tammi, 1987). Now, it is understood that most load bearing tissues remodel in response to mechanical stimuli, by which they gain the ideal characteristics of the tissue. Therefore, it is beneficial to consider the mechanical stimuli on these tissues while cultivating *in vitro*.

Integrin molecules transmit mechanical stresses to the cells, which may change with the rigidity in ECMs. On floating collagen gel, epithelial cells tend to differentiate, whereas collagen-coated culture dish promotes proliferation (Figure 12.9). Detachment of collagen gel from solid surface allows their contraction by the cells into small rafts, which causes a release in tension, thereby promoting differentiation of epithelial cells in place of proliferation. It has been hypothesized that rigid surface promotes high contractility, leading to accumulation of focal adhesion proteins at the site of attachment. This results in increase of phosphorylation of *focal adhesion kinase* (FAK) and other focal adhesion proteins, and corresponding stimulation of growth factor-induced activation of MAP kinase pathway, leading to the proliferation of the cells (Schoenwaelder and Burridge, 1999; Schwartz and Ginsberg, 2002). In case of malleable surface. MAP kinase activation does not increase beyond the level with growth factors, thus differentiation process is favored over proliferation.

12.4 BIOREACTORS IN TISSUE ENGINEERING

The culture of tissue in plastic plate or in 3-D scaffold does not ensure that the growing tissue will be analogous to that found in mammals. A functional tissue means that it possesses all specialized properties of the tissue. Each cell of the tissue constantly experiences mechanical stimuli; even in static culture they are exposed to the gravitational forces. Physical stimuli such as shear stress, fluid-flow, compression and tensile not only alter the organization and distribution of structural elements and organelles within cells, but also transduce signal inputs that modulate signaling pathways within or between two cells. For some tissues, exposure to mechanical stresses, while growing, is necessary to accomplish the functional properties, otherwise they become useless in tissue reconstruction. Similarly, not all types of forces are effective for the maturation of tissue, e.g., gravitational force affects the growth and maturity of cartilage.

In manufacturing porous scofflods, solvent casting followed by particulate-leaching technology does not have accurate control over pore size, their geometry and distribution and inter-connection. It has been shown that in gas foaming technique, only 10-30% of the pores are interconnected. Besides above limitations there are many drawbacks that are associated with the present forms of scaffold, these are toxicity of residual solvent, non-uniform distribution of cells, limited diffusion of oxygen and other nutrients, and toxic effects of trapped metabolic wastes. *In vivo* studies in animal have revealed that any tissue, beyond 500 μm away from the vascular system, suffer from the limitation of oxygen and other nutrients. This problems become critical in case of tissues growing *in vitro*.

The problem of diffusion of oxygen through the scaffold has been explained in Figure 12.11. The porous scaffold is open on all sides, through which oxygen and nutrients can easily diffuse. In general, cells are seeded on the top of the scaffold, on which they adhere and proliferate and finally migrate through the pores to the interier of the scaffold. The cells on the top layers consume most of the oxygen and nutrients; again these cells secrete ECMs. The ECMs secreted by cells may prevents diffusion of oxygen from the upper surface.

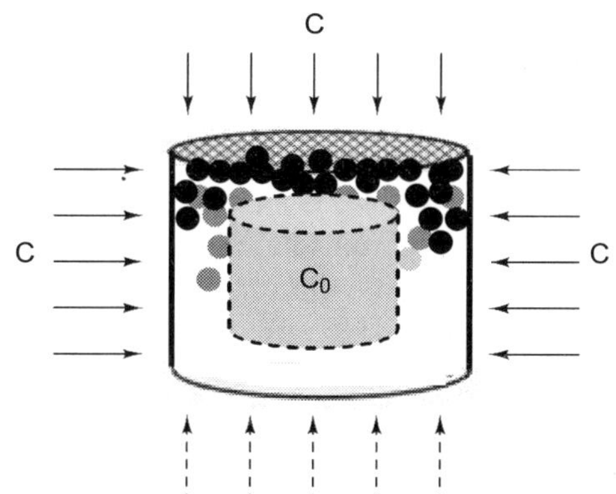

Figure 12.11 Transfer of oxygen across the thickness of the scaffold. In static culture, the oxygen concentration at the solid-liquid interface of the scaffold surface (except bottom) is assumed 'C', which is at equilibrium with that in the Liquid. Oxygen and other nutrients can diffuse inside the scaffold up to certain length. Due to diffusional limitation, a hypoxic zone is developed in the center of the scaffold, which could be as low as zero, 'C_0'. The black solid circles are cells growing in the presence of adequate oxygen level.

Eventually, the migration of cells is also stopped, and those already invaded into the matrix will die due to hypoxia. Oxygen and nutrients may diffuse from other sides of the scaffold, but very few cells actually utilize this, as cells cannot migrate from the top surface. The diffusion problem is amplified in a static culture. As a result, nutrient limited, no-cell zone is developed in the center of the scaffold, the volume of which increases with the thickness of the scaffold. The oxygen concentration may become zero (C_0) at the no-cell zone of the scaffold (Figure 12.11). Since the diffusion of oxygen in medium is generally low and oxygen is consumed at a high rate, very soon it becomes the limiting factor for survival of the growing cells within the scaffold. On the other hand, nutrients like glucose, amino acids can diffuse easily. The overall effect will be no growth of cells beyond 200 μm distances from the surface they are applied. In bone tissue engineering, the high rates of nutrient and oxygen consumption at the surface of the scaffold promote mineralization of cells, which further limits the mass transfer to the interior of the scaffold (Martin *et al.*, 1998). Osteoblasts seeded on porous scaffold form a viable tissue of thickness within 200 μm (Martinv *et al.*, 1998). Similarly, the islet cells undergo necrotic death when oxygen diffusion path exceeds 100 μm. Cartilage is an exception, which can maintain viability in avascular area up to 1000 μm thickness (Guyton and Hall, 1996). This is possible, as chondrocytes require less oxygen than osteoblasts and islet cells. Transport of oxygen and other nutrients inside the scaffold can be improved by increasing the convective flow of medium across the scaffold. In summary, the diffusion limitation of oxygen and inadequate mechanical force (e.g., shear stress, fluid-flow, compression and tension, gravitational effect) on tissue growing in static culture may yield constructs different from the natural sources.

Over the past few years, several bioreactors have been developed to overcome these limitations of growing cells/tissues in culture dish. The bioreactors have been broadly divided into two groups—rotating and non-rotating types. In rotating bioreactors, the culture chamber rotates permanently in which tissue matrices are either kept fixed or in floating condition. In case of floating tissue matrices, the rotation speed of the bioreactor is adjusted to mimic the free-falling state of the growing tissues. Free-falling state protects fragile tissues by lowering shear stress and avoiding contact between cells and the vessel wall. Since the vessel rotates continuously, the diffusional limitation of nutrient will be minimized. The culture medium is exchanged batchwise at definite intervals to avoid the limitation of key nutrient and accumulation of toxic metabolites. Some of the rotating bioreactors used for tissue-type growth of cells on scaffolds are shown in Figure 12.12. NASA microgravity bioreactor is one such system where cartilage and mammary epithelial cells have been grown as 3-D constructs on scaffold (Fig. 12.12B, D, E).

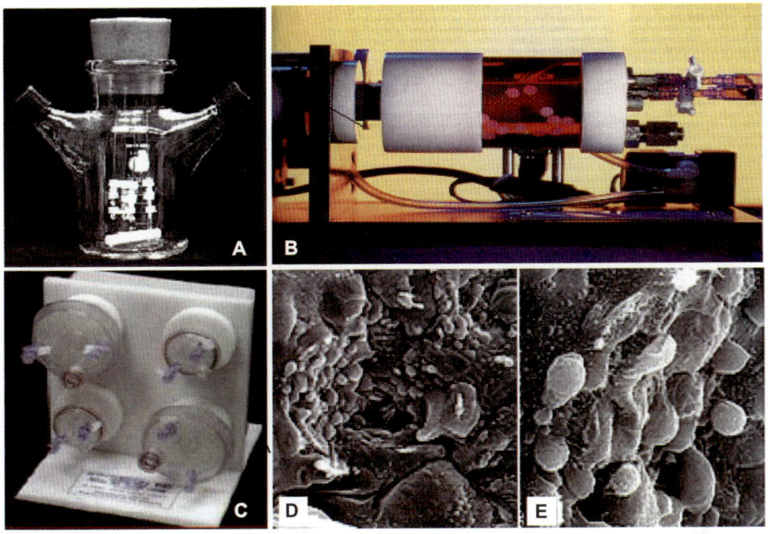

Figure 12.12 Rotating bioreactors for tissue engineering application. **A.** Spinner flask bioreactor; **B.** NASA μ-gravity bioreactor; **C.** Synthecon[R] rotating wall vessel bioreactor [*Source*: Synthecon Inc, Houston, Texas]; **D & E.** Human mammary epithelial cells grown in 3-D constructions on attachment beads in the NASA Bioreactor. [*Source*: Dr. Robert Richmond, NASA/Marshall Space Flight Center]

Non-rotating bioreactors are developed based on perfusion columns or chambers, where the tissue matrices are kept in fixed positions and the medium is continuously circulated through the tissue (Fig. 12.13A). The physical conditioning of some of the engineered tissue constructs relies on hydrodynamic shear forces. It is believed that hydrodynamic force affects on growing tissue via pressure fluctuations that stretches the cell membrane. Hydrodynamic shear force induced bioreactor has been used to grow tissues up to 5 mm of thickness (Freed *et al.*, 2000). Another kind of bioreactor has been developed in which both hydrodynamic and mechanical stimulation can be applied on growing tissues (Fig. 12.13B). The EnduraTEC ITEMS (Intelligent Tissue Engineering via Mechanical Stimulation) Bioreactor is such a system that has been used for culturing blood vessels and heart valves. The same *multi-station bioreactor* has the capability of directly stimulating six scaffolds simultaneously for cartilage tissues. A novel pulsatile perfusion bioreactor has been developed to grow small-caliber arteries on PGA scaffold (Fig. 12.13C). Here, the vascular graft is constructed in two stages—the tubular PGA scaffold, secured in the bioreactor, is first inoculated with smooth muscle cells. The bioreactor operates under *dynamic pulsatile radial force* through distensible silicon tube inserted in the lumen of the PGA tube. After 8 weeks of culture, a smooth muscle cells layer replaces the PGA tube, which looks identical to that of a natural artery (Niklason *et al.*, 1999). In the second stage, the silicon tube is removed, endothelial cells are added in the lumen and further cultured for another 2-3 weeks to obtain an artificial artery. In the second stage, the luminal medium flow rate is gradually raised to 0.1 ml/s, and maintained throughout the culture period. The laboratory-grown artery has a burst pressure of 2000 mmHg and exhibits serotonin and endothelin-1 vasoactivity.

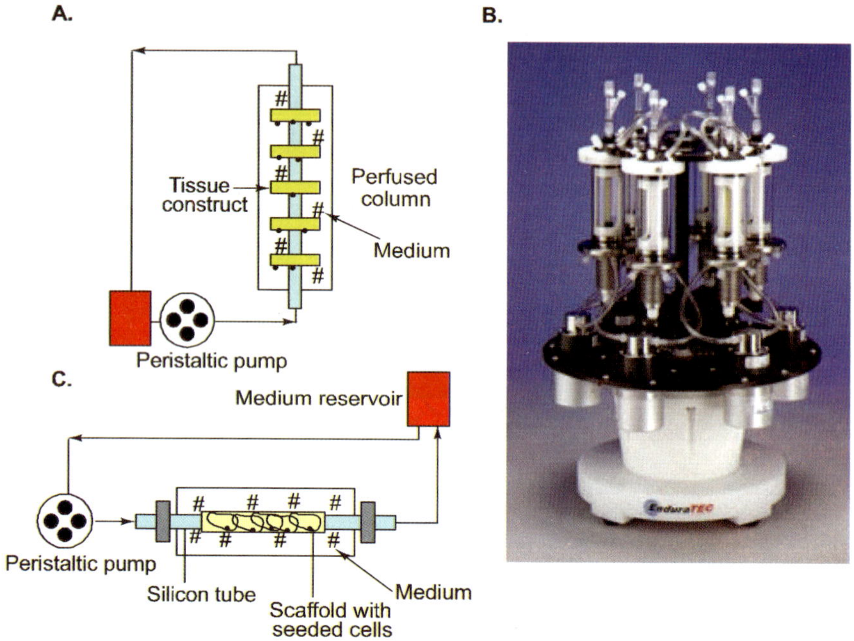

Figure 12.13 Non-rotating bioreactors for tissue engineering application. **A.** Perfusion column bioreactor [Adopted from Ref. 7]; **B.** EnduraTEC ITEMS bioreactor [*Source*: Reed Elsevier Inc., 2004]; **C.** Pulsatile radial flow bioreactor. [Adapted from Ref. 32]

The comparative merits and demerits of rotating and non-rotating bioreactors for different applications of tissue construction are shown in Table 12.3.

Table 12.3 Merits and demerits of rotating and non-rotating bioreactors

Type of tissue	*Rotating bioreactor*	*Non-rotating bioreactor*
Cartilage	Nutrient and gas exchange is optimum, but mechanical stress on growing tissue will be inadequate	Nutrient and gas exchange is optimum, adequate mechanical stress is possible, complex shape of the tissue can be achieved
Cardiac tissue	Free-falling environment protects fragile tissue	Excessive mechanical stress
Blood vessel	Insufficient mechanical stress and strain	Mechanical stress and strain (pulsatile nature) can be imparted in the tissue, complex tissue shape possible
Heart valve	Inadequate mechanical stress	Complex shape possible

12.5 CURRENT APPROACHES IN TISSUE ENGINEERING

12.5.1 Chemical modification of surface

Two important considerations are kept in mind while designing new biomaterials—cells should be uniformly distributed in the scaffold and able to migrate inside the scaffold. For tissue grafting, the time required for cells to spread and migrate inside the scaffold should match with the ingrowth of new blood vessels to support the growth of new tissues and their survival. It is known that the surface chemistry of the material plays a significant role in adhesion and spreading of cells. Another factor that affects the behavior of cells is presence of serum proteins on scaffold surface. Mammalian cells are inherently adhered on the serum proteins via cell surface integrin receptors. Pre-adsorbed serum proteins on PLGA, and collagen on synthetic scaffold make them more conducive for adhesion and migration of epithelial cells. Exogenous collagen not only improves adhesion and migration of cells, but also allows cells to secrete their own ECMs (e.g., fibronectin), which in turn play role in regulating adhesion and migration (Seal et al., 2001).

The modification of biomaterial surface has been widely studied in the context of regeneration of different tissues. It is possible to modify the surface of a material favourable for adhesion of a specific cell type the mechanical properties of the material can also be improved. For example, cell adhesion peptide, RGD (-arginine-glycine-aspartic acid-) and a heparin binding domain, FHRRIKA (-phenylalanin-histidine-arginine-arginine-isoleucine-lysine-alanine-) have been grafted on metal oxide and polymer surface to make them more biocompatible (Healy *et al.*, 1999). RGD improves endothelial cell adhesion and its migration. Again, incorporation of different bioactive molecules with scaffold polymer improves the growth and differentiation of cells. One such molecule is bone morphogenic protein-2 (BMP-2) involves in bone formation. The formation of bone is found to be significantly improved when BMP-2 is incorporated with type I collagen-coated porous PLA scaffold (Winn *et al.*, 1998). Similarly, local release of acidic fibroblast growth factors (aFGF) in vascular grafts, initially grown on Dacron or ePTFE, stimulates the proliferation of endothelial cells.

12.5.2 Nanotechnology in fabricating scaffold

Proteins present in ECMs of mammalian tissues are nano-structured. For example, the diameter of fibrillar collagen fiber ranges from 50 to 500 nm; laminin is about 70 nm; vitronectin and fibronectin are further lower in dimensions. Due to the presence of numerous nano-structured materials in the body, it is believed that cells in culture will also behave naturally if they interact with surfaces that have a large degree of nanometer roughness. It has been hypothesized that nano-structured scaffolds in tissue engineering applications will behave superior than the presently used macro level structures (>100 μm). Again, the nanophase materials are shown to improve surface, mechanical, optical, and electrical properties compared with the existing biomedical implants. Attempts have been made to design scaffolds in subcellular (0.1-10 μm) and nanoscale (1-100 nm) levels for better realization of cellular control.

Three different ways have been tried to fabricate nanoscale scaffolds. These are known as atom-by-atom, molecule-by-molecule (also known as top-down), or self-assembly (also know as self-organization or bottom-up). The first two techniques will not be discussed in this book; students are advised to consult references. The principle of *self-assembly technique* for fabricating nanoscale scaffold has been briefly explained in this section. Self-assembly method is based on the principle of self-organization of predefined components into an ordered superstructure, as the living cells synthesize nucleic acids or proteins. How a self-organization structure is formed can be better explained by the formation of bone. Bone is an inorganic-bioorganic composite material consisting of mainly collagen proteins and hydroxyapatite (a crystalline form of calcium phosphate). The first attempt to mimic the formation of natural bone has been through self-assembly in which collagen induces calcium, phosphate, and hydroxide to form hydroxyapatite crystallites (Hartqeriak *et al.*, 2001; Hartqeriak *et al.*, 2002). In order to obtain the composite of collagen and hydroxyapatite, a small self-assembling molecule with a hydrophobic alkyl tail and a hydrophilic oligopeptide head was designed. The amphiphilicity of these molecules creates micelle structures *in situ.* A row of four cysteine residues on adjacent molecules forms disulfide bridge at low pH, leading to the formation of crystals of size 200 kDa polymeric nanofibrils. On the surface of the fibrils, flexible glycine residues efficiently present both phosphoserine residue, which induces nucleation and growth of mineral, and a peptide sequence (RGD), which promotes cell adhesion. The resultant mineralized matrix (10–50 nm in length) is structurally similar to the mature bone matrix. The collagen matrix recognizes undifferentiated bone-marrow stem cells, which differentiate into bone-forming osteoblasts in response to signals from bone-specific proteins present in the matrix.

12.6 ORGAN CULTURE

Organ culture means the cultivation of whole organs or parts, by which the structural organization of individual cell types and their organization as a whole, as they are maintained *in vivo*, are restored. For example, glandular structures are maintained in case of glands, small bronchi in case of lung tissues, etc. Due to that organ culture study has several advantages over the classical cell culture method where tissue architecture cannot be maintained, cell-to-cell interaction is reduced and extracellular matrix is not adequately synthesized by the cells. Thus, the organ culture is able to closely model the functions of an organ in various states to perform studies like (a) morphogenesis and differentiation during embryonic development, (b) effects of chemical and physical agents (*e.g.*, to hormones, drugs, or radiation), (c) growth and behavior, and (d) organ replacement.

12.6.1 Alternate techniques of organ culture

There are many alternate techniques that are developed for organ culture, these are: (a) plasma clot or watch glass method, (b) agar gel method, (c) raft method, (d) grid method and (e) cyclic exposure to medium/gas phase method. Plasma clot is the oldest technique for organ culture where the explant is placed on a mixture of plasma and chicken embryo extract, kept in a watch-glass. Watch-glass can be kept in a Petri plate placed on a humidified CO_2 incubator. This method of organ culture allows microscopic tissue growth and cellular differentiation. There are limitations in plasma clot method, like low of oxygen tension, exhaustion of nutrients, etc. In agar gel method, the explants is cultured on 1% agar gel in physiological salt solution containing serum and defined mixture of amino acids and vitamins. Like plasma clot it is also close culture technique, so the explant is sub-cultured in fresh agar gel. This method was adopted to examine many aspects of organ development and tumor growth. In third method, the explant is placed on a raft of rayon acetate, which is floated on serum in a watch-glass. The modified form of this is known as grid method, where a raft containing the explant is placed on a metal grid. The grid is immersed in culture medium inside of a petry plate. The plate is then incubated in a CO_2 incubator. This system can be adopted to study growth and differentiation of embryonic tissues. Better method of organ culture is cyclic exposure to medium and gas, where the limitation of oxygen and nutrient supply is taken care by rocking the culture flask where the explants is attached.

12.6.1.1 Fetal thymus culture

Fetal thymus is the organ where functional T cells are formed from the progenitor $CD4^+CD8^+$ population by negative/positive selection through a series of developmental events. In this regards, other than progenitor T cells, thymic stromal cells play most critical role. Thus, fetal thymus culture is considered as an important tool to study interactions between these two cell types, which otherwise difficult to proceed *in vivo*. The major steps for initiating reaggregate thymus organ culture (RTOC) are described below; students are advised to consult the reference section: (Andersoon & Jenkinson, 2007)

Step I: (dissection of thymus lobes)

1. Asceptically removes the uterus from pregnant mice (E14 – E16) and transfer to a 90-mm Petri plate. Surgically remove the embryos and wash in PBS.
2. Transfer the embryos in a new Petri plate containing RF10-H medium, decapitate the embryos and remove entire thoracic tree (heart, lungs, trachea, thymus lobes) under a dissecting microscope, place them in a new Petri plate containing RF10-H medium. Each thoracic tree will have two thymus lobes, remove them by the help of a forceps.
3. Endogenous thymocytes from the lobes are eliminated by culturing the lobes in the presence of 1.35 mM 2′-deoxyguanosine (dGuo) at 37°C for 5-7 days. After that the thymocytes free lobes, containing thymus stroma

ECM components, are transferred to a dish containing prewarmed RF10-H medium.

Step II: (initiation of RTOC)

4. About 20 lobes are taken in a 1.5 ml microcentrifuge and washed properly with PBS. Resuspend the lobes in 0.5 ml prewarmed 1 x trypsin solution and incubate at 37°C for 10 min. After complete dispersion of the lobes, neutralise the trypsin by adding equal volume of RF-10H medium supplemented with 10% FBS.
5. Centrifuge cell suspension, resuspend in the same medium and enumerate cell number.
6. Mix equal number of $CD4^+CD8^+$ thymocytes and stromal cells in a 1.5 ml microcentrifuge tube (total cell number 1.5 to 2.0 × 10^6), centrifuge and resuspend in prewarmed DMEM.
7. Transfer 30 µl cell suspension to a single Terasaki well, invert the plate so that the medium forms a hanging drop. Incubate the culture at 37°C in humidified 10% CO_2 incubator. Cohorts of $CD4^+CD8^-$ and $CD4^-CD8^+$ cells are first appeared and around in 5 days of culture.

12.6.1.2 Organ culture on decellularized matrices

It is now well accepted that complex processes of differentiation, proliferation, tissue-specific gene expression, morphogenesis and cell migration are controlled by interactions between cells and ECM and maintained the tissue architecture. Three-dimensional native structural features with proper tissue assembly are critical for an organ to function like naturally grown organ. Recent studies have shown that decellularization of whole organs, like kidney, liver, heart and lung are possible in rodent, which following recellularization are expected to perform like a complete organ. The main objective of organ culture on acellular scaffolds is to manufacture complete organ *in vitro* to fill up the demand and supply gap. The basic steps are as follows:

12.6.1.2.1 Step I: Decellularization of organ

Complex structures of various organs cannot be entirely created on artificial scaffold ECM, therefore researchers have attempted to recover natural scaffolds from organs of small or large animals. The first step of organ tissue engineering is development of a successful decellularization process. The objective of effective decellularization is to minimize undesirable effects that are associated with cell removal techniques on the three dimensional (3D) architecture, while ensuring that all immunogenic cells are removed including DNA, cellular proteins and even membrane fragments. Nonetheless, left-over decellularizing chemicals may cause harm to the cells during cellularization step, so entire scaffold requires through cleaning. Alternative techniques followed in decellularization process have been documented in the literature, students are advised to consult references. In this book, the method of organ culture has been restricted to rats and mice, whereas decellularization process is restricted to the use of detergent only. Following precautions are to be taken during decellularization of an organ.

1. The effectiveness of decellularization process is vastly depend on nature of the tissue. The degree of success of a decellularization process then depend on many other parameters, such as preservation of vasculature within the organ under treatment, maintenance of mechanical properties and 3D shape of the organ, and restoration of the chemical compositions of decellularized membrane same as native membrane.
2. Most commonly used detergent for decellularization process are SDS (ionic detergent) and Triton X-100 (non-ionic), besides reagent like sodium deoxycholate and CHAPS. The concentrations of these reagents and the duration of perfusion changes from tissue to tissue. It is essential that all solution is prepared in either sterile tissue culture grade water or filtered sterilized before use. The decellularization process should be conducted in side of a biosafety cabinate.
3. As harsh treatment may be detrimental, the decellularized organ scaffold is evaluated by various techniques like DNA content and PCR analysis to know remnant cellular genomic materials, histological and immuno-histochemical analysis to know ECM components of the scaffold, surface structure analysis by scanning electron microscopy. Soluble collagen, elastin, and glycosaminoglycans are quantified by colorimetric analysis. It is also necessary to determine the trances of decellularizing chemicals, if any, present in the scaffolds. For example, residual SDS is quantified by Stains-All assay. It is also important to test the patency in vasculature by injecting dye through the main supply vein of the organ. The basic steps of the decellularization processes for few organs are described below:

A. Decellularization of liver (taken from Basak et al., 2011; Hiroshi et al., 2013)

Follow steps 1 & 2 of 15.X to perfuse the liver through portal vein leaving inferior and superior vena cava open to remove blood. Harvest the liver and place on a 10 cm petri plate; continue perfusion at a rate of 1ml/min with 0.01% SDS for 12 h, followed by 0.1% DSD for next 12 h. Then after the perfusion Fig. 12.14 is continued in 2 spells of 3 h each, using 0.2% and 0.5% SDS. Washed our residual SDS by perfusing tissue culture grade water for 1 h, followed by 1% Triton X-10 for 30 min to remove bound SDS and nucleic acids. The liver scaffold is then perfuse with PBS for 3 h. The liver scaffold is sterilized in 0.1%

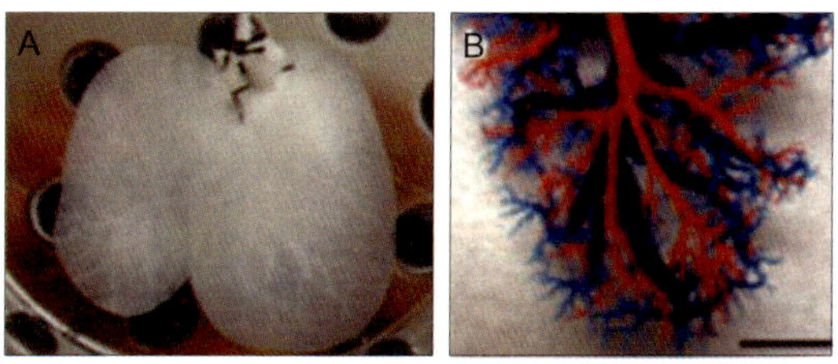

Figure 12.14 A. Decellularized liver matrix at the end of the decellularization process. B. The microvascular structure is retained in the decellularized liver matrix. red, portal ; blue, venous vasculature. [Taken from Ref. 4 & 17]

peracetic acid in PBS for 3 h, followed by another 3 h in 70% ethanol. Finally, the scaffold is thoroughly washed in sterile PBS and 2% penicillin-streptomycin solution in PBS and stored in the same solution at 4°C till used. Decellularized liver scaffold and inside vasculatures are shown in Fig. 12.14.

B. Decellularization of kidney (taken from Jeremy et al., 2013)

Anesthetize rat/mouse, a median incision expose the abdomen and carryout heparinization through the intrahepatic inferior vena cava. Remove the kidney from the abdomen by transecting the renal artery, vein and ureter. The renal artery is cannulated with heparinized PBS prefilled 25-guage cannula to allow antegrade arterial perfusion at 0.5 to 0.75 ml/min flow for 15 min to remove residual blood through the ureter which is previously cannulated with another 25-gauge cannula. The decellularization solutions is then administered at the same flow though renal artery in the following sequence: 1% SDS solution for 12 h, followed by 15 min washing with water and 1% Triton X-100 perfusion for 30 min. Finally, the decellularized kidney scaffolds is throughly washed with PBS containing 10,000 U/ml penicillin G, 10 mg/ml streptomycin and 25 μg/ml amphotericin B at 1.5 ml per min perfusion for 4 h. The decellularized kidney is stored at 4°C under asceptic condition in the presence of antibiotics solution. The decellularized kidney at different stages of decellurization process is shown in Figure 12.15.

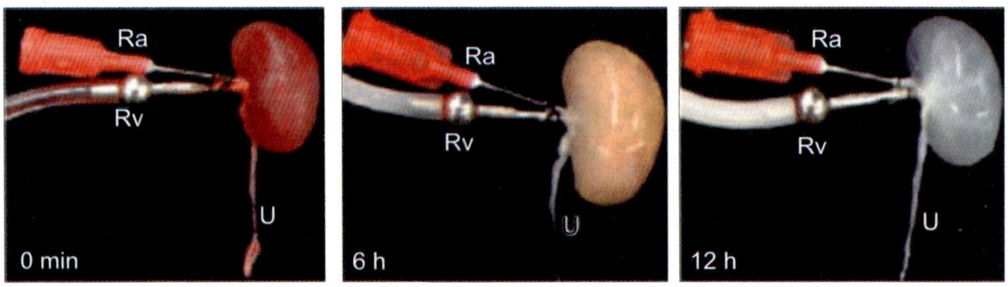

Fig. 12.15: Decellularization of whole rat kidney. Time-lapse photographs of a rat kidney undergoing antegrade renal arterial perfusion-based decellularization are shown. At the beginning, freshly isolated kidney (left) and the same kidney after 6 h (middle) and 12 h (right) of SDS perfusion depict that after complete decellularization kidney turns into transparent scaffold. Ra, renal artery; Rv, renal vein; U, ureter. [Taken from Ref. 28]

C. Decellularization of lungs (taken from Andrew et al, 2010; Gilpin et al, 2013)

Expose thoracic cavity of an euthanize mouse, perfuse the lungs by injecting 3 ml of sterile water containing 5X penicillin-streptomycin mixture into the right ventricle of the heart and the same amount of solution is transferred through trachea to inflate lungs. Isolate lungs from other connecting organs and rinse with same solution for 1h. Thoroughly perfuse the lungs by repeated injection of antibiotics solution through trachea and the right ventricle. Inject 3 ml each of 0.1% triton X-100 solution through the trachea and right ventricle and incubate at 48°C for

24 h. Wash-off lungs by perfusion of antibiotics solution through the trachea and right ventricle, then inject 2% sodium deoxycholate through the same routes and further incubate for 24 h. Wash-off deoxycholate from lungs by perfusion of antibiotics solution as before. The lung pair is then treated with 1 M NaCl and 30 mg/ml DNase in two independent steps. Finally, lung pair is washed thoroughly with PBS before initiation of scaffold bioreactor system.

12.6.1.2.2 Step II: Recellularization of scaffold

Recellularization is the most important step for successfully creation of an organ in laboratory. The viability of cells to be used for recellularization process, flow rate they are pumped inside decellularized scaffold, and the medium used for culture of cells once recellularization process is completed are important. As the whole process is susceptible to contamination, it is needless to mention that biosafety cabinet is the only working area where recellularization process is conducted. For demonstration purpose in animal model, cells may be harvested from an identical organ of freshly euthanized animal, but in clinical application, due to shortage of donor organ, it is targeted that the cells are derived from ES or iPS sources. Another important issue is that all organ are made of multiple cell types arranged in a specific manner thus, it is a difficult task to attain such cellular arrangement in ex vivo grown decellularized scaffolds. In this part, I have explained recellularization of liver using primary hepatocytes, by this process physiologically functional liver is not obtained as liver is multi-cellular organ composes of hepatocytes, liver sinusoid endothelial cells, hepatic stellate cells, Kupffer cells, NK/NKT cells, etc.

A. Recellularization of liver

Place the decellularized in perfusion chamber containing culture medium (DMEM, supplemented with 10% FBS, 100 U/ml penicillin and 100 μg/ml streptomycin. Connect the same to the peristaltic pump through the portal vein cannula and perfuse medium at 5 ml/min through the liver scaffold for 30 min. In the mean time keep ready of rat primary hepatocytes by liver perfusion (16.3), stop medium perfusion, instead inject 50×10^6 hepatocytes in the perfusion chamber and recirculate cell suspension through the decellularized (dc) liver at 10 ml/min flow rate for 10 min. Follow the above step in three more spells so that 200×10^6 cells are introduced in the dc-liver. Collect the perfusate and enumerate the cell number to determine seeding efficiency. Fill up a fresh perfusion chamber with 50 ml medium (Williams' E, 5% FBS, 0.5 U/ml insulin, 20 ng/ml EGF, 14 ng/ml glucagon, 7.5 μg/ml hydrocortisone, 100 U/ml penicillin and 100 μg/ml streptomycin). Connect the rc-liver to perfusion system through the portal vein cannula and perfuse at 5 ml/min. The whole system is placed in a CO_2 incubator, culture is continued for 10 days with daily replacement of oxygenated medium. The culture medium may be sampled at regular intervals of time to monitor the metabolic products, such as urea, albumin, etc. (Fig. 12.16)

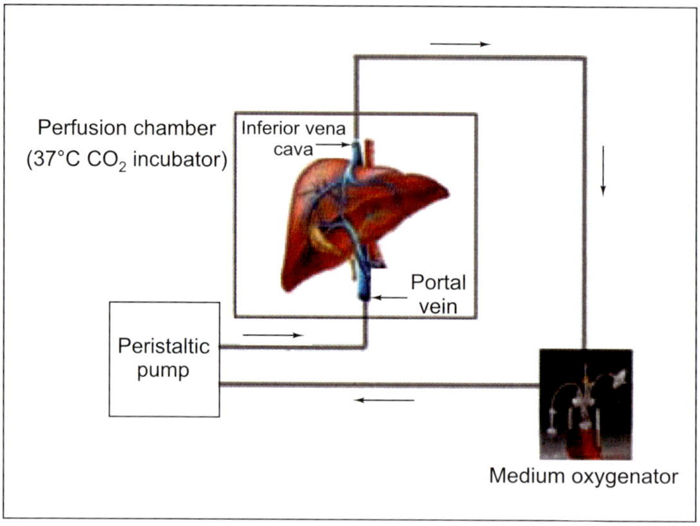

Figure 12.16 Perfusion setup for recellularization of liver is shown.

12.7 BIOARTIFICIAL ORGANS

The concept of bioartificial organs has been developed due to acute shortage of donor organs/cells in case of patients with terminally diseased organs. The failure of vital organs the heart, kidney and liver requires their replacement. Until properly matched heart and kidney is available, patients are kept in the assistance of heat-lung machine and dialysis, respectively. However, in case of liver failure, similar device is not available; rather patients are kept in assistance of artificial liver. We shall now discuss on artificial liver and pancreas. The purpose of these artificial organs is to provide organ specific functions to the potients, avoiding immune rejection.

12.7.1 Bioartificial liver (BAL)

In artificial liver, tissue engineering may contribute in two ways: creation of a liver support device to keep the patient alive and creation of a functional implantable liver replacement in place of *orthotopic* transplantation. The BAL, the extra-corporeal liver support systems, involves culturing hepatocytes (isolated from unmatched human or porcine liver) in hollow fibers or capsules. Thus, hepatocytes are separated from the human blood (immune) cells by semi-permeable membranes. Small molecules from blood plasma are able to diffuse through the pores but not the immune cells, thus hepatocytes are able to perform their detoxification function. In this way, immune-rejection can be avoided. BAL functions reasonably well, but there is a problem due to the short functional life span of the hepatocytes. Under the most optimum condition of the culture, hepatocytes cannot be kept viable for more than 7 days in culture, which raises issues like cost and feasibility of the device for long-term support. The working principle of BAL has been described in Figure 12.17A. Initially, the blood cells are isolated, if not fully, using a plasma-separating machine. The isolated blood cells are returned back to the patient, and plasma is oxygenated before passing through the hollow fibers. Hepatocytes grow on the extracapillary space (ECS) of the hollow fibers; plasma carrying the toxins enters into ECS, gets detoxified by hepatocytes, and the detoxified plasma returns back to the lumen of the fibers. Finally, the detoxified plasma is collected in a chamber and circulated back to the patients. The future development of BAL depends upon how the viability of hepatocytes is improved for longer duration. This would be possible if hepatocytes are cultured on ECMs (e.g., collagen I, laminin, heparin sulfate proteogycan), and they are intermittently incubated (regenerated) in the presence of a specialized medium.

A. B.

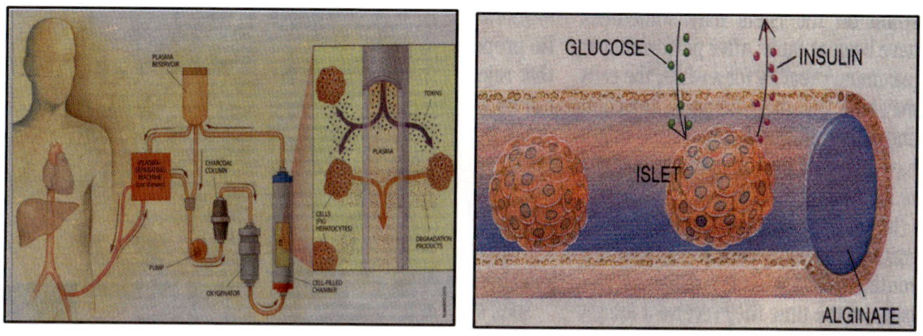

Figure 12.17 Bio-artificial organs. **A.** Bio-artificial liver: The artificial liver hepatocytes is kept out side the body, through which blood plasma is circulated for detoxification of toxins. The pig or cadaver hepatocytes are grown on the extracapillary space of the hollow fiber. The pores of the fiber allows diffusion of impure plasma from inner lumen of the fibers to the extracapillary space and returns back to the lumen, but do not allow cross flow of cells from the extracapillary space to the lumen and vies-a-versa; **B.** Bio-artificial pancreas: The pancreatic cells (pig or cadaver) encapsulated in a gel are implanted inside body. The semi-permeable membrane of the capsules allows small molecules (e.g. glucose, insulin) for the diffusion in either directions, but not the cells, thus immune rejected is avoided.

12.7.2 Bioartificial pancreas

Several groups have been trying to develop artificial pancreas, which can produce insulin for longer duration and is protected from the host immune system. One such technique is *encapsulating* islet cells in a semi-permeable membrane. The materials used for manufacturing capsules must be biologically and chemically inert, i.e., their chemical composition should not induce inflammation or any other adverse reaction in the body, and they must resist auto-degradation. Again, the bioartificial pancreas works on the basis of the same principle as used in BAL, which has selective permeability of soluble substances (Fig. 12.17B).

Thus, the capsules are designed such that the pores are small enough to prevent entry of immune system (e.g., macrophages and lymphocytes), but large enough to allow the release of insulin by the islet cells in response to blood glucose levels. Several methods have been tried to encapsulate islet cells, which include alginate capsules, hydrogels, and semi-permeable membranes. These encapsulation techniques protect the cells from host immune recognition, but are found to generate a foreign body reaction, which coat the capsular, thus inhibits oxygen and nutrient diffusion, thereby reduces islet function and viability. Among these three methods, encapsulating islet cells in hydrogel is relatively simple. The hydrogel compound, N-isopropylacrylamide (NiPAA), mixed with acrylic acid, exhibits thermosensitive reversible gelation property at temperatures between 30-34°C. By suspending islet cell clusters in NiPAA matrix (liquified at lower temperature), it is possible to inject the suspension as subcutaneous implant, which forms gel at body temperature as artificial pancreas. When insulin secretion by artificial pancreas is stopped or reduced, the hydrogel is liquefied and removed by aspiration, and at the same time new islet NiPAA suspension is reinjected. Diabetic monkeys and dogs have become insulin independent using the bioartificial pancreas for more than two years. The FDA has approved clinical trials in humans using microencapsulated porcine cells.

12.8 SUMMARY

1. Three different ways by which a tissue or an organ can regenerates: these are, compensatory hyperplasia, activation of adult tissue specific progenitor stem cells, and dedifferentiation.
2. The failure of tissue regeneration after damage could be due to three main reasons: (a) lack in regeneration-competent cells, (b) absence of inductive microenvironment, and (c) synthesis of inhibitory factors.
3. Tissue engineering has basic four steps: (a) isolation of target cells to be transplanted, (b) seeding target cells on scaffold, (c) culture cells on suitable designed bioreactor, and (d) transplantation in the affected area.
4. Tissue like growth is not possible, unless target cells are cultured on porous 3-D scaffolds.
5. Ideal biodegradable scaffolds are highly porous with interconnected channels, high surface in unit volume, impart no toxic effect to the cells, physically and chemically stable untill enough cell growth occur, and induce cells to produce own ECM components.
6. ECM components have multifaceted role in reconstructing tissue.
7. Mechanical stimuli is essential in tissue remodelling *in vitro*.
8. Two general classes of bioreactors are used in tissue reconstruction, these are either rotating or non-rotating type. The bioreactor used for tissue engineeering applications are specially designed.
9. ECM proteins are nano-structured, thus better realization of cellular control can be achieved in nanoscale (1-100 nm) scaffolds.

ABBREVIATIONS

ASC	Adult stem cell
2-AAF	2-acetylaminofluorene
Akt/PKB	Protein kinase B
BAL	Bioartificial liver
CAM	Cell adhesion molecule
ECM	Extracellular matrix
EC	Endothelial cell
FDA	Food Drugs Administration
FAK	Focal adhesion kinase
GAG	Glycosaminoglycan
ICAM-1/2/3	Intracellular adhesion molecule-1/2/3
NiPAA	N-isopropylacrylamide
PDGF	Platelet derived growth factor
PGA	Polyglycolic acid
PLLA	Poly L-lactic acid
PLGA	Poly lactic glycolic acid
SMC	Smooth muscle cell
TRAIL	TNF related apoptosis inducing ligand
TSP	Thrombospondin
VCAM-1	Vascular cell adhesion molecule-1
vWF	von Willebrand factor

REFERENCES

1. Andermn G, Jenkinson EJ (2007) Cold Spring Harbor Protocol. doi:10.1101/pdb.prot4808
2. Andrew PP, *et al.* (2010) Tissue Engg (Part A) 16, 2581 Volume 16, Number 8, 2010, 2581
3. Aoudjit F, Vuori K (2001) *J Cell Sci.* 152, 633.
4. Basak EU, *et al.* (2011). Journal of Visualized Experiments (DOI: 10.3791/2394)
5. Buckwalter JA, Mankin HJ (1997) *J Bone-Am* 79, 612.
6. Bettencourt-Diaz M, Mittnacht S, Brockes JP (2003) *J Cell Sci* 116, 4001.
7. Frisch SM, and Francis HJ (1994) *J Cell Biol.* 124, 619.
8. Folkman J, Moscona A (1978) *Nature*, 273, 345.
9. Fausto N, Campbell JS (2003) *Mech. Dev.* 120, 117.
10. Freed LE, Vunjak-Novakovic G (2000). In: Principles of Tissue Engineering (RP Lanza, RP Langer, J Vacanti, eds.). Academic Press, New York, p.143.
11. Gumbiner BM (1996) *Cell* 84, 345.
12. Goss RJ (1969) *Principles of Regeneration*. New York: Academic Press.
13. Gevers W (1884) *J Mol. Cell Cardiol.* 16, 3.
14. Geraudie J, Akimenko M-A, Smith MM (1998) *In: Cellular and Molecular Basis for Regeneration*. (Ferretti P, Geraudie J, eds.) New York: John Wiley & Sons, Ltd, p. 167.

15. Guyton AC, Hall JE (1996) In: *Text book of Medical Physiology*, 6th edition, WB Saunders, Philadelphia, p. 183.
16. Gilpin SE, *et al.* (2013) The Journal of Heart and Lung Transplantation 32, S69.
17. Hiroshi Y, *et al.* (2013) Cell Transplant 22, 231.
18. Huang S. Ingber DE (1999) *Nat. Cell Biol.* 1, E131.
19. Hudson BG, Gunwar S, Chung A, Burgeson RE. (1995). *In*: Extracellular Matrix- A practical Approach (MA Haralson, JR Hassell, eds.) IRL Press, Oxford, p. 99.
20. Heinegard D, Oldberg A (1993) In: Connective tissue and its heritable disorders. Molecular, genetic and medical aspects (PM Royce, B Steinmann, eds.) Wiely-Liss, New York, p. 189.
21. Hay ED (1981) *J. Cell Biol.,* 91, 205.
22. Hartqeriak JD, Beniash E, Stupp SI (2001) *Science* 294, 1684.
23. Hartqeriak JD, Beniash E, Stupp SI (2001) *Proc. Natl. Acad. Sci., USA* 99, 5133.
24. Healy KE, Rezania A, Stile RA (1999) *Ann. New York Acad. Sci.* 875, 24.
25. Ingber D. (1993) In: *Tissue Engineering- Current Perspectives* (E Bell, ed.) Birkhauser, Boston. p. 69.
26. Ingber DE, Madri JA, Folkman J (1987) *In Vitro Cell Dev Biol.* 23, 387.
27. Jenkinson EJ, *et al.* (1982) Effect of deoxyguanosine on lymphopoiesis in the developing thymus rudiment *in vitro*: Application in the production of chimeric thymus rudiments. *Eur. J. Immunol.* 12:583–587.
28. Jeremy JS, *et al.* (2013) Nature Medicine (doi:10.1038/nm.3154)
29. Komuro I, Katoh Y, Kaida T, *et al.* (1991) *J Biol. Chem.* 266, 1265.
30. Kratochwil K (1972) In: General properties in tissue interaction in carcinogenesis (Tarin D., ed.). Academic Press, New York, p. 1.
31. Kamiya A, Togowa T (1980) *Am J Pathol.* 239, H14.
32. Li ML, Aggeler J, Farson DA, Hatier C, Hassell J, Bissel MJ (1987). *Proc. Natl. Acd. Sci. USA.* 84, 136.
33. Levesque MJ, Sprague EA, Nerem RM (1990) *Biomaterials* 11, 702.
34. Mayne R, Var der Rest M, Bruckner P, Schmid TM (1995). *In*: Extracellular Matrix- A practical Approach (MA Haralson, JR Hassell, eds.) IRL Press, Oxford, p. 73.
35. Martin I, *et al.* (1998) *J. orthopaedic Res.* 16, 181.
36. Michalopoulos G, Pitot HC (1975) *Exp. Cell Res.* 94, 70.
37. Molecular, genetic and medical aspects (PM Royce, B Steinmann, eds.) Wiely-Liss, New York, p. 211.
38. Netter Frank MD (1987) The Ciba collection of medical illustrations, vol. 8.
39. Niklason LE, Gao J, Abbott WM, *et al.* (1999) *Science* 284, 489.
40. O'Brien FJ, Harley BA, Yannas IV, Ginson LJ. (2005). *Biomaterials* 26, 433.
41. Reid LM (1990) *Curr. Opin. Cell Biol.* 2, 121.
42. Rafii S, Lyden D (2003) *Nature Medicine* 9, 702.
43. Rytomaa M, Martinis LM, Downward J (1999) *J Curr. Biol.* 9, 1043.
44. Sage EH, Bornstein P (1995) *In*: Extracellular Matrix- A practical Approach (MA Haralson, JR Hassell, eds.) IRL Press, Oxford, p. 131.
45. Seal BL, Olero TC, Paritch A (2001). Mat. Sci. Engg: Reports, 34, 197.
46. Sell S, Leffert HL (1982) *Hepatology* 2, 77.
47. Schwartz MA, Assoian RK (2001) *J Cell Sci.* 114, 2553.
48. Schoenwaelder SM, Burridge K (1999) *Curr. Opin. Cell Biol.* 11, 274.

49. Schwartz MA, Ginsberg MH (2002) *Nat. Cell Biol.* 4, E65.
50. Trembly JH, Steer CJ (1998) Perspectives on liver regeneration. *J Minn Acad Sci* 63, 37.
51. Timpl R, Aumailley M (1993) In: Molecular and Cellular aspects of basement membranes (DR Rohrbach, R Timpl, eds.). Academic Press, San Diego, p. 211.
52. Tammi M, Paukkonen M, Kiviranta I *et al.* (1987) In: *Joint Loading: Biology and Health of Articular Structures* (HJ Helmenin *et al.*, eds.) Bristol, UK Wright-Butterworth Scientific.
53. Yannas I V (2001) Tissue and Organ Regeneration in Adults. Springer, New York.
54. Von der Mark K, Goodman S (1993) In: Connective tissue and its heritable disorders. Molecular, genetic and medical aspects (PM Royce and B Steinmann, eds.) Wiley-List, New York, p. 211.
55. Weissman IL (2000) *Cell* 100, 157.
56. Winn SR, Schmitt JM, Buck D, Hu Y, *et al.* (1998) *J Biomed. Mater. Res.* 45, 414.
57. Zerwes H-G, Risau W (1987) *J Cell Biol* 105, 2037.
58. Zhang Z, Vuori K, Reed JC, Ruoslahti E (1995) *Proc. Natl. Aca. Sci., USA* 92, 6161.

SUGGESTION FOR FURTHER READING

1. Bell E, ed. *Tissue Engineering- Current Perspectives,* Boston.
2. Chae SY, Kim SW, Bae YH (2001) *J Drug Target* 9, 473.
3. Griffith LG, Swartz MA (2006). *Nature Rev. Mol. Cell Biol.* 7, 211.
4. Korossis SA, Bolland F, Kearney JN, Fisher J, Ingham E (2005). In: *Topics in Tissue Engineering* (N Ashammakhi, RL Reis, eds.) vol 2.
5. Kim SS, Oberpenning F, Meng J, *et al.* (1999) *Nat Biotechnol.* 17, 149.
6. Miranti CK. Brugge JS (2002). *Nat. Cell Biol.* 4, E83.
7. Mikos AG, Temenoff JS (2000). *EJB Elec. J Biotechnol.* 3, 1.
8. Portner R, Stephanie N-H, Christiane G, *et al.* (2005) *J. Biosci. Bioengg.* 100, 235.
9. Utsunomiya H, Koski JA, et al. (1998) *Ann Surg* 2287, 8.
10. Wen X, Shi D, Zhang N (2005). In: *Handbook of Nanostructured Biomaterials and Their Applications in Nanobiotechnology* (HS Nalwa, ed.). Americal Scientific Publishers. vol. 1, p. 1.

13

Regenerative Medicine: Cellular Therapy

In previous chapter I have introduced regenerative medicine with reference to tissue engineering. In past 20 years tissue engineering has progressed to a new pick. Scientists have understood how to grow tissue, if not complex one, in 3D conformation. The progress has been made particularly toward regeneration of bone, cartilage and skin. Besides these, tissue engineering has been applied in repair/reconstruction of wind pipe of the lungs, uninary bladder and lymbus of the eye. In many cases, at least in some of the above areas, tissue engineered reconstructs have become a regular practice in the clinic. However, the vital organs in human, such as heart, liver, kidney, pancreas, etc. are more complex in nature, not only due to functional reasons, but also due to multi-cellular type, special arrangement of the cells and highly vasculature. Complexity of these organs has become the main obstacle to develop tissue engineered organ for human applications, it may take a decade to create such functional organs in the laboratory. The present chapter will cover another aspect of 'regenerative medicine' that is 'cell therapy'. It is believed that the damaged cells leave the organ creating an empty space on the scaffolds. If the tissue specific cells are delivered targeting to these empty spaces, cells will engraft, become mature and integrated with the existing host tissue, in turn the diseased organ become functionally active. Various cell types that have been used or proposed for the therapy are depicted in the Figure 13.1. It may be remembered that all these cells not used to deliver cells per se, they are also used as the vehicle for delivering genes or editing (correcting) genetic defects. In some cases, the donor cells do not engraft in the host environment for longer period, but secrete trophic factors to rejuvenate the week tissue for their proliferation and normal functions. Different sources of cells, their manipulations for clinical applications have been described in the chapter.

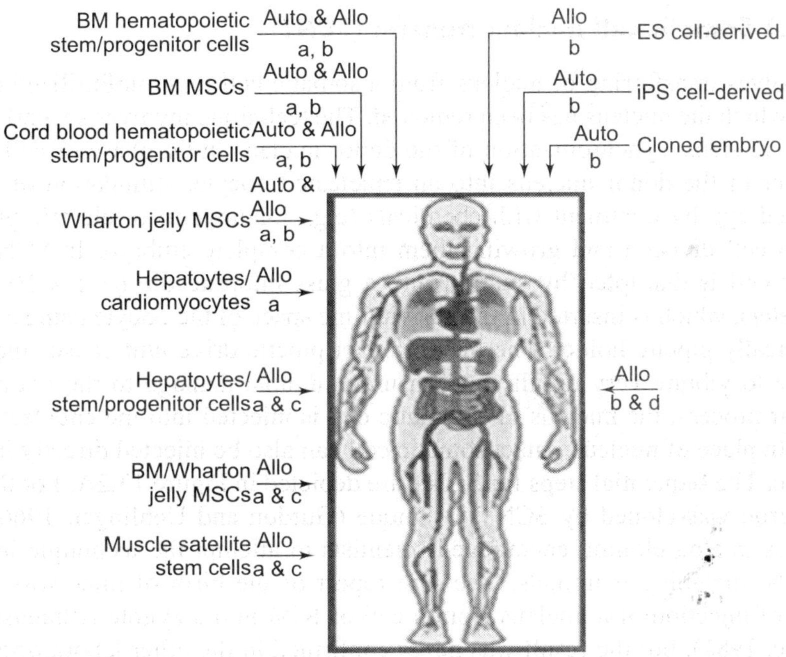

Figure 13.1 Different approaches of cell therapy. Auto: Autologous; Allo: Allogeneic; **a.** Original tissue replacement; **b.** Differentiation into target tissue; **c.** Gene therapy; **d.** gene editing.

13.1 CLONING

The process of cloning is general refers to the change in entire genome of an oocyte. Cloning is highly controversial in ethical reason; it involves prediction of defects in an organism and correcting that before birth, or creating normal tissues for the adult life. Through cloning, it is expected to create intact human cells, outside the body. By cloning it might be possible to avoid or even arrest the progress of diseases like Parkinson's, leukemia, diabetes, etc. At molecular level cloning means copying of DNA fragments containing the genes of interest and amplifying them in a suitable host cell. In similar analogy, copying of somatic cells by growing in culture means cellular cloning, thus creating an identical copy of an organism. The same process occurs naturally in single cell organisms (e.g. bacteria and yeast), as they reproduce by cell division. In human natural cloning occurs in identical twins due to formation of multiple blastocysts from a zygote. How cloning is performed for multicellular organisms has been described in this section. In view of applications cloning in mammals are of two types – reproductive and therapeutic. Both types of cloning are based on transfer of entire genetic materials from one somatic cell to an enucleated oocyte to generate an embryo in which the characteristics of the donor nuclear materials are restored.

13.1.1 Somatic cell nuclear transfer (SCNT)

It involves transferring of nucleus from a somatic cell to an unfertilized oocyte from which the nucleus has been removed. The technique involves several highly sensitive steps: synchronization of the donor nucleus in G_0 phase of cell cycle, transfer of the donor nucleus into an enucleated oocyte, stimulation of reconstructed egg by treatment with chemicals (e.g., ionomycin) or electric pulse to obtain cell division and growing them into a complete embryo. In SCNT, the donor cell is disrupted by suction into a glass microneedle (> 2×10^{-4} inch diameter), which is inserted in the perivitelline space of the oocyte using a piezo-electrically pipette holder. The piezo-impact pipette drive unit causes the glass needle to vibrate very rapidly and ensures minimal damage to the oocyte. By similar process, the nucleus of a somatic cell is injected into the enucleated oocyte. In place of nucleus, intact somatic cell can also be injected directly into the oocyte. The sequential steps for SCNT are depicted in Figure 13.2A. For the first time frog was cloned by SCNT technique (Gurdon and Uehlinger, 1966). The success in frog cloning encouraged scientists to attempt the technique in other species, including mammals. The first report of the birth of mice was on the basis of injection of a nucleus from a cell of ICM into a zygote (Illmensee and Hoppe, 1981), but the result was never confirmed in the other laboratories. The reproducible result in SCNT in mammal was obtained after extensive studies, which lead to the recognition of three major physiclogical factors—the biological state of recipient cell, the developmental status of donor cell, and the relative cell cycle position of donor nucleus and recipient cytoplasm.

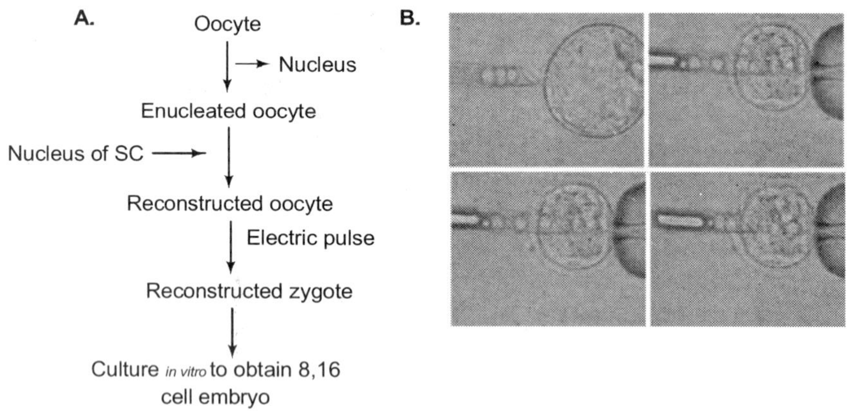

Figure 13.2 **A.** Basic steps in somatic cell nucleus transfer. **B.** Various stages of injection of nuclear materials into an enucleated oocyte. [*Source*: Harvard Stem Cell Institute]

13.1.2 Possible applications of SCNT

There are many areas in which SCNT can be applied. These are as follows:

(A) ***Production of farm animals***—it can be used to create an enumerable number of clones of high productive farm animals such as cows, pigs, lambs, etc. For a successful program of cloning in farm animals, SCNT is integrated with animal breeding facility.

(B) ***Production of therapeutic proteins***—human therapeutic proteins can be produced in milk of transgenic sheep or goat at a very high titer, as compared to that produced in mammalian cell culture process.

(C) ***Cell based therapies***—by therapeutic cloning it would be possible to cure diseases like Parkinson's, stroke and diabetes without immune rejection.

(D) ***Xenotransplantation***—nuclear transfer will allow the deletion of genes from pigs that are highly immunogenic and responsible for rejection of an organ/tissue of pig by human. Thus, genetically modified pigs could be the alternate source of organs like heart, liver, pancreas, etc.

(E) ***Gene therapy***—by nuclear transfer it would be possible to develop new therapies to treat and overcome diseases like cancer, genetic disorders and to control normal aging process.

(F) ***Reproductive cloning***—artificial embryo twinning can be achieved by properly cloning or genetic copying of an individual.

13.1.3 Reproductive and therapeutic cloning

In artificial embryo twinning, an early embryo is separated from mother's womb, which is then splitted and cultured *in vitro* before transferring into a surrogate mother. Since all the embryos are derived from the same zygote, they are genetically identical. SCNT is an entirely *in vitro* technique in which the nucleus from a somatic cell is transferred into an enucleated egg. The egg cell is then stimulated to develop as an embryo. The embryo is implanted into a surrogate mother for the birth of cloned individual. The basic difference between fertilization and SCNT is that in the former case both haploid cells, sperm and egg, are involved for the formation of zygote having two sets of chromosomes, whereas, in SCNT, enucleated egg receives two sets of chromosomes from a somatic cell, resulting in the formation of hybrid egg or zygote. The comparison between normal fertilization and *reproductive cloning* is described in Figure 13.3. Reproductive cloning means genetic reprogramming of the nucleus of a fully differentiated somatic cell to produce a totipotent cell clone that develops into an identical new copy of the organism from which the somatic cell had been taken. The birth of Dolly was the result of above procedure (Wilmut *et al.*, 1997).

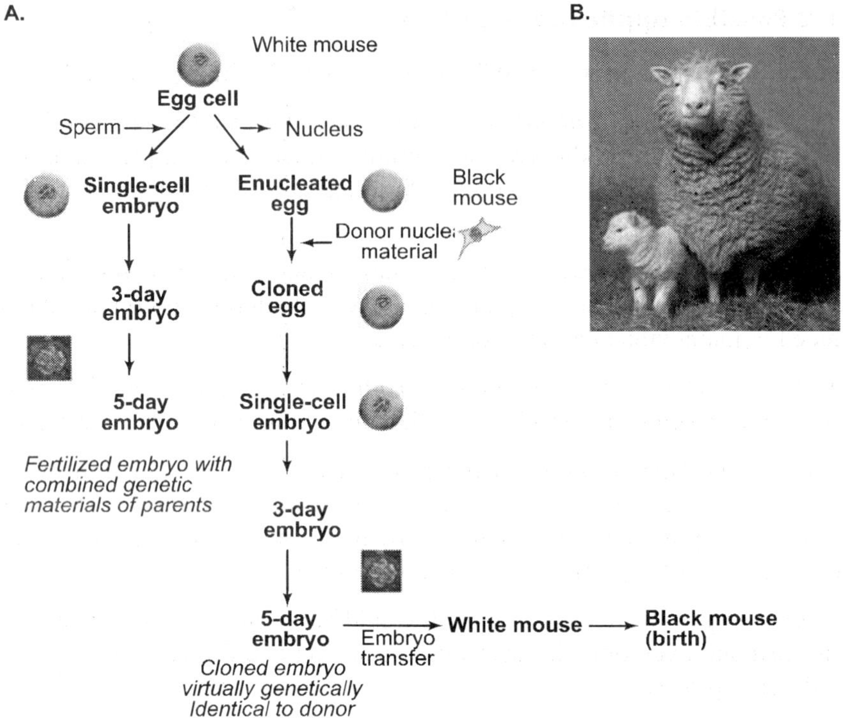

Figure 13.3 A. Comparison between normal fertilization and reproductive cloning. Normal fertilization is based on the formation of zygote with the help of two haploid cells (sperm and egg). The reproductive cloning is based on the transfer of nuclear materials of a somatic cell to an enucleated egg; B. The sheep Dolly and her first-born lamb, Bonnie.

In therapeutic cloning or embryo cloming, the embryos so formed are used for research and application in regenerative biomedicine. Thus in therapeutic cloning stem cells are harvested from embryos to study mammalian development and to treat various diseases. Many human diseases involve irreversible cell degeneration, which include Parkinson's disease, some heart failure conditions, diabetes, etc. If a method is available to create normal human cells and inject them into the patients, it might arrest the progress of such diseases. The current allogeneic cell therapy practice to treat above diseases is limited due to immune rejection. Similar is the situation in case of organ transplantation (despite perfect MHC match), as the donor organ is recognized by the host immune system as a foreign object. Thus, host immune system targets donor cells/organs for destruction. The target of host immune system towards the donor cells is termed as host versus graft disease (HVGD), which is normally suppressed by using immunosuppressive drugs (e.g., cyclosporin). However, this drug has debilitating side effects including high risks of infections due to immuno-compromise and malignancies. In therapeutic cloning, ES cells are created from the nucleus of patient's own cell, thus genetically identical cells (equivalent to autologous) are formed, avoiding the risks of immune rejection and the requirement of immunosuppressive drugs. Furthermore, ES cells can become an uninterrupted source of cells to replace tissue and allowing repeated therapy. The scheme of *therapeutic cloning* has been described in Figure 13.4. The basic difference between therapeutic and reproductive cloning is that in case of former, the cloned blastocyst is not transferred to uterus, rather, ICM is harvested and expanded in culture and induced to tissue specific differentiation prior to transplant into the patients.

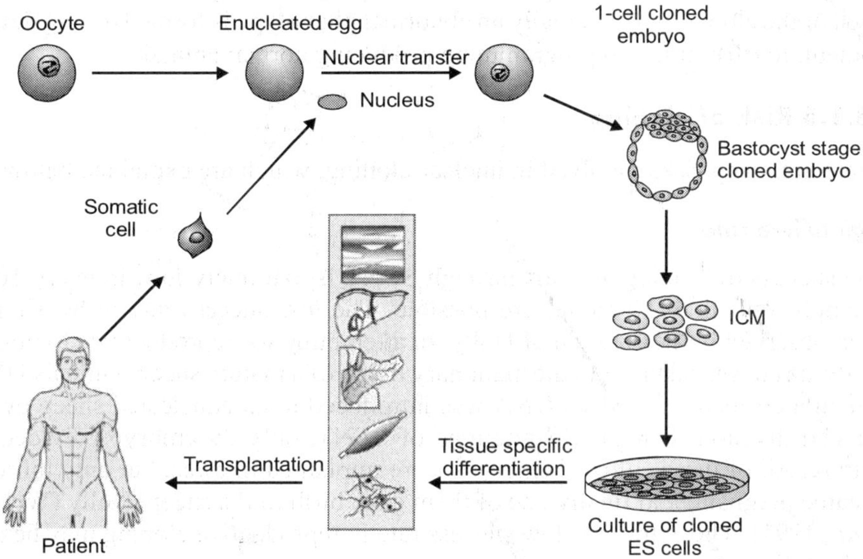

Figure 13.4 Various steps involved in therapeutic cloning. In therapeutic cloning, the ICM is expanded, differentiated to specific tissue before they are transplanted to the patients.

13.1.4 Biology of somatic nuclear cloning

In normal fertilization, the combination of genomes of the two mature gametes leads to faithfully activation of genetic program to ensure normal embryonic development. Due to reprogramming of genome in gametogenesis, the mature gametes are formed. This reprogramming process is initiated right at the time of gastrulation, when PGCs are formed (Rideout *et al*., 2001). In somatic nuclear transfer, the genome of the donor cell is not in the same state as that in the fertilized egg. In order to begin the developmental process, many genes (normally expressed during embryogenesis), which usually remain silent in the somatic cell, are reactivated in somatic cloned embryo. The complex process of egg and sparm maturation, in which extensive genomic reprogramming is normally taken place are simplified by SCNT. In normal embryogenesis, epignetic remodeling results in shut down of some of the genes, whereas other keep expressing. For example, epigenetic state of mammary gland cells assures expression of genes important for mammary gland function, such as the milk production to be active where irrelevent genes are to be inactivated. At the same time, inactivation of one X chromosome in female cells is done by random and stable silencing of one of the two X chromosomes early in embryogenesis. In embryo cloning from female somatic cells, the inactive X chromosome may be further activated.

The cytoplasm of egg contains reprogramming factors that are believed to convert the epigenetic state of the donor nucleus to one that is similar to an embryonic cell. This radical differences in reprogramming during somatic nuclear cloning may lead to: (i) no reprogramming of the genome, leading to death of the newly formed embryo; through (ii) partial reprogramming, allowing initial survival of the clones, but eventually an abnormal phenotype is formed during development; to (iii) faithful reprogramming producing normal animals.

13.1.5 Risk of cloning

There are many risks involved in nuclear cloning, which are explained below:

High failure rate

The success of cloning animals through SCNT is extremely low, in every 1000 attempts, only 1 to 30 clones are possible. The low success rate in SCNT has been observed in the creation of Dolly. In attempting for reproductive cloning of Dolly, a cell was taken from the mammary tissue of a mature sheep while its DNA was in a dormant state. This DNA was introduced to an enucleated sheep ovum for cell division. Out of 277 attempts of SCNT, only 29 embryos divided in culture. All of these cultured embryos were implanted in ewes, but only thirteen became pregnant, and finally one of them gave birth to the sheep, Dolly (Wilmut *et al.*, 1997). The extremely low success rate in reproductive cloning may be due to the following reasons:

(a) Lack of compatibility between enucleated egg and the transferred nucleus.
(b) Exogenous source of nucleus may not be properly activated by recipient (egg) cytoplasmic factors.
(c) Unsuccessful implantation of the embryo into the surrogated mother.
(d) Fail of pregnancy or abortion.

Problems in post-development

Cloned animals were found to be much larger at birth than their natural counterparts, this abnormality is known as *large offspring syndrome* (LOS). The offspring may have large organs, which may lead to breathing, blood flow and other problems. Abnormal kidney, brain and impaired immune functions are also evidenced in care of LOI.

Abnormal gene expression patterns

It has been found that the genomes of cloned mice are compromised, 4% of the liver and placenter cells genes showed abnormalities. It was felt that above abnormalities did not arise due to mutations in the genes but from changes in the normal activation or expression of certain genes, which lead to errors in programming of the genetic material of the donor nucleus. In natural fertilization the embryo receives copies of most genes from both parents. '*Genetic imprinting*' marks the DNA from the mother and father, so that only one copy of a gene (either the maternal or paternal gene) is activated. Defects in the genetic imprinting, single donor nuclear may lead to some developmental abnormalities of the cloned embryos.

Telomeric differences

It is not clear why cloned animals show differences in *telomere* length than the normal counterparts. Shorting telomere length is a natural process of aging, as cells divide many times in lifespan. What happens to the clone animals if nucleus of an old donor is transferred? Does shortened telomere length affect on development or lifespan of the offspring? When the telomere lengths of cloned animals were examined, no clear cut answers were obtained in this regard. Chromosomes from cloned mice had longer telomeres than normal mice. In cloned cows, each cell showed extended lifespan compared with cells obtained from a naturally conceived cow. On the other hand, Dolly's chromosomes had shorter telomere length than the normal sheep. This implies that Dolly's cells had aged faster than the cells of a normal sheep.

13.1.6 Ethical issues in reproductive cloning

Therapeutic cloning is less controversial, whereas the ethical issues are just about reproductive cloning. Should reproductive cloning be continued? The contemplation against such cloning has originated from many social and cultural considerations. The most crucial ethical, legal and social issues about cloning arise in the context and process that may lead to the birth of the first human clone. Like plant, human reproduction does not occur by the replication of DNA from one generation to another. Only in case of twinning, simultaneous birth of siblings does occur by the replication of DNA. The debate against cloning of humans has become stronger due to high failure rates (< 90%) and high morbidity in cloned animals. Furthermore, cloned animals seem to suffer from high deformity and disability rates. All these social and cultural issues culminate an important question—do we have appropriate technology to perform reproductive cloning. It is also important that proper quality controls should to make this technology safe, less riskful to those using it. Thus, the present ethical issues regarding human reproductive cloning are pertaining to the following issues:

(a) Availability of a safe technology for reproductive cloning,
(b) If reproductive cloning becomes successful, it may abate the concept of reproduction and family.
(c) Most of us have a strong interest in the notion of individuality; the clone individual will lost personal identity.
(d) An ambiguous relationship may develop between the cloned child and the progenitor (biological parents).
(e) It is contrary to the human dignity.

13.2 GENE THERAPY

Gene therapy is specific, it involves supplementing the genome of the target tissue with wild (correct) type gene that has been found mutated due to any reason. Genes are specific sequences of bases that encode instructions on how to make proteins, involve for most biological functions and even make up the majority of cellular structures. Most of us do not suffer from the harmful effects of the defective genes, as our genome carries two copies of nearly all genes, each derived from the parents. The problem is associated mainly with the gene present in male sex chromosomes. Each cell in male carries single copy genes of X and Y chromosomes, as that are sufficient to take care of all the symptoms of the diseases linked with sex chromosomes. If one of the recessive genes not belonging to sex chromosomes is mutated the counterpart will carry out the normal function. If both copies of recessive gene are defective, it will lead to the development of the disease. In case of male, the defect in the recessive genes of sex chromosomes results in the development of a genetic disease. The only exception to this rule is in case the gene is dominant in nature, it alone can cause the disease, even the counterpart is normal. By gene therapy, it is possible to complement the defective gene that is believed to be responsible for the development of the disease. Gene therapy is likely to have the greatest success with diseases that are caused by single gene defects; many clinical trials are under progress to correct genetic defects. This section will cover the therapeutic molecular approaches for treating diseases. At the molecular levels, three major approaches are followed to introduce changes in the defective genes, these are: (a) inserting normal gene into a nonspecific location within the genome to supplement the loss function of the defective gene, (b) replacing abnormal gene with the normal gene through homologous recombination, and (c) by selective reverse mutation of abnormal gene to derive its normal function. The common approach of gene therapy has been the insertion of trans-gene in nonspecific location of the genome.

13.2.1 Candidate disease for gene therapy

Gene therapy is found to be more successful method in cases where the diseases are caused by the defect of single gene. The first clinical trials on gene therapy were approved in 1994. Presently, there are more than twenty diseases, where gene therapy trials are at various phases. The best examples of the clinical trials are hemophilia A and B, cystic fibrosis, severe combined immune deficiency, Gaucher's disease, muscular dystrophy, leukemia, cardiac disorder, and Parkinson's disease. According to Eve Nichols (1988), the selection criteria of candidate disease for gene therapy should be based on:

(1) Incurable and life-threatening disease.
(2) Diseases in which the normal counterpart of the defective gene has been isolated and cloned.
(3) Cell types (organs) affected by the disease have been identified (e.g., muscle cells in case of muscular dystrophy).

(4) Gene can be introduced in the substantial fraction of the affected cells/ tissues.
(5) Adequate expression of gene product is possible to reverse the disease.
(6) Adverse effect of the gene therapy, if any, can be determined in patients receiving the therapy.

13.2.2 The gene delivery system

The purpose of gene therapy is to introduce genetic modification in cells in order to make them produce a therapeutic effect to rescue patients from the disease. As mentioned above, in most gene therapy, a normal gene is inserted into the genome of the recipient cells to replace an abnormal, disease-causing gene. It has been explained in earlier chapter that in any genetic modification, a carrier molecule, known as vector is used to deliver the target gene. In gene therapy, the therapeutic gene is also delivered to the diseased cells with the help of a vector. The most common vectors are viral-based that have been genetically altered to carry the transgene. Virus naturally evolves a way of *packaging* and delivering their genes to normal human cells. A similar approach has been exploited to deliver the target gene through the viruses, the only difference is that its genome is manipulated to remove the disease-causing genes. The viral vectors commonly used for gene therapy are:

1. ***Retroviral vector***—a class of viruses (e.g., human immunodeficiency) that can create double-stranded DNA copies of its RNA genomes. It has been extensively used for gene therapy applications due to easy manipulation, integration in the host chromosome, and sustained expression of the desired protein. Retroviral vector is able to accommodate foreign genes ranging from 1 to 7 kb, and can infect wide variety of cell types. There are two major limitations of retroviral vectors:

(a) They cannot infect postmitotic cells, and are thus limited to *ex vivo* gene therapy applications.
(b) There is a risk of integration in the vicinity of the proto-oncogene, which may cause neoplasia.

2. ***Adenoviral vector***—a class of viruses (Ad2 and Ad5) with double-stranded DNA genomes of 36 kb that may cause respiratory, intestinal, etc. infections in humans. This virus is much safer than retroviruses. The advantage of adenoviral vector is that they can infect postmitotic cells, hence both *in vivo* and *ex vivo* gene therapy is possible. It is able to infect wide variety of cell types, produces high titer (10^{11} to 10^{12} virus particle/ml) of recombinant viruses, and can accommodate up to 10 kb of foreign DNA. The limitations of adenoviral vectors are lack of sustained expression (as the viral DNA does not integrate into the host chromosomes), toxic at high doses of virus, antigenic response against viral proteins, and the preexisting immunity in the host.

3. ***Adeno-associated viral (AAV) vector***—it is a defective human parvovirus, a class of small and single-stranded DNA virus, which can insert its genetic material at a specific site of chromosome 19. AAV vector offers unique advantages which include:

(a) Ability to infect a wide variety of cells.
(b) It can infect postmitotic cells, hence both *in vivo* and *ex vivo* gene therapy is possible.
(c) It has no immunogenic response like adenoviral vectors.
(d) Viral particles are remarkably stable.

The major limitations of AAV vector are that it accommodates relatively small size of foreign DNA (4 to 4.5 kb), requires Rep protein to integrate in a defined chromosome locus, and produces low titers (10^5 to 10^6 virus particle/ml) due to lack of suitable packaging cell lines.

4. ***Herpes simplex viral vector***—a class of double-stranded DNA viruses that infect a particular cell type, e.g., neurons. Herpes simplex virus type 1 genome can carry multiple inserts, up to total of 30 kb DNA.

5. ***Lentiviral vector***—it is a type of retrovirus, e.g., HIV-1, that can infect both dividing and nondividing mammalian cells because their preintegration complex can penitrate the intact nuclear membrane of the target cell. Lentivirus based system has wide applications in gene therapy, as it can change the target cell's gene expression for quite sometime. Most importantly lentivirus can be used for nondividing cells such as neurons, macrophages, muscles and liver cells. Cells at quiescent cells (G_0 state) cannot be infected with lentiviruses, because reverse transcription is blocked in these cells (Amado and Chen, 1999).

Beside virus-mediated gene-delivery, several nonviral gene delivery techniques have been tried. The simplest method is direct introduction of therapeutic DNA into target cells by physico-chemical methods such as DNA transfection with DEAE dextran, calcium phosphate, electroporation, optoporation, etc. Each approach has its own limitation, as it is used in specific cell types, requires large amounts of DNA for transfection, and low transfection efficiency. Another nonviral approach is of using an artificial lipid sphere with an aqueous core (liposome). The liposome-trapped therapeutic DNA can be easily delivered into the target cells through the cell membrane. Therapeutic DNA can also be delivered inside the target cells by chemically linking DNA to a molecule that binds to a specific cell surface receptor. Once attached with the receptor(s), the therapeutic DNA construct enters inside the cells by a process known as receptor-mediated endocytosis. The most effective method of nonviral gene therapy is known as particle bombardment or microprojectile or gene gun. This technology was initially developed for the transformation of plant cells, but later found suitable in case of wide ranges of mammalian cells. Microscopic particles, coated with the gene of interest, are accelerated for penetrating cells and deliver the DNA.

13.2.3 Development of retroviral vector for DNA transfer

Retrovirus is surrounded by a lipid bilayer with proteins specific for cell surface (Figure 13.5A). One of these membrane proteins is gp120, which recognizes CD4 receptor of the host T cell. It is also binds to CXCR4 present on T cells and CCR5 on macrophages, and induces a membrane fusion with gp41 transmembrane protein. This allows the entry of virus particles inside the cells and the release of genetic materials in the cytoplasm. Replication-competent retroviruses have three protein coding genes *gag*, pol, and *env*. Gag protein is involved in viral encapsidation and assembly, Pol is involved in the synthesis of viral DNA and integration, whereas, Env protein is responsible for viral budding and entrance. In addition, retroviruses have *cis*-acting sequences for reverse transcription, integration and encapsidation. As genetic information of retroviruses is integrated into the host cell DNA, its RNA is converted into DNA by reverse transcription. Reverse transcriptase synthesizes the first strand of DNA from the RNA template, and RNA dependent DNA polymerase activity of reverse transcriptase synthesizes

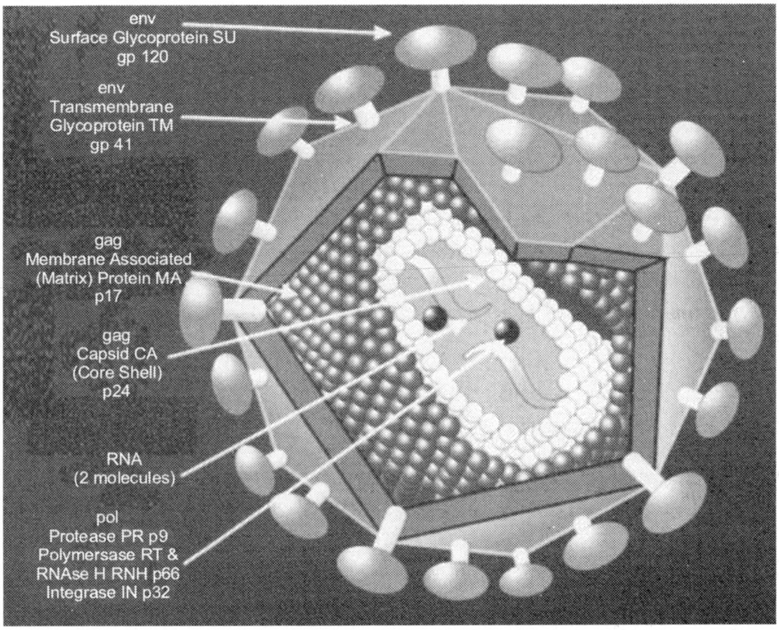

Figure 13.5 Retrovirus based vector development for DNA transfer. **A.** Schematic drawing of a retrovirus virion. The outer envelope contains products of the *env* gene. The inner hexagon contains products of the *gag* gene as structural components, and products of the *pol* gene are mostly different enzyme of the virion; **B.** RNA and DNA genomes of C-type retroviruses. The viral RNA contains a copy of the DNA missing long terminal repeats (LTR), attachment sequences (*att*L, *att*R), and a 3' RNA processing site of the LTR. The additional genes present in the viral RNA are small repeat (r) and polyadenylic acid (poly (A)). The viral RNA and DNA contains encapsidation sequences (E), a primer binding site (PBS), splice donor sequences (sd), viral coading genes (gag, pol, env), splice acceptor sequences (sa), a polypurine tract (PPT), enhancer and promoter sequences.

the second strand to produce dsDNA. The DNA copy contains the genes *gag*, *env*, and *pol*, are inserted by integrase into the host genome (Adler et al., 1999). Integrase ligates attachment sequence (*att*) of viral DNA to the host chromosome. There is no apparent specificity of retrovirus integration with respect to cellular DNA, although there is a strict sequence specificity as far as the attachment sequence of viral DNA is considered.

As mentioned above, the major proteins expressed by all retroviruses are necessary for self-replication, which are encoded by viral genes *gag*, *pol*, and *env*. In retroviral vectors, these 3 genes are deleted and the viral genome is modified with transgene(s) and a marker gene. The DNA copy of the retroviral vector genome contains transgene, 2LTRs and the packaging sequence (Ψ), which are required to package the recombinant genome into virions. In addition to the transgene, three more genes are separately introduced in the packaging cells for the expression of proteins Gag, Pol, and Env. The resulting retroviral vectors release from the packaging cells are infectious, but replication-defective. They can be used to infect other packaging cells for one cycle, or suitable target cells for gene therapy purpose.

13.2.4 Lentiviral vector

As on today, the viral vectors are the best choice as a delivery vehicle for both *in vivo* and *ex vivo* modification of target cells in gene therapy applications. Among the five viral vectors used in gene therapy, lentiviral vector is most popular, as it has distinct advantages over other vectors these are: stable expression of transgene is possible and virus can infect non-dividing cells. The following portion of this section has been restrict to lentiviral vector system.

13.2.4.1 Construction of lentiviral vector

The lentiviral vector containing transgene is produced in a transient transfection system in which the packaging cell line is transfected with the following three separate plasmid expression systems:

1. ***Transfer vector plasmid***—it contains the CMV promoter directed transfer vector genome (between two asterisks), as show in Figure 13.5A. The therapeutic gene is cloned into a vector, which is flanked by LTRs and the Y-sequence of HIV. The LTRs allow to integration of the therapeutic gene into the genome of the target cell. The Ψ-sequence acts as a signal sequence and is necessary for packaging RNA with the therapeutic gene in virions. The Rev responsive element (RRE), in the presence of Rev, facilitates nuclear export of the vector genome. The expression vector contains internal promoter and the transgene. The 3′ region contains the self-inactivating (SIN) deletion, ΔU3. The R region contains the vector polyadenylation signal. The viral proteins that make virus shells are provided in the packaging cell line, but are not in context of the LTRs and Ψ-sequence. Thus, the virus particles produced are replication deficient (Naldini *et al.*, 1996).

2. ***Packaging plasmid***—it contains the backbone of the virus system, such as the elements required for vector packaging e.g., structural proteins, HIV genes, and the enzymes that generate vector particles (Amado and Chen, 1999). The plasmid is CMV promoter directed, which is responsible for the expression of capsid and matrix as polyprotein Gag, for packaging the viral genome. Protease, reverse transcriptase, and integrase, expressed as polyprotein Pol, cleave the structural elements of the vector particles (Figure 13.5B). Here the packaging and their adjacent signals are removed, such that the packaging sequences will not be incorporated into the viral genome and virus will not multiply in the infected host cells (Naldini *et al.*, 1996).

3. ***Envelop vector plasmid***—this plasmid contains genes of envelope proteins of different viruses that infect broad range of possible target cells, other than helper T-cells as in case of normal HIV (Amado and Chen, 1999). Two types of heterologous envelope proteins are used. The one most commonly used is G glycoproteins of the vesicular stomatitis virus known as VSV-G (Figure 13.6C). This provides stability to the vector by bringing together the particles that are made by the packaging plasmid (Naldini *et al.*, 1996).

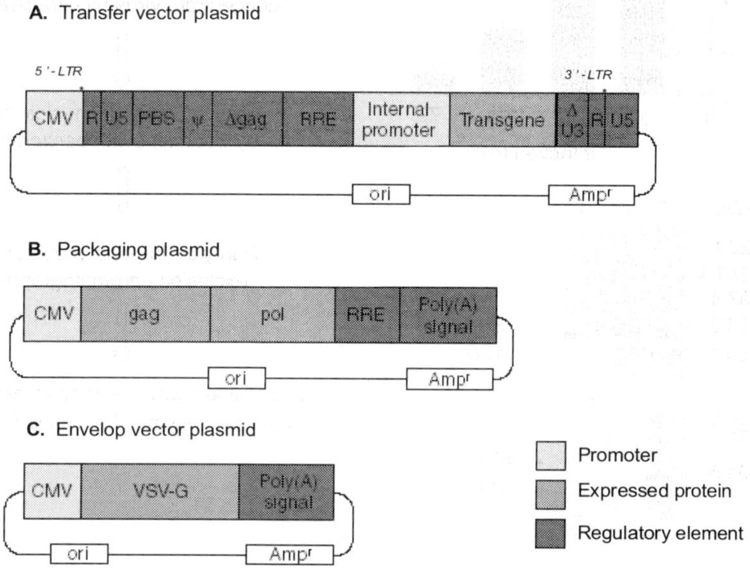

Figure 13.6 Construct of a third generation lentivirus vector. It has three separate plasmid expression systems: transfer vector plasmid, packaging plasmid, and envelope vector plasmid. [Taken from Ref. 9]

13.2.4.2 Production of lentiviral vector

The popular packaging cell line for the expansion of lentiviral vector is 293T (human embryonic kidney). Embryonic kidney cells may be cultured on a 10-layer Nunc® cell factory or in a roller bottle. For optimum transfection, it is important to optimize the ratio of plasmid DNA (combination of above three plasmids) to transfection reagent, and the time of transfection. It has also been found that 293T cell passage up to 60 times do not cause any significant difference in the yield of final product. The maximum virus titer is normally obtained between 30 and 45 h of transfection (Figure 13.7A), and thus considered the optimum time for the production of lentiviral vector (Coleman, 2003).

The plasmid DNA and transfection reagent are mixed in an optimum ratio in plain DMEM and incubated for 10 min at room temperature. Prior to transfection the spent medium is removed and the culture is incubated with plasmid mixture for 4-5 h. Following transfection, the culture medium is replaced with growth medium and allowed to complete the production of vector. Medium containing the first batch of virus particles is harvested between 25-30 h of transfection.

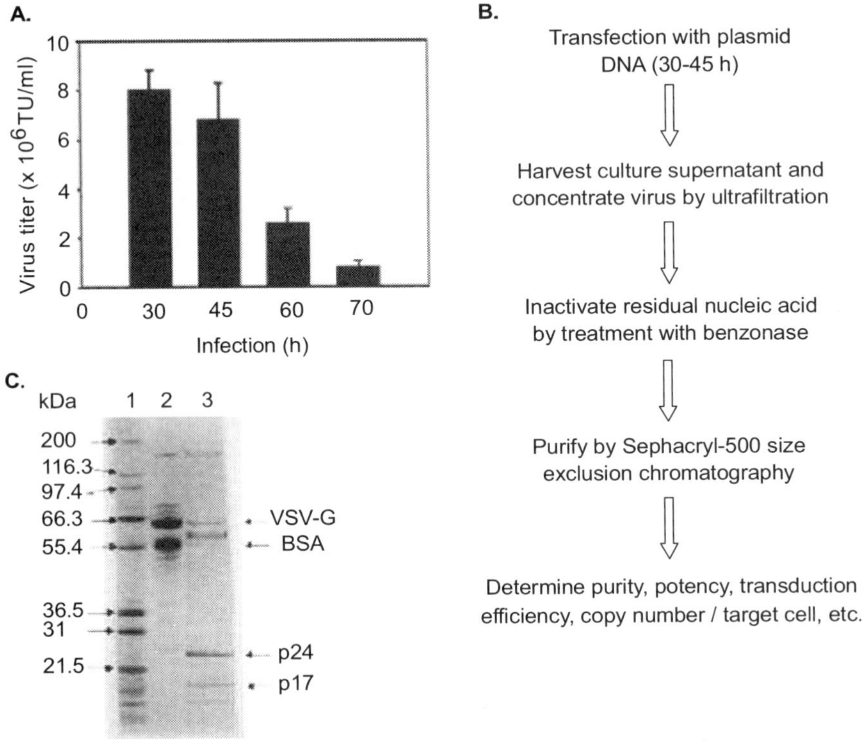

Figure 13.7 Clinical grade lentiviral vector production procedure; **A.** Production of lentivirus particles by transfection in 293T cells is a function of time (Taken from Ref. 2); **B.** Basic steps for production and purification of lentiviral vector. **C.** Electrophoretic profile of vector product. [Taken from Ref. 13]

The culture is continued with fresh medium for another 15-20 h, and finally terminated. The pooled harvest is first clarified in microfilter, and the product is concentrated (30 to 40-folds) by removing spent medium using hollow fiber cartridge in *diafiltration* mode. The buffer (pH 8.0) used for diafiltration contains benzonase. The residual nucleic acid in vector is destroyed by treating the solution with benzonase (500 U/ml) for 30 min at 37°C. The benzonase-treated material is purified by gelfillration chromatography using a Sephacryl S-500 packed column. Alternately, the purification of vector can be performed by Mustang Q-ion exchange chromatography (Slepushkin *et al.*, 2003). Finally, the purified vector is sterile-filtered and stored at –80°C. The flow sheet for the production of lentiviral vector is shown in Figure 13.7B.

The purified vector is tested in terms of purity, titer, and transduction efficiency. The purity of vector is determined by examining electrophoretic mobility of proteins in SDS-PAGE gel (Figure 13.7C). The gel shows the major bands of VSV-G envelope protein, the p24 core protein and the p17 matrix protein. Contaminating protein as BSA is observed, which may not be a problem in case of *ex vivo* modification, as the target cells are cultured in the presence of FCS. Prior to inject into patient, it is necessary to check the BSA level by ELISA. The maximum limit of BSA in cellular product is 1 μg/ml (Schonely *et al.*, 2003). The overall yield of the purification of clinical grade lentiviral vector following size exclusion chromatography is about 30% (Slepushkin *et al.*, 2003). Besides SDS-PAGE analysis, another way to determine the purity of the vector is measuring the ratio of p24/total protein in pre- and post-purification step. This ratio increases greater than 1000-fold in the purification step, indicating the selective purification of p24-containing particles (Slepushkin *et al.*, 2003). Finally, the quality of clinical grade vectors are determined by their transduction efficiency and potency. The purified vector is titrated in primary $CD8^+$ T cells, which may be as high as 2.19×10^8 TU/ml (Slepushkin *et al.*, 2003). Low copy number vector is normally maintained in the transfected cells, which usually varies from 0.5 to 5.0.

13.2.5 Strategy for gene therapy

Two strategies are followed during the delivery of gene to the target cells (Kaji and Leiden, 2001). The first strategy is based on direct infusion of gene to the subjects (also known as *in vivo* modification of cells). Viral vectors are used as vehicle for delivering genes, assuming that the delivery of viral genes into the mammalian cells follows the same mechanism as seen in case of viral infection. This strategy is limited to few viral vectors, as other vectors cannot be used in case of non-dividing cells. For example, it is difficult to deliver genes in heart and brain tissues by *in vivo* modification of cells, as these organs are largely composed of non-dividing cells. The illustration (Figure 13.8, left) shows the delivery of gene into liver via a catheter through the portal vein.

The second strategy for delivering therapeutic genes into recipient involves the use of live cells (also known as *ex vivo* modification of cells). In this case, the target cells are isolated from the body, therapeutic gene is introduced, and the modified cells are expanded before infusing back to the defective organ (Figure 13.8, right). *Ex vivo* modified cells for gene therapy offer several unique advantages over *in vivo* modification:

(a) It is possible to test the quality of the *ex vivo* modified cells in terms of expression of transgene product and chromosomal integration.

(b) The conditional modification is possible in *ex vivo* strategy, in which the expression of therapeutic agent can be regulated in response to certain signals.

(c) In case DNA transfection is not possible in non-dividing cells, they can be treated *ex vivo* with growth factor to make them exit from G_0-phase before transduction.

(d) DNA transfection efficiency is more than *in vivo* modification process, and toxic effects to the non-target cells can be avoided.

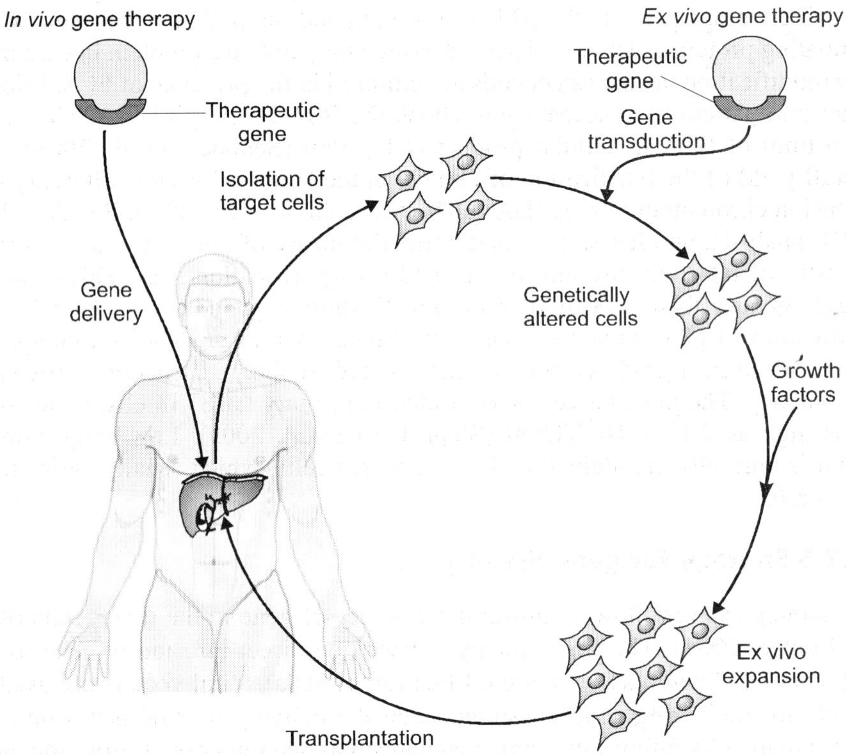

Figure 13.8 Alternate strategies for delivering therapeutic transgenes in the patients. An *in vivo* approach (left), the DNA is directly applied to the target cells of the patient. *Ex vivo* approach (right) involves harvesting cells from the tissue of interest, transducing them with the gene, and readministrating genetically modified cells into the patient.

13.2.6 Somatic cell gene therapy

The current research on gene therapy in humans deals with somatic cells. There are a number of somatic tissues that have been tried to transfer genes, but the precise choice of the tissue is depending on the nature of the disease and the vector used. The most commonly used somatic cells for gene therapy are hematopoietic stem cells (HSCs), hepatocytes, lung epithelium, fibroblasts, muscle tissue, and endothelial cells. Hematopoietic stem cells have been used as model somatic cells for gene therapy due to various reasons as described below:

Hematopoietic stem cells are the largest single somatic cell type that have been used for gene therapy. About 40% of gene therapy clinical trials conducted in USA are based on hematopoietic stem cells (NIH: Stem cells, 2002). The major advantages of using hematopoietic stem cells in gene therapy are as follows:

(a) Self-renewing nature, so long lasting effect is possible without repeated administrations of the gene.
(b) Easily available compared to other stem cells, and also easily identifiable and manipulative.
(c) Ability to give rise to different blood cells, which offer wide applications in blood and immune system disorders.
(d) Strategic delivery of gene is possible due to homing of cells in different organs.

Retroviral and lentiviral vectors are useful tools for the genetic modification of HSCs as they have the ability of stable integration in the host genome. The successful integration of transgene in the genome may result in the expression of same gene in all the progeny of the HSCs. Various lines of evidence suggest that both retroviral and lentiviral vectors do not efficiently transduce primitive human HSCs ($CD34^+CD38^{+/-}Thy\text{-}1^+Lin^-$) that belong to G_0-phase. The primitive HSCs in G_1 or $S+G_2/M$ phase are three to four times more transducible by lentiviral vectors than HSCs in the G_0-phase (Sutton *et al.*, 1999). Culture of $CD34^+$ cells with Flt-3 ligand, SCF and Tpo has been found to enhance gene transfer, compared to the unstimulated cells used as control (Logan *et al.*, 2002).

13.2.7 Germline gene therapy

Germline gene therapy refers to the introduction of genetic changes in germ cells, that incorporated in all cell types of the offspring and passed into the future generations. Germline gene transfer opens new possibility for the treatment of rare genetic disorders in humans. Two stages of the germline are suitable for genetic engineering—one the released oocyte (before or after fertilization) and another, the cells in the blastomere stage. The blastomere stage is especially convenient for genetic alteration, as the derived cell line (ES) can be easily manipulated in the culture and propagated indefinitely. The basic steps involved in the hypothetical germline gene therapy in humans are described in Figure 13.9. So far, germline gene therapy has been restricted to produce only specially altered animals for experiments. For example, in knock-out mice with specific genes are inactivated for determing their functional role and transgenic sheep for synthesizing valuable hormone or enzyme.

Isolate totipotent stem cells from
preimplanted blastocyst (8-cells stage)

↓

Diagnose any genetic abnormality (disease)

↓

Expand ES cell in culture

↓

Transfer genetic materials to the ES cells
with respect to the defective gene

↓

Select cells transduced by the transgene

↓

Target gene replacement

↓

Marker removal

↓

Confirm genomic integrity

↓

GCNT

↓

Reimplant into mother

Figure 13.9 Sequential steps in germ line gene therapy.

13.3 ADULT STEM CELLS THERAPY

Adult stem cells therapy could be a part of gene therapy where these cells are used a vehicle to deliver the trans gene. In this subsection, adult stem cells therapy, without any genetic manipulation of the cells will be discussed. Adult stem cells mean any somatic stem cells or progenitor cells, but those are popular in clinical applications are mentioned.

13.3.1 Hematopoietic stem and progenitor cells

Hematopoietic stem cells transplantation is carried out for the purpose of reconstituting bone marrow with functional cells that can support patients for the reasonable period of time. Details of the characterization of hematopoietic stem/progenitor cells are described in Chapter 11. As mentioned earlier these cells are present in the bone marrow and in cord blood. For allogeneic transplantation, healthy donors are usually the sources of these cells harvested either directly from bone marrow or from the peripheral blood as cytokine (G-CSF) mobilized cells, which are preferred in these days. In case of autologous therapy, these cells are harvested either from patient's bone marrow or peripheral blood. In this sub-section, I will provide one example of logistics for the hematopoietic stem cell transplantation in case of a cancer patient. It is important to not that in case nonmalignant hematological diseases (e.g., aplastic anemia, thalassemia beta, etc.) similar steps are followed except that the patients are not subjected to radiation or chemotherapy. The entire cycles of the procedure for transplantation of bone marrow or peripheral blood stem cells are shown in Figure 13.10. The steps are as follows:

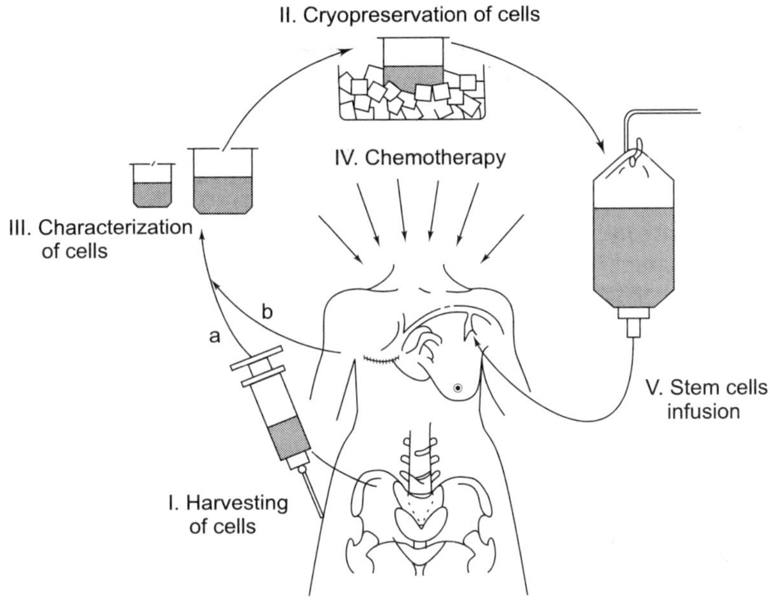

Figure 13.10 Transplantation of bone marrow (a) or peripheral blood (b) stem cells are shown. Cells are harvested from the cancer patient, processed and cryopreserved at -80°C (for 1 to 3 months) or in LN2 (more than 3 months). A small aliquot of cells is used for laboratory characterization. In the mean time, the patient is subjected to chemotherapy and/or radiation therapy. Post chemotherapy tumor burden is analyzed, when it is reached to expected low level, cryopreserved cells are infused to the patient.

I & II Harvesting and cryopreservation of stem cells: The donor, in this case patient, is anesthetized and with the help of a special needle about 1 liter marrow is harvested from different sites of iliac crest bone. To avoid clotting, marrow is mixed with sterile normal saline containing heparin. Marrow is filtered through 500 μm nylon filter to remove bone particles and other particulate matters, and collected in infusion bags for cryopreservation (see section 3.3). Every attempt is made to harvest rich bone marrow in which at least 2×10^6 $CD34^+$ cells/kg body weight can be obtained. In alternate, G-CSF stimulates mobilization of stem cells by decreasing SDF-1α expression and increasing proteases in the marrow, leading to mobilization of cells in the peripheral blood. G-CSF is subcutaneously injected at a dose of 10 mg/kg b.w/day for a period of 10 to 15 day, on 7th days CD34$^+$ cells start accumulation in the peripheral blood. Once 5 to 20 CD34$^+$ cells µl blood are accumulated, the patient is schedules for apheresis and 250 to 300 ml of stem/progenitor cells enrich suspension (devoid of erythrocytes and other white blood cells) are collected in infusion bags for cryopreservation.

III. Characterization of cells: At this stage a small fraction of bone marrow or G-CSF mobilized peripheral blood cells are characterized in three ways: (a) number of nucleated cells per µl or ml, (b) number of CD34$^+$ cells or CD34$^+$CD38$^-$ cells per 10^6 nucleated cells, and (c) number of CFC (GEMM, G, E) per 10^5 nucleated cells.

IV. Chemotherapy: Patients are treated with chemotherapy and/or radiation therapy to destroy unwanted tumour cells and develop bone marrow failure. The nature of chemotherapeutic agent and its dose is depending upon the type of cancer and its' stage. The most commonly used anti-cancer agents are hydroxylurea and 5-Flurouracil (antimetabolites), vincristine and vinblastine (antimitotics), daunor-ubicin and doxorubinin (antitumor antibiotics), carboplatin and cyclophosphamide (DNA-damaging agents), etc.

V. Stem cells infusion: Stem cells are transferred to patients by infusion, a procedure similar to a blood transfusion. Blood is delivered through a catheter — a thin flexible tube — into a large blood vessel, usually in your chest. Infusing the stem cells usually takes several hours. Post transplantation period is critical for the patients for proper engraftment, which is indirectly assessed by counting platelets and neutrophils in the peripheral blood.

13.3.2 MSCs

MSCs have been considered as promising therapeutic tool in regenerative medicine due to several advantages including self-renewal potential, immunomodulation property, engraftment capacity, secretion of trophic factors and multi-lineages differentiation potential. Moreover, the current evidences have shown that MSCs can be isolated from various tissues, promising among them are bone marrow, adipose, Wharton jelly of umbilical cord and lesser extent from dental pulp are important. Due to these unique advantages, since after the first trial with MSCs in 1995, the database http://clinicaltrials.gov show 206 clinical trials have been pursued between 2011 to 2012 for wide range of applications (Figure 13.11). Since single major application of MSCs has been found to be in case of myocardium infarction (MI), this sub-section will cover elaborately on that. In MI cardiomyocytes undergo rapid degeneration in the infarcted area due to inadequate supply of oxygen and other nutrients as a result of blockage in one of the major arteries. Not only cardiomyocytes die, along with entire vasculature in that part cripple down to death leading to aberrant regeneration of the fibroblasts. Therefore, three major cell types are required for complete regeneration of infarcted cardiac tissue, one parencymal cardiomyocytes, and other two nonparencymal endothelial and smooth muscle cells. Different cell types have been tested in experimental animals for the regeneration of infarcted cardiac tissue; except for ES cells none of them produces all three cell types. Use of ES cells in human has been prohibited due to teratogenic in nature. The best source of cardiomyocytes is cardiac progenitor cells, isolated from primary adult source, but it is difficult to obtain due to shortage of cadaver organs and difficulty of on time retrieval of the cells; moreover in absence of new blood vessels tissue regeneration could not occurs properly. Endothelial progenitor ($CD133^+$) cells, present in the bone marrow, are found to be involved in the formation of new vasculature in regenerating heart. Thus, scientists and clinicians have anticipated transplanting hematopoietic cells in infarcted area will form new vasculature result in the regeneration of damaged tissue with the help of existing cardiac progenitor cells present in the heart. In the present clinical trials, MSCs have been considered the best candidate to treat MI as they can differentiate into desired cellular phenotypes (Table 13.1). The characteristics of stem cells to be used for therapeutic application in infracted heart will depend up on the differentiation ability to form cardiac of and other cells, at the same time will be capable of inducing growth and maturation of cardiac progenitor, protecting their apoptosis and forming vasculature (Figure 13.12). In therapeutic application, better improvement of heart has been observed in cases when stem cells are directly injected in the heart rather that infusing through peripheral intravenous route. This is believed to be due to better engraft of cells on infarcted heart rather than distributing cells to the entire body in case of intravenous route. Presently, stem cells are applied to the infarcted heart via three alternate routes: transepicardial, transendocardial and intracoronary (Figure 13.13). The number of MSCs transplanted in clinical trials for MI patients have been widely varied from 0.5×10^6 to 6×10^6 cells/kg body weight.

Table 13.1 Different cell sources have tendencies to form important cells in regenerating MI heart of experimental animals

Cell types generated	ES cells	Skeletal muscle	Cardiac progenitor cells	Hemato-poietic cells	MSCs
Cardiomyocytes	++	+	++	–	+
Endothelial cells	++	–	–	++	+
Smooth muscle cells	++	–	–	–	+

++, very high; +, moderate; -, nil

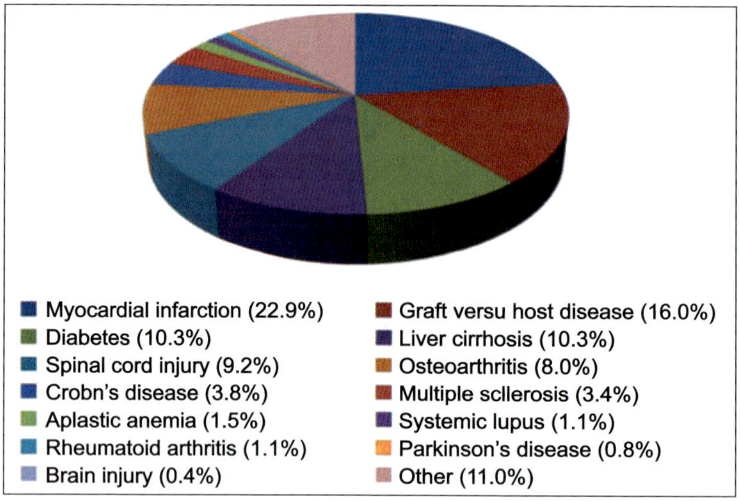

Figure 13.11 Percentage distribution of the undergoing clinical trials of MSCs on various diseases. [*Source*: https://clinicaltrails.gov/]

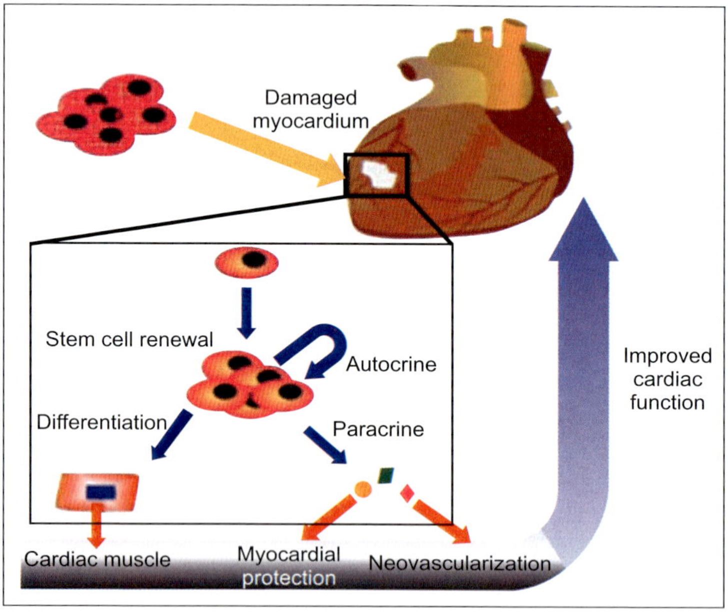

Figure 13.12 Schematic representation of the proposed mechanisms by which stem cells are expected to be involved in repair of ischemic heart. The cells are capable of proliferation into cardiomyocytes and activate existing cardiac and endothelial progenitors for repair and revascularization. [Taken from Ref. 6]

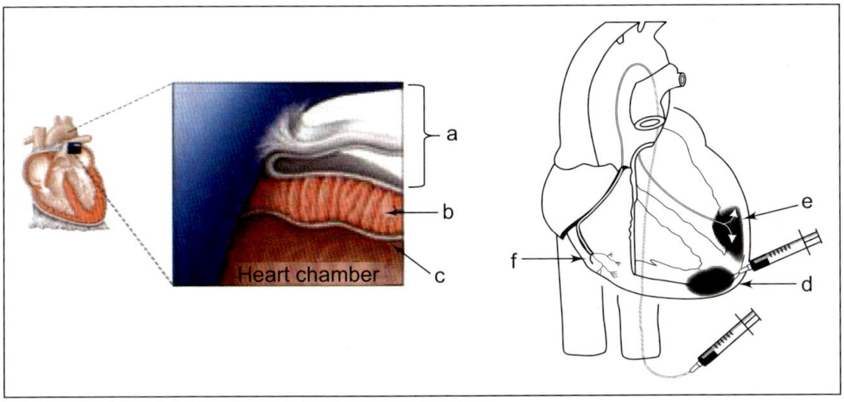

Figure 13.13 Layers of the heart wall and current strategies of direct stem cell application to the heart: epicardium (pericardium)- outer layer; (b) myocardium- cardiac muscle, (c) endocardium-endothelium lines, (d) transepicardial injection, (e) catheter-based transendocardial transplantation, (f) intracoronary catheter-based infusion. [Taken from Ref. 19]

13.4 PLURIPOTENT STEM CELLS THERAPY

In this category, ES cells (embryonic nature) and iPS cells (embryonic nature) are most important. As both the cell types are pluripotent in nature, they can differentiate into all cell types belonging to 3 germ layers. It has been mentioned earlier that due to embryonic in nature these cells have a tendency to form teratoma, so it is not recommended to use in undifferentiated form. Various steps followed in ES and iPS cells based therapy are depicted in Figure 13.14. It is interesting to note that both these cell therapies follow common steps. So far, cell therapeutic using pluripotent stem cells has been restricted only in case of experimental animals, except for one application where ES cell-derived neurons were introduced in clinical trial. In this sub-section, I will discuss more detail in iPS cells based applications.

In cell therapy for treatment of age related degeneration, targeted differentiation of iPS cells will be adequate for transplantation in order to restore the normal function of the organ. Allogeneic cell therapy, such as in case of ES cells, has become complicated by the limited availability of matched tissues. This requires life-long treatment with immune-suppressive drugs to prevent rejection of the graft, which may cause serious side effects in the other organs. iPS technology offer the advantage to circumvent this problem as patient's own cells could be coaxed into the desired cell types. The most exciting application is *custom-tailored* cell therapy in which the current transplantation approaches is combined with repairing mutation, causes the disease, by homologous recombination. Hanna et al (2007) by an elegant study have shown that iPS technology can be

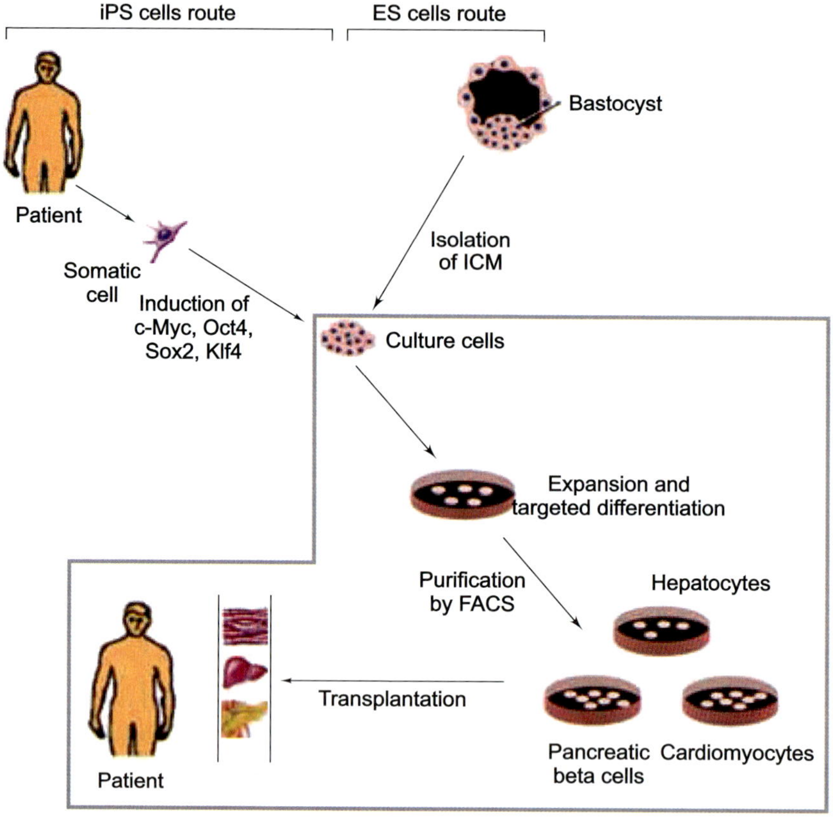

Figure 13.14 Cross-talk between ES and iPS cells based therapy. The common parts of both the routes have been enclosed in the box. ES cells based therapy is essentially allogeneic, whereas iPS cells based therapy could be autologous and allogeneic.

used to rescue defects in animals suffering from sickle cell anemia. The disease is caused by single point mutation in the hemoglobin gene rendering red blood cells nonfunctional. In the proof-of-concept study, skin cells of the diseased mouse were converted into iPS cells, subsequently mutation was corrected in the cells by gene targeting (gene editing), prior to coaxed into blood-forming progenitors. The healthy progenitors were transplanted back into anemic mice, where they produced normal red blood cells (Figure 13.15). In principle, the same method can be applied for any human diseases causing due to mutation of a gene and which is known.

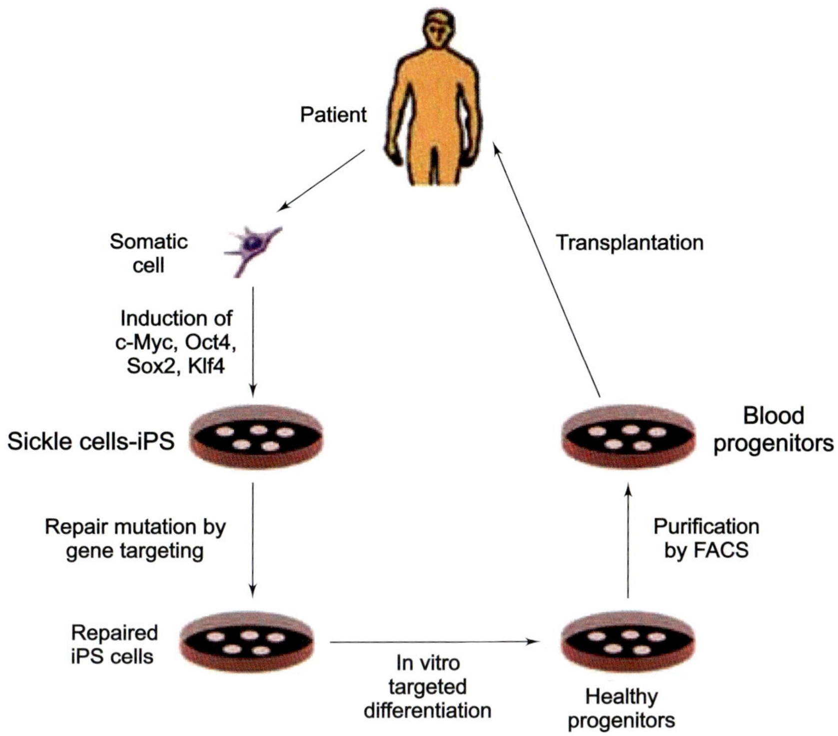

Figure 13.15 Potential applications of iPS technology in sickle cells anemia. Sickle cell anaemia-specific iPSC cells are subjected to gene targeting to repair the mutation. After that coaxed blood progenitors are transplanted back into the diseases subject. [Adapted from Ref. 18]

13.5 SUMMARY

1. Cloning targets the replacement of entire genome of an unfertilized oocyte, whereas gene therapy refers to the introduction of the correct form of the mutated gene responsible for the disease.
2. Therapeutic cloning is linked with the regenerative biomedicine, by which patient's own body cells can be produced to replace the degenerated cells avoiding immune rejection.
3. Gene therapy is more promising in targeting defect in single gene. Viral vectors consisting the transgene are introduced to the target cells through *in vivo* or *ex vivo* techniques.
4. The viral vectors used for gene therapy are one cycle infective, but are replication defective.
5. Adult somatic or pluripolent stem cells can be used for regeneration of tissue/organ. In can of pluripotent stem cells (ES/iPS), targeted differention into desired cell type in must before transplantation.

ABBREVIATIONS

AAV	Adeno-associated virus
BSA	Bovine serum albumin
CMV	Cytomegalo virus
DEAE	N,N-diethylaminoethyl
dsDNA	Double stranded DNA
ELISA	Enzyme linked immunosorbent assay
HVGD	Host versus graft disease
HIV	Human immunodeficiency virus
LOS	Large offspring syndrome
LTR	Long terminal repeat
MHC	Major histocompatibility complex
PAGE	Polyacrylamide gel electrophoresis
RRE	Rev responsive element
SCNT	Somatic cell nuclear transfer
SIN	Self-inactivating
VSV-G	Vesicular stomatitis virus-G
SDS	Sodium dodecyl sulphate

REFERENCES

1. Amado RG, Chen YIS (1999) *Science* 285, 674.
2. Coleman JE (2003) *Efficient transduction and targeted expression of lentiviral vector transgenes in the developing retina.* Ph.D. Thesis Submitted to University of Florida, Florida.
3. Copelan EA. N Engl J Med 2006, 354, 1813
4. Dull T, Zufferey R, Kelly M, Mandel RJ, Nguyen M, Trono D, Naldini L (1998) *J Virol.* 72, 8463.
5. Gurdon JB, Uehlinger V (1966) *Nature* 18, 210.
6. Feng Y, Wang Y, Cao N, *et al.* Ann Pallial Med. 2012, 1, 65.
7. Hanna J, Werning M, Markoulaki S, *et al.* Science 2007, 318, 1920.
8. Illmensee K, Hoppe PC (1981) *Cell* 23, 9.
9. Kaji EH, Leiden JM (2001) *JAMA* 285, 545.
10. Logan AC, Lutzko C, Kohn DB (2002) *Curr Opn Biotechnol.* 13, 429.
11. Nielsen TO (1997) *MJM* 3, 126.
12. Naldini *et al.* (1996) *Science* 272, 263.
13. Nichols E (1988) Institute of Medicine: National Academy of Sciences. *Human Gene Therapy.* Harvard University Press, Cambridge, MA.
14. NIH stem cells—Scientific progress and future research direcions, 2002.
15. Rideout WM III, Eggank, Jaenisch R (2001) *Science* 293, 1093.
16. Slepushkin V, Chang N, Cohen R, *et al.* (2003) *BioProcessing J*, 1, 89.
17. Schonely K, Afable C, Slepushkin V, *et al.* (2003) *BioProcessing J* 2, 39.
18. Stadtfeld M, Hochedlinger K, Geves & Dev 2010, 24, 2239
19. Steinholf G, Choi Y-H, Stamm C. Eu Heart J Supp 2006, H32 [dio:10.1039/earhearty/sub065.

20. Sutton RE, Reitsma MJ, Uchida N, Brown PO (1999) *J Virol* 73, 3649.
21. Wilmut I, Schnieke AE, McWhier J, *et al.* (1997). *Nature* 385, 810.

SUGGESTION FOR FURTHER READING

1. Adler K, Gifford J, Summar R. HIV as a vector in gene theragy (LHP://wwwpp.wwrf.elu%7Ekk00/hirvector/ivvector.htm).
2. Change PL, ed. (1994) *Somatic gene therapy*. CRC Press, Inc. Boca Raton, Florida.
3. Kucherlapati R, ed. (1986) *Gene transfer*. Plenum Press, New York.
4. Suter DM, Dubois-Dauphin M, Krause Karl-Heinz (2006) Genetic engineering of ES cells. *Swiss Med Wkly* 136, 413.
5. Wolff JA, ed. (1994) *Gene therapeutics- Methods and applications of direct gene transfer*. Birkhauser, Boston.
6. Hollis RP, Nightingale SJ, Wang X, *et al.* (2006) *Exp. Hematol.* 34, 1333.
7. Dixit P, Katare R. Stem Cell Research & Therapy 2005, 6, 26
8. Wei X, Yang X, Ha Z-P, *et al.* Acta Pharmacologica Simica 2013, 34, 797

14

Present Trend and Future Perspective

The major driving force behind scientific improvement in healthcare is innovative research in bioscience and medical fields. Understanding the causes of many diseases and development of treatment modalities, followed by the growth in biopharmaceuticals have created an unprecedented progress in cell culture based and allied industries. The first cell culture based product, the rabies vaccine, was introduced about 50 years back. After that cell culture technology and the relevant industries have been reached to a new peak with the advent of recombinant DNA and cell hybridization technologies. Cell culture industry is now a major segment of the biopharmaceutical market. The old products and technologies are gradually entering into saturation, while newer products and technologies have created new market place. The development of animal cell based products is the result of combined efforts of basic scientists, physician doctors and engineers, while the end-users of the products are exclusively patients, who require medical supports. Like any other products, the market force for the innovation depends on the following factors:

1. Who needs the product and specific requirements?
2. What are the choices of the competing products that the patients have or want?
3. Whether the product is the result of best research and innovation?
4. Who can provide or manufacture the products?

On the basis of end-users, the future perspective of animal cell culture services is broadly divided into two classes. First, those who utilize animal cell technology-the manufacturers, and the second, those who consume the final products of animal cell technology- the patients. Therefore, the driving forces behind adopting the new trend among respective classes of end-users are different. Manufacturers rely on the available technology and the method of production; so, their deciding factors to adopt a new innovative technology are cost of the production, market volume and profit. On the other hand, patients have very little choices as merely few alternate products are at their disposal, and moreover products (drugs) cannot be consumed unless prescribed by a physician doctor.

14.1 SHIFT OF GOAL IN ENGINEERING AND PRODUCTION TECHNOLOGIES

The technological achievement in modern animal cell culture facility during past twenty years is the fruits of continuous effort in development of bioreactors, culture media, supplements and manufacturing processes. The progress has been guided by few goals: (a) maintenance of culture sterility for longer duration, (b) gentle handling and safe culture harvest, (c) increase of product yield, and (d) improvements in process design. During past more than two decades, tremendous improvement have been registared both in scientific and technical aspects of animal cell culture. To maintain such trend in the future, scientific and technical efforts have been made in many areas; some of them are mentioned in the Table 14.1.

Table 14.1 Present and future trends in technology and engineering aspects cell culture

Technology/Engineering improvement	*Present status*	*Future trends*
1. Host cells	(a) Type culture (host cells) and some improvement in protein glycosylation pattern. (b) Human ES cell lines have been established (1998 to 2001). (c) iPS cell line (2006 onwords)	(a) Host cell engineering to make custom made glycosylated proteins, matching with the native molecules. (b) Development of growth factor independent cell lines. (c) Development of apoptosis resistance cell lines. (d) Derivation of cGMP cerlified ES cell lines with proper characterization. (e) Simple technology to derived iPS cells, genetic correction, etc. (f) Development of universal donor therapeutically important cells (ES & iPSC based) (g) Use of CCL (non-tumorigenic) for viral vaccine production.
2. Technology for expansion and differentiation of transplantable cells	(a) Un-manipulated/ un-expanded HSCs are used for transplantation.	(a) Expansion of HSCs with maintaining long-term multilineage marrow engraftment potential.

(*Contd.*)

(*Contd.*)

Technology/Engineering improvement	*Present status*	*Future trends*
	(b) GMP certified processes for expansion of MSCs, ES/iPS cell lines.	(b) *Ex vivo* differentiation of HSCs into hepatocytes, cardiomyocytes, and other useful cells. (c) Expansion of human ES/iPS cells without feeder layer. (d) Targeted differentiation of ES/iPS cells.
3. Cell culture medium	(a) Standard formulation and commercially available culture media. (b) High cost of media.	(a) To develop medium that produce less metabolic toxic products. (b) To develop low cost medium for serum free formulation. (c) Media that induce self-renewal of stem cells but prevent their differentiation. (d) Media that induce targeted differentiation of stem cells.
4. Bioreactor and control systems	(a) Presently used systems are designed on the basis of experiences from microbial culture process with some modification to facilitate animal cell growth.	(a) High mixing, low shear, high perfusion rate bioreactors need to be developed. (b) Development of bioreactor that can facilitate high oxygen transfer in low shear field. (c) Development of bioreactor for high cell-density culture of anchorage-dependent cells. (d) Development of niche bioreactors for expansion of different stem cells. (e) Development of bioreactor for tissue engineering purposes: pulsatile, hydrodynamic shear force induced, etc.
5. Matrices for supporting growth of cells: tissue engineering	(a) Several classes of microcarriers have been developed, but very few of them exist in the market.	(a) Development of cheap microcarriers for routine cell culture applications for manufacturing vaccines and therapentic proteins.

(*Contd.*)

(*Contd.*)

Technology/Engineering improvement	*Present status*	*Future trends*
and other cell culture applications.	(b) Commercially available microcarriers are high price, which causes cell culture based processes cost-intensive. (c) Due to high cost of microcarriers, the trend for the past many years has been adopting free suspension culture instead of microcarrier based culture. (d) Synthetic and natural polymer based matrices are developed in various laboratories for use in tissue enginee-ring field.	(b) Development of customized biocompatible 3D matrices for tissue engineering applications. (c) Development of culture process that can support high cell density culture in suspension.

14.2 PARADIGM SHIFT IN TREND OF FUTURE PRODUCTS

The broad future perspective of cell culture related products are depicted in the Table 14.2; few of them have been elaborated in this section. The first group of products are from the *transgenic organisms*. A transgenic organism carries additional genetic information in its genome. Besides transgenic proteins, production in milk, the future medicines are expected to be the life cells. In analogous to the recombinant protein produced in cell line using bioreactor, these transgenic cells could be directly transplanted to the patients suffering from the same disease. A relatively new and promising application of the transgenic production technology is in the use of animals (e.g., porcine) as a donor of genetically modified solid organs for human transplantation. This would be possible with the advancement of our knowledge on therapeutic cloning, transplant immunology, and prevention of transmission of pathogens from animal to human. The impetus behind the progress in *xenografts* has been the acute shortage of donor organs: cadaver or live organ donation.

Another aspect of future development is in the area of tissue engineering for reconstructing every possible tissues and organs in our body. Tissue engineering incorporates the fields of cell transplantation, material science and engineering. Individuals who have mastered the techniques of cell harvest, culture, expansion, transplantation, as well as polymer design have important roles for successful application of this technology. At present different engineered tissues are at several stages of the development; some are already used clinically, a few are in preclinical trials, and many are in the invention stages. Recent progress suggests that engineered tissues may have major clinical applicability in the future as they represent a viable therapeutic option for that require tissue replacement. Advances in stem cell biology, tissue engineering and iPS cell technology have made it possible to combine above and to create the comprehensive discipline of regenerative medicine.

The development of cancer vaccine is another most interesting area on which continuous progress has be made. The term *cancer vaccine* is often used to describe a process whereby a person's immune system is activated for recognizing and destroying malignant cells without harming normal cells. The treatment with cancer vaccine may be considered as *immunotherapy*. Immunotherapy against cancer refers to antigen-specific adoptive immunotherapy with antibodies and/or T cells, and antigen-specific active immunization (vaccination). The limitation of using antibodies is that the target proteins need to be expressed specifically at the surface of cancer cells. Identification of few defined tumor antigens in humans has catalyzed the development of adoptive T-cell therapy. Vaccines against cancer aim

Table 14.2 Current and future perspective in mammalian cell derived products

Category of products based on health disorder	*Products available in the market*	*Future trend*
1. Cell culture based vaccines against virus and other diseases.	(a) Polio, (b) Rabies, (c) Measles, (d) Mumps, (e) Rubella, (f) Smallpox, (g) Chickenpox, (h) Hepatitis A, etc.	(a) Development of next generation seasonal influenza and pandemic vaccines. (b) Discovery of therapeutic vaccines against HIV that will induce virus-specific cytotoxic T cells destroys latently infected cells. Other important vaccines of similar kind are against human papilloma virus, herpes, diabetes, and multiple sclerosis. (c) Immunotherapy for treatment of cancer patients. (d) Subunit vaccines.

(*Contd.*)

(*Contd.*)

Category of products based on health disorder	*Products available in the market*	*Future trend*
2. Therapeutic proteins against blood disorders, viral infection, and other health disorders.	(a) EPO, (b) t-PA, (c) Factor VIII & IX, (d) Interferon β, etc. (e) Gene therapy to introduce normal target gene in the diseased organ/body	(a) Development of next generation immuno-tolerated proteins for humans (more effective and less side effect). (b) Product derived from mammary gland of transgenic goat and sheep: antithrombin III, t-PA, EPO, Factor VIII, etc. Antibodies and hemoglobin from transgenic goat and swine, respectively. (c) Construction of efficient vectors for gene therapy, which are less immunogenic, stable, and where gene expression can be regulated. (d) Cell replacement therapy to supplement with right kind of cells making the protein, such as, pancreatic islet for insulin production, etc.
3. Therapeutic monoclonal antibodies against immunological rejection and cancer	Chimeric antibodies against (a) Rheumatoid arthritis, (b) Transplantation rejection, (c) Crohn's disease, (d) Breast cancer, (e) Colon cancer, and (f) Other malignant disorders	(a) Chimeric and humanized (CDR technology) antibodies against malignant and non-malignant disorders (highly specific and less side effect). (b) Development of catalytic antibodies.
4. Prosthesis devise, cells/ tissues, organ transplantation (HLA matched donors)	(a) Bone marrow reconstitution, (b) Skin grafting, (c) Liver, kidney, heart-lung transplantation, (d) Cornea grafting.	(a) *In vitro* organogenesis (tissue engineering) for artificial liver, skin, etc. (b) Cell therapy to repair damaged organs. (c) Xenotransplantation using barrier technology

(*Contd.*)

(*Contd.*)

Category of products based on health disorder	*Products available in the market*	*Future trend*
		(d) Development of universal donor stem cells. (e) Development of extracorporeal devices to support kidney and liver.
5. Cells/tissues, organ transp-lantation (allogenic donors, using immuno-compromising drugs)	(a) Bone marrow reconstitution, (b) Skin grafting, (c) Liver, kidney, heart-lung transplantation	(a) Therapeutic cloning followed by tissue engineering (b) Xenotransplantation of organs.
6. Biomedical/drug testing	2D culture of sensitive cells for: (a) Testing anti-cancer drug (b) Testing of cosmetics, (c) Drug metabolism study for hepatocytes.	(a) Development of 3D culture for drug sensitive cell lines (b) Development of artificial organ like liver, kidney, etc.

to induce tumor-specific effector T cells that can reduce the tumor mass, as well as maintain tumor-specific memory T cells that can control tumor relapse. Owing to their capacity to regulate T-cell immunity, dendritic cells are increasingly used as an adjuvant for vaccination, and the immunogenicity of antigens deliver by dendritic cells has now been shown in patients with cancer. Dendritic-cell vaccines have shown some promise against melanoma, prostate cancer, and lymphoma.

SUGGESTION FOR FURTHER READING

1. Kumar R (1994) In: *Animal cell technology: Basic and applied aspects* (T. Kobayashi, Y. Kitagawa, K. Okumura, eds.). Kluwer Academic Publishers, Dordrecht, The Netharlands, p.87.
2. Gregory AP, Dennis M, Ruben B-G (2002) *Br. J Med.* 324, 1315.
3. Banchereau J and Palucka AK (2005) Nat. Rev. Immunol. 5, 296.

Section V
Analytical Techniques and Cell Isolation

15

Quantitative Analysis

Quantitative analysis of each experiment and interpretation of the results are two most important tasks for all studies in biology and engineering. Analysis of results help to optimize process variables, select right clone, optimize medium composition, determine the efficiency of a drug, and many others. Quantitative analysis of any experiment is carried out on the basis of certain readouts or results of the assays/tests, by which one can compare the outcome of an experimental events with that of the other. Some of the important analytical techniques by which animal cell culture experiments are quantitativety interpreted has explained in this chapter.

15.1 ENUMERATION OF CELL NUMBER

Enumeration of cell number is most common in any cell culture experiments. It is not possible to determine the consistency of culture processes and to know the optimum time for harvesting a culture or to subculture unless cell number is known. Without knowing the exact cell number, it is not possible to evaluate cytotoxicity of a drug or to determine plating efficiency of a batch of serum. In microbial culture, cell number is indirectly determined by measuring the *optical density* (OD) of medium or estimating dry cell weight; however, these methods are not popular in animal cell culture. The following two popular techniques can be adopted for enumerating cell number of a culture.

15.1.1 Counting Chamber

It is the simplest device to enumerate cell number in unit volume of culture. The most widely used type is known as hemocytometer (improved version of Neubauer chamber), as it was originally designed for counting blood cells. A hemocytometer is a thick glass mirror-like polished surface having grids on the surface, as shown in Figure 15.1. It is extremely important for careful handling of hemocytometer in highpower objectives, as the counting chamber is much thicker than the conventional glass slide. One entire grid on a standard hemocytometer with all rulings can be seen at 40 ×. The main divisions separate the grid into 9 large squares. Conventionally, four large squares are utilized to count the number of cells in each application. The volume of one such large square (seen as × in Fig. 15.1) is 0.0001 cm^3 (ml) (length × width × height; i.e., 0.1 cm × 0.1 cm × 0.01 cm).

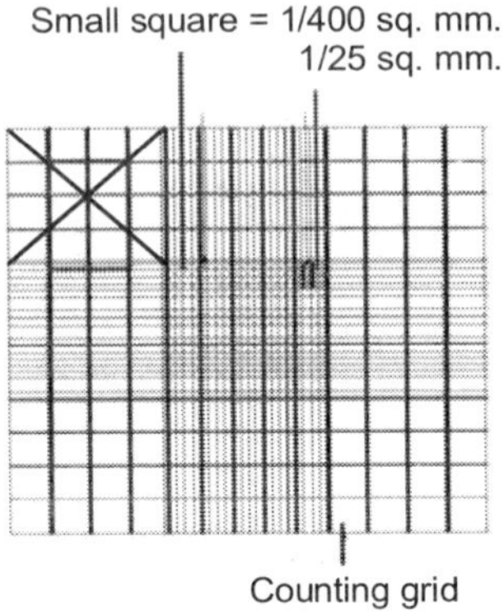

Figure 15.1 Hemocytometer for enumerating cell number.

For counting cell number, the chamber and the cover slip are properly cleaned to remove oily substances. The cover slips used in counting chamber are thicker than those used for mounting in conventional microscopy to overcome the surface tension of liquid droplet. Prior to apply cell suspension, the cover slip is placed over the counting surface, and then the sample is introduced into one of the V-shaped wells with the help of a Pasteur pipette or micro-tip. The area under the cover slip become loaded with cell suspension by capillary action. Enough liquid (~ 10 μl) should be introduced so that the mirrored surface is just covered. The loaded counting chamber is then placed on the microscope stage and then counting is performed at low power (4× or 10× objectives). For accurate counting, cell surpenion is diluted enough, so that they are uniformly distributed within the grid. The optimum dilution of cell suspension should be such that the cell number in each large square is between 25 and 50. As a convention, the cells that overlap the top or right ruling are counted, and those that overlap the bottom or left ruling are excluded. Since viable cell counting is the main function of chamber-based counting technique, cell suspension is mixed with 0.4% (w/v) trypan blue solution (1:1) prior to applying to the counting chamber. Trypan blue is a vital dye; intact plasma membrane of the viable cells excludes the dye, whereas damaged membrane of dead cells retains the blue color. By counting unstained bright cells one can determine viable cell number.

Single cell suspension is preferred for accurate counting of cell number. If cellular aggregates are unavoidable, it is recommended to follow alternate technique for counting cells. Aggregated cells are lysed by incubating them in presence of 0.1 M citric acid containing 0.1% crystal violet at 37°C for 1 h. Citric acid disintegrates cell membrane, thus the nucleus is stained with violet dye. By counting the number of violet color nuclei, one can enumerate total cell number.

The same technique is followed for enumerating cells growing on solid microcarriers. The disadvantage of crystal violet method is that nuclei of both viable and dead cells are stained with the dye. Cell density is calculated as follows:

$$\text{Cell number per milliliter} = [C_1 + C_2 + C_3 + C_4] /4 \times \text{dilution factor} \times 10^4 \qquad 15.1$$

where, C_1, C_2, etc. are cells in four large squares, and volume of each large square is 10^{-4} ml.

15.1.2 Coulter particle counter

Coulter particle counter is based on the principle that the current-flow between two electrodes is interrupted when a particle moves through a small aperture separating the two electrodes (Figure 15.2A). In addition to that, each particle also displaces its own volume of conducting fluid, resulting in momentarily increase of impedance of the aperture. This change in impedance produces a tiny but proportional current flow, which is converted into voltage pulse and is registered in the recorder.

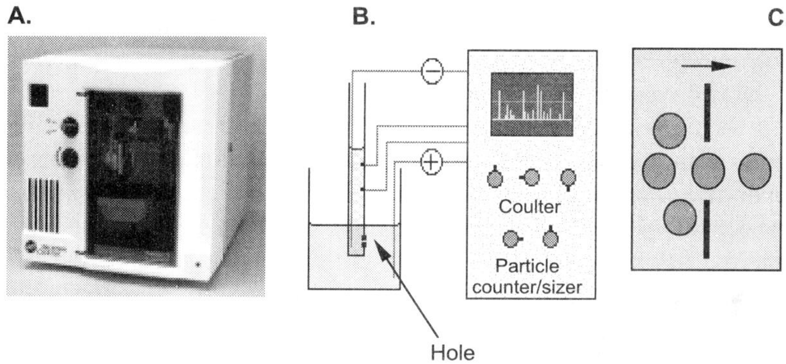

Figure 15.2 Coulter Multisizer^{TM} 3 particle analyzer and the function (Beckman Coulter). **A**. The basic Coulter counter; **B**. Narrow tube in which sample is loaded; **C**. Flow of particles on the basis of electrical gradient.

The basic system of Coulter particle counter consists a reservoir into which a narrow tube is partly immersed (Figure 15.2B). The tube is filled with liquid up to the point above the level of the outside reservoir. A small hole of slightly larger diameter than the average size of the cells is present near the lower part of the tube. The content of the tube is slowly flow through this hole until the liquid levels of both lower and upper chambers are equalized. Cathode is attached to the tube and anode to the reservoir. Thus, with the flow of liquid an electrical current will also flow through the hole, which is measured by an oscilloscope. When a particle (cell) flows through the hole (Figure 15.2C), the electronic system detects a sudden and momentary interruption of current flow, followed by an increase in resistance. Two more electrodes are placed in the upper part of the tube to detect inside fluid level. These electrodes are placed in such a way that if the liquid meniscus breaks contact with the upper sensor, the counter starts counting, and when the meniscus breaks contact with the lower sensor, the counter stops counting. This process counts the number of particles present in 0.1 ml liquid volume. The quantity of suspension is precisely drawn by the system to allow counting of particles in reproducible manner. The oscilloscope is able to detect about 300,000 partial interruptions per second. For accurate particle counting, properly diluted sample should be applied. It is needless to mention that Coulter counter does not differentiate between the live and dead cells. In other words, this instrument measures total cell number, and the detection of particle is independent of its shape, color and density.

All suspensions, to be used in culture counter, must be made in prefiltered and degassed buffer as the intrument is extremely sensitive to air bubbles and anything that might clog the fine hole. Coulter counter has small preset microscopes that look at the hole, so that one can see if it is plugged, and with the help of a miniature brush, the blockage is removed. The method for enumerating cell number is as follows:

1. Switch-on the machine 5-10 minutes before the actual counting process starts.
2. Prepare single cell suspension of properly diluted sample in Tris buffer (note down the dilution factor, say 20), and apply the sample under the tip of the orifice tube.
3. Turn the glass valve 'A' directly above the cells to vertical (open) till a green baseline in the sizing monitor represents cells. Turn the valve 'A' back horizontal (closed) to begin the counting.
4. The length of the count should be approximately 13 sec, it is continued from the first number of cell appears till the green display goes out.
5. The count displays the number of cells in 0.5 ml. If 0.2 ml of cells is diluted to 10 ml, multiply the displayed number with dilution factor 100 [(10/0.2) × (1/0.5)] to determine the number of cells per ml of the original culture. At high cell densities, the probability of two or more cells passing together through the aperture is increased, which will lead to false negative counting. Therefore, cell counting is performed at 2-3 different dilutions and then compound.
6. After counting, entire system including electrode is flashed with double distilled water, and the machine is turned-off.

15.2 ANALYSIS OF CULTURE MEDIUM

15.2.1 Estimation of glucose (off-line)

Glucose is the major carbon source in all cell culture media, the label of which influences the metabolism and final yield of cells. Thus, it is necessary to know the exact concentration of glucose in the culture medium. Knowing glucose concentration, it is possible to optimize medium feeding strategy in perfusion culture. The concentration of glucose can be analyzed by HPLC method using a C18 chromatography column for carbohydrates; however, enzyme-based assay is most popular due to specific, reproducible, sensitive, and less time consuming. Due to high specificity and sensitivity, enzyme assays can be performed on crude materials with little or no sample preparation. Hexokinase (HK)-based glucose assay (Sigma Diagnostics) has been described in the literature. The assay is based on the principle that nicotinamide adenine dinucleotide (NAD) is reduced of form equimolar amount of reduced NAD (NADH), resulting in the increase in absorbance at 340 nm. The increase in absorption is proportional to the NAD utilized, to the glucose concentration in the sample. The enzymatic reactions take place as follows:

First, glucose is phosphorylated by adenine triphosphate (ATP) in the presence of HK. In the second step, glucose-6-phosphate (G6P) is oxidized to 6-phosphogluconate in the presence of NAD in a reaction catalyzed by glucose-6-phosphate dehydrogenase (G6PDH).

$$\text{Glucose} + \text{ATP} \xrightarrow{\text{HK}} \text{Glucose-6-Phosphate} + \text{ADP}$$

$$\text{Glucose-6-Phosphate} + \text{NAD} \xrightarrow{\text{G6PDH}} \text{6-Phosphogluconate} + \text{NADH}$$

The detailed assay is described below:

1. Mark three test tubes as sample blank, reagent blank, and test. Add 1 ml distilled water and 0.1 ml of sample in the tube marked as sample blank. Add 1 ml reagent and 0.1 ml of distilled water in the tube marked as reagent blank, and 1 ml reagent and 0.1 ml of sample in the tube marked as test.
2. Mix the contents and incubate at room temperature for 15 min and measure the absorbance at 340 nm against distilled water as blank. The reading should be taken within 1 h of incubation.

Suppose, the absorbance of the sample blank, reagent blank, and test samples is A_{SB}, A_{RB}, and A_T, respectively, the absorbance of total blank (A_{TB}) = A_{SB} + A_{RB}, and the net increase in absorbance due to sample (A_N) = $A_T - A_{TB}$.

Glucose concentration (mg/ml)

$$= (A_N)(\text{Assay volume})(\text{Mw of glucose})(\text{Dilution factor})]/ [(\varepsilon)(\text{light path})(\text{Sample volume})(\text{conversion factor from } \mu\text{g to mg})]$$

$$= [(A_N)(1.1\ \text{ml})(180\ \mu\text{g/mmoles})(\text{Dilution factor})]/ [(6.22\ \text{ml}/\mu\text{moles.cm})(1\ \text{cm})(0.1\ \text{ml})(1000)] \qquad 15.2$$

where, ε: molar extinction coefficient for NADH at 340.

15.2.2 Estimation of ammonia and glutamine

Both ammonia and glutamine in the spent medium can be determined by measuring dissolved ammonia gas with the help of a gas-sensing electrode (Ozturk *et al.*, 1989). The assay is based on the principle that ammonia is produced by metabolism of glutamine by the cells. Glutamine is converted into glutamate and ammonia by the action of glutaminase. A gas-sensing electrode (Model EA 940, Orion Research Inc., USA) is used to determine initial level of ammonia in each sample. The unused glutamine concentration in the same sample is then reacted with glutaminase and liberated ammonia concentration is further estimated. The difference between two ammonia concentrations is the amount produced from equivalent quantity of unused glutamine, as one mole of ammonia is liberated from one mole of glutamine. The ammonia calibration curve is prepared by measuring mV readings from log dilutions of ammonium chloride standards (0.1, 1.0, 10.0 mM). Using this calibration curve and the mV reading, the concentrations of ammonia and glutamine in the samples are estimated.

15.2.3 Estimation of lactate

In the presence of excess NAD, lactate is converted into pyruvate by lactate dehydrogenase (LDH), and one mole of NADH is produced in the reaction. The increase in absorbance at 340 nm due to NADH is the measure of lactate originally present in the medium, as the reaction involves one mole of lactate with one mole of NAD with the formation of one mole of NADH, as shown below:

$$\text{Lactate} + \text{NAD} \xrightarrow{\text{LDH}} \text{Pyruvate} + \text{NADH}$$

Fifty micro-liter of properly diluted sample (0.1 – 0.5 lactate g/l) is incubated with 150 μl of reagent (Sigma) in a microtiter plate at 37°C for 15 minutes. Calibration curve is established using standard L-lactic acid (0 – 0.5 g/l). The concentrations of lactate in unknown samples are measured by comparing absorbance reading with the standard curve.

15.2.4 Estimation of amino acids

Amino acids are analyzed using PICO.TAG method (M/s Waters Associates, Inc.). Diluted culture supernatant is vacuum evaporated and derivatized using phenyl isothiocyanate (PITC) solution (ethanol: water: TEA: PITC :: 7:1:1:1) at room temperature for 10 min. Derivatized amino acid mixture is again dried under vacuum to remove excess reagents. The dried powder of each sample is dissolved in 200 μl of 4 mM Na_2HPO_4 (pH 7.4) containing 5% acetonitrile. Five to ten microliter sample is injected in C-18 amino acids column (Water Associates) and resolved via reversed phase HPLC. Gradient of acetonitrile is pumped using gradient control program. To determine the calibration curve, standard amino acid mixture and an internal standard aminohexanoic acid (AHA) are used.

15.3 STUDY OF GROWTH AND METABOLIC PARAMETERS

15.3.1 Establishing growth curve

Like microbial systems, animal cells in culture initially exhibit a initial lag phase, which is followed by a period of exponential growth, and finally a period of reduced or no growth. The duration of each period is the characteristic of each cell type and its nutritional requirement for growth. A typical batch growth curve of hybridoma is shown in Figure 15.3. The growth of cells usually means the increase in cell number; however, an increase in cell mass is often used to establish the growth curve. If cell mass is used to establish growth curve, two errors are automatically incorporated in the final results. These are due to dead and *hypertrophy* of the cells. Thus, by all practical reasons, growth curve should be established on the basis of viable cell number. The growth of cells can be quantified in terms of following relationships:

Specific growth rate (h^{-1}), μ (i.e., the rate of cell growth in unit number of cell)

$$\mu = [(1/N)(dN/dt)] \qquad 15.3$$

where 'N' is the cell number, and 'dN' is the increase in cell number in the time interval of 'dt'. Assuming that the growth rate does not change in the exponential period,

$$\ln (N_t) = \ln (N_0) + \mu t \qquad 15.4$$

where, N_0 is the number of cell at time t_0.

Doubling time (h), t_d (i.e., the time required to double the number of cells)

$$t_d = \ln 2/\mu = 0.693/\mu \qquad 15.5$$

Degree of multiplication, x (i.e., the number of times the inoculated cells double their initial number)

$$x = 3.32 \log (N_t/N_0) \qquad 15.6$$

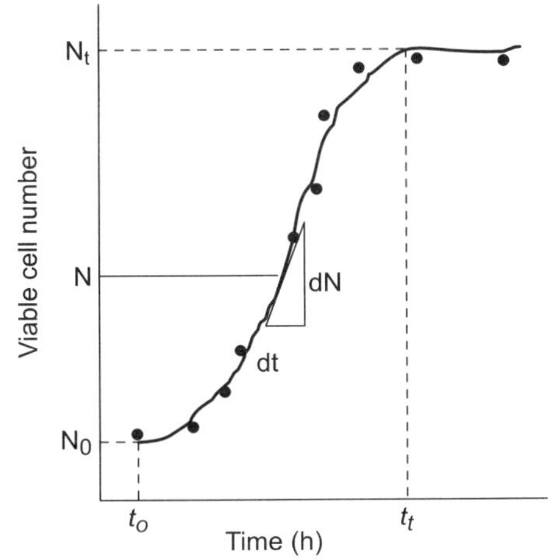

Figure 15.3 Growth curve of animal cells. Viable cell number is plotted against the time of the culture. Slope is determined at the linear (exponential) portion of the curve.

15.3.2 Specific oxygen uptake rate

Oxygen transfer rate (OTR) in microbial and animal cell culture is proportional to (i) *oxygen transfer coefficient* (K_L), (ii) area of interface (a) across which oxygen is transferred, and (iii) driving force of oxygen transfer, i.e., difference between gas phase and liquid phase oxygen concentration ($C_G - C_L$), thus

$$OTR = K_La\,(C_G - C_L) \qquad 15.7$$

where, K_La is volumetric oxygen transfer coefficient (h^{-1}).

The optimum OTR is the amount of oxygen transferred to the culture to satisfy *oxygen uptake rate* (OUR) by the culture at steady dissolved oxygen concentration, thus

$$OUR = Q_{O2}.X \qquad 15.8$$

where, 'Q_{O2}' is *specific oxygen consumption rate* (mmol/10^6 cells.h), and 'X' cell density (number/l). In steady state, when oxygen transfer and consumption is equal (i.e., $dC_L/dt = 0$), the C_L will remain constant (assuming that X is not change during that time), then

$$K_La\,(C_G - C_L) = Q_{O2}.X$$

or,

$$Q_{O2} = [K_La / X]\,[(C_G - C_L)] \qquad 15.9$$

knowing the values of K_La, X, C_G, and C_L, one can calculate Q_{O2}.

Taguchi and Humphrey (1966) introduced dynamic technique to calculate Q_{O2} of microorganisms, which can be applied in case of animal cells also. According to the dynamic technique, if oxygen supply is turned-off, the dissolved oxygen level in the culture medium is declined in the following rate:

$$dC_L/dt = Q_{O2}.X \qquad 15.10$$

Therefore, by monitoring dissolved oxygen concentration with time, a plot can be drawn for 'C_L' versus 't', the slope of which is the measure of Q_{O2}.X. Specific oxygen uptake rate (Q_{O2}) can be calculated from Figure 15.4, if 'X' is known.

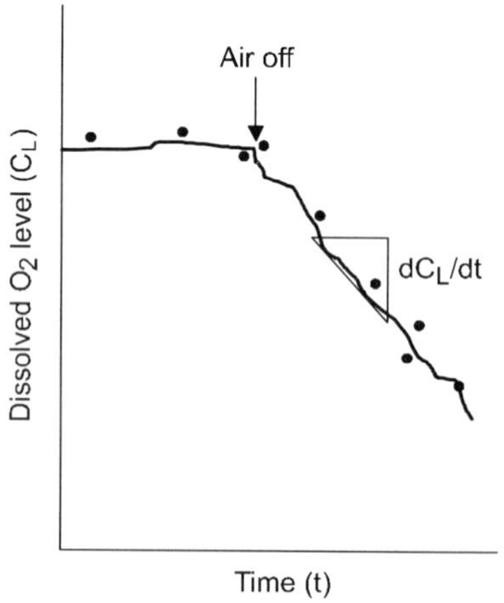

Figure 15.4 Determination of sp. oxygen uptake rate. Air supply is turned-off, and the dissolved oxygen level (C_L) in the medium is allowed to fall and recorded through a paramagnetic oxygen analyzer. The slope determines the drop of oxygen in unit time.

15.3.3 Determination of metabolic quotients

The *metabolic quotients* of animal cells are the important state variables to be studied to monitor and control cell culture processes. Metabolic quotients mean specific substrate consumption rate, or specific metabolite formation rate. The most common metabolic quotients in animal cell culture are specific glucose and glutamine consumption rates (Q_G and Q_{GL}, respectively) in mM/10^6 cells.h, and specific lactate and ammonia formation rates (Q_L and Q_A, respectively) in mM/ 10^6 cells.h. The metabolic quotients of cells during culture are determined from the slope of the plots of metabolic consumption of substrate or production of metabolites versus ($e^{\mu t}$–1), as suggested by Smiley *et al.* (1989). If the metabolic quotients are known, it becomes easy to design the supply of carbon source (feeding of media) and removal of toxic metabolic end products (e.g., ammonia and lactic acid). The metabolic quotients of animal cells are calculated by estimating concentrations of glucose at zero time (G_0) and at any time (G); glutamine at zero time (GL_0) and at any time (GL); lactate at any time (L); ammonia at any time (A); cell number at zero time (N_0); and specific growth rate (μ). All parameters are customarily determined in exponential growth phase, as metabolic quotients and specific growth rate are assumed constant in this phase, thus

$$G_0 - G = [(Q_G.N_0)/\mu][e^{\mu t}-1] \qquad 15.11$$

$$GL_0 - GL = [(Q_{GL}.N_0)/\mu][e^{\mu t}-1] \qquad 15.12$$

$$I_L = [(Q_L.N_0)/\mu][e^{\mu t}-1] \qquad 15.13$$

$$I_A = [(Q_A.N_0)/\mu][e^{\mu t}-1] \qquad 15.14$$

15.4 ASSAY OF PROTEINS/ANTIGENS

15.4.1 Enzyme-linked Immunosorbent Assay (ELISA)

The assay combines the specificity of antibodies with the sensitivity of enzyme assays by using antibodies or antigens coupled to an easily assayed enzyme that possesses a high turnover number. ELISA can be used to measure antigen or antibody concentration present in serum or in culture supernatant. If a mixture of antibody (Ab), antigen (Ag), and labeled antigen (Ag*) is allowed to equilibrate, it ends up with a mixture of both free and bound antigen with the antibody. As the quantity of both 'Ab' and 'Ag*' is constant, the amount of 'Ag*' bound (AbAg*) is dependent on the amount of 'Ag' present. In other words, if 'Ag' is high, the amount of 'Ag*' bound (AbAg*) will be low and vice versa. By measuring the amount of 'Ag*' in the free and bound fractions, it is possible to determine the amount of 'Ag' present in the sample. With the help of standard antigen, a dose response curve can be established to determine the concentration of unknown 'Ag'. Two types of ELISA techniques are commonly used for quantitative determination of antigens or antibodies in any samples.

These are as follows:

Sandwich ELISA—This is a two-antibody *sandwich* ELISA assay by which unknown antigen concentration is determined. This ELISA consumes less time and accurate, and if a purified antigen standard is available, the assay can be used to determine the absolute quantity of antigen in any unknown sample. The sandwich ELISA requires two antibodies that bind two different epitopes on the same antigen. In sandwich ELISA, the purified 'captured' antibody is generally coated on micro-titer plate. Antigen is allowed to make complex with the bound antibody. The labeled 'detection' antibody (second antibody) is then allowed to bind with the antigen, captured on the first antibody (capture antibody). Thus, a sandwich of antigen between two antibodies is formed. The assay is quantified by measuring the amount of labeled second antibody bound to the antigen through the use of a colorimetric substrate. The major advantages of sandwich ELISA are that it is very specific, and unpurified antigen at a very low concentration can be used. However, the assay has a limitation—unless the monoclonal antibody combinations are "matched pairs", which means they recognize separate epitopes on the same antigen without hindering each other's binding sites, the assay is not

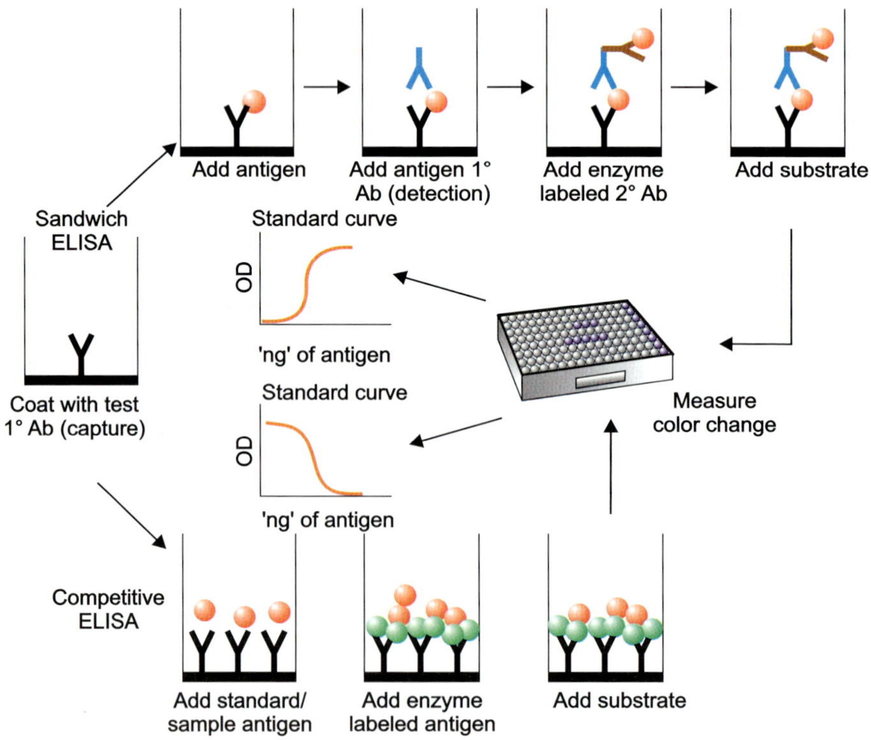

Figure 15.5 Enzyme linked immunosorbent assay: the methods for sandwich and competitive ELISA are shown. 1°, 2°Ab- primary, secondary antibody; brown dot—sample (antigen); Red dot—enzyme; Green dot— standard enzyme, conjugated with enzyme; Black (Y)— Capture antibody; Indiyoblne (Y)—primary detection antibody; Merun (Y)— secondary detection antibody.

possible. The schematic presentation of sandwich ELISA is shown in Figure 15.5. The sensitivity of the sandwich ELISA depends on following: (1) amount of the first antibody that bounds to the solid phase, (2) the *avidity* of the first antibody for the antigen, (3) the avidity of the second antibody for the antigen, and (4) the specific activity of the second antibody. The amount of the capture antibody that is bound to the solid phase can be adjusted by using dilute or concentrated antibody solution. The general procedure for sandwich ELISA is shown below:

1. Coat each well of micro-titer plate by incubating with 100 μl of unlabeled antibody (10 μg/ml in PBS) at 37°C for 2 h or overnight at 4°C. For optimum coating, approximately 1 μg/well antibody is required. The binding of antibody depends upon the active surface on the well.
2. Remove the unbound antibody by flicking the contents of the plate, and wash the well 3-time with PBST (PBS containing 0.5% Tween-20).
3. The unbound sites for protein on micro-titer plate are blocked by incubating with 200 μl of blocking solution containing either 5% nonfat milk or 1% BSA solution in PBST. The blocking of wells is carried out by incubating the plate with above solution for 1-2 h in an incubator set at 37°C.
4. Wash the wells thrice with PBST solution and add 50 μl of the antigen of different dilutions. All dilutions should be done in PBST. Incubate the plate for 1 h at 37°C incubator.
5. Wash the plate 3-times with PBST, add the labeled [with alkaline phosphatase (AP) or horseradish peroxidase (HRP)] second (or detecting) antibody, and incubate further at 37°C for 1 h. The amount of secondary antibody to be added is determined in preliminary experiments, however, always second antibody is added in excess. Sodium azide is an inhibitor of HRP, so it is to be avoided in buffers or in wash solutions if an HRP-labeled antibody is used for detection.
6. Wash the plate 3 times with PBST, and add substrate as per the enzyme labeled with the second antibody. After incubation for a period of 2 to 5 min, dilute H_2SO_4 is added to stop the reaction. The optical densities of the reaction product are measured in an ELISA plate reader. For quantitative analysis, compare the signal of unknown samples against standard curve established using known concentration of the antigen.

ELISA procedures utilize substrates that produce soluble products. Ideally, the enzyme substrates should be stable, safe and inexpensive. Popular enzymes are those that convert a colorless substrate to a colored product, e.g., para-nitrophenylphosphate (pNPP), which is converted to yellow p-nitrophenol (measured at 405-410 nm) by alkaline phosphatase. Substrates used for HRP include 2,2′-azo-bis(3-ethylbenzthiazoline-6-sulfonic acid) (ABTS) (measured at 405-410 nm), o-phenylenediamine (OPD) (measured at 490 nm), and 3,3′5,5′-tetramethylbenzidine base (TMB) (measured at 450 nm), and 3,3′-Diaminobenzidine (DAB) (measured at 510 nm), which yield green, orange, blue, and brown color product, respectively.

Competitive ELISA— When two matched pair antibodies are not available for a target antigen the second option for the analysis is competitive ELISA. There are several forms of *competitive* ELISA, one of which has been described below. In order to perform competitive ELISA, either the immunogen or the primary antibody is conjugated to a detection enzyme, HRP. The protocol shown here is based on labeled immunogen as the competitor. In short, unlabeled purified primary antibody is coated onto the wells of a 96 well microtiter plate. This primary antibody is then incubated with different dilutions of unlabeled standard immunogen or unknown samples. Once the equilibrium is established, HRP-conjugated immunogen is added to the same plate. This conjugate will bind to the free binding sites of the primary antibody. Thus, more the immunogen present in the sample or the standard, lower will be the binding of conjugated immunogen to the primary antibody. The plate is then developed with substrate, and the color change is measured using an ELISA reader. The general procedure for competitive ELISA is shown below:

1. Coat each well of micro-titer plate by incubating with 100 μl of purified primary capture antibody (10 μg/ml in PBS) at 37°C for 2 h or overnight at 4°C. The amount of antibody required may change with the assay, but if maximum binding is required, at least 1 μg/well should be taken. The binding of antibody also depends upon the capacity of the well.
2. Remove the unbound antibody by flicking the contents of the plate, and wash each well 3-times with PBST (PBS containing 0.5% Tween-20, PBST).
3. The unbound sites for protein on micro-titer plate are blocked by incubating with 200 μl of blocking solution containing either 5% nonfat milk or 2% BSA solution in PBST. The blocking of wells is carried out by incubating the plate with above solution for 1-2 h in an incubator set at 37°C.
4. Wash the wells thrice with PBST solution, and add 50 μl of the standard immunogen or sample to the wells. All dilutions should be done in PBST. Incubate the plate for 1 h at 37°C in an incubator.
5. Remove the blocking buffer and add 50 μl of the HRP-conjugate immunogen solution of different dilutions to the wells. Incubate the plate for 1 h at 37°C incubator.
6. Wash the plate four times with PBST. Add the required substrate and incubate the plate for 5-10 min. The optical densities of the reaction product are measured in an ELISA plate reader at a wavelength, depending upon the substrate.

15.4.2 Radio immuno assay (RIA)

It is a competitive binding assay in which fixed amounts of antibody (Ab) and radiolabeled antigen (Ag^*) are allowed to react in the presence of known (standard antigen) or unknown (sample antigen) antigen. The binding of labeled antigen with antibody is competitively inhibited by increasing the concentration of standard antigen or antigen in the unknown sample, and hence a dose response curve can be established. In RIA, I_GG subclasses are quantified as radioactive labeled immune complexes are bound with specific antibody. In case of a direct technique, a radioactive labeled anti-I_GG subclass-specific antibody is used.

Whereas, an anti-I$_G$G subclass-specific antibody directed against the first antibody is used in indirect technique. Since working with radioactive labeled reagents the may requires special precaution and is relatively costly.

15.4.3 Bioassay

Bioassay commonly means biological assay with the live system *in vitro*. Bioassays are typically conducted to measure the effects of a compound/drug on a living organism or on a biological process, and the results are either directly or indirectly measured. In direct readout (result), the effects of test compound on biological system (cell type or certain biological event) are determined by straight forward methods. For example, the effect of growth factors on proliferation of cells (e.g., proliferation of human TF-1 cells in the presence of different growth factors), or the effect of death ligand on apoptosis of cells (e.g., apoptotic death of mouse NIH 3T3 cell line by TNFα), or the degradation of fibrill clot by tissue type plasminogen activator, etc. Bioassay could lead to indirect readouts also, such as the effect of human chorionic gonadotropin hormone or mouse Leydig cells to stimulate testosterone production. Bioassays may be qualitative or quantitative, the latter often involves an estimation of the concentration or potency of a substance by measuring the biological responses that it produces. The importance of bioassay can be conceived if one understands the complicacy of the manufacturing process. For example, the therapeutic proteins are expressed in bacterial or mammalian systems. Depending upon the expression system used, the proteins are expressed either in native or non-native conformation. By successive purification steps, the target proteins are separated from contaminating substances to obtain biological activity similar to natural counterpart. The biological activity may lost, if the protein is not produced in the correct environment, and/or purified in harsh condition. In other words, despite the protein is 99.5% pure in SDS-PAGE or GPC-HPLC, the specific activity may become much lower than the native protein. Using bioassay, one can determine the specific activity (also expressed as biological activity) of the product in terms of ED_{50}/ng protein or IU/ng protein. The use of bioassays include: (a) measurement of the pharmacological activity, (b) determination of the side effect profile, and (c) investigation of the function of endogenous mediators.

15.5 PLATING EFFICIENCY

Plating efficiency is a measure of the number of colonies originated from fixed number of cells (Mather and Roberts, 1998). It is a very sensitive test and is often used for determining nutritional requirements by the cells, testing serum lots, measuring the effects of growth factors, and toxicity study of some inhibitors. The assay is based on the major assumption that each colony has been originated from individual cells. Cells at very low densities (2-50 cells/cm^2) are normally used in determining the plating efficiency. The procedure followed to determine plating efficiency is described below:

1. Count the exact viable cell number and dilute the cell suspension in medium in such a way that a known volume can dispense cell densities at 2, 10, and 20 cells/cm^2 of surface area of culture plate.
2. Maintain equal volume of the culture at each density in triplicate plates.
3. Incubate the cells for approximately 10 days. The time of the culture may vary on cell type. Distinct isolated colonies (20-50 cells) are normally visible by naked eye within the test period.
4. Wash the plate with PBS fixed with 10% formaldehyde for 10 minutes.
5. Remove formaldehyde and add 0.2% crystal violet in PBS to cover cells for 10 minutes.
6. Rinse the cells with tap water until the water becomes colorless; count the number of colonies by visual inspection.

$$\text{Plating efficiency (\%)} = [\text{Number of colonies/Number of cells plated}] \times 100 \quad 5.15$$

15.6 CYTOTOXICITY ASSAY

Out of 10 candidates, only 1 drug qualifies to enter into the development stage. One of the main reasons for failure is abnormal toxicity of in the candidate drugs. *Cytotoxicity* assay is a rapid and cost-effective tool to select such compounds before they enter into the costly development and then in the pre-clinical evaluation process. There are many goals for cytotoxicity assay—(a) identification of potentially active compounds, (b) determination of the toxic concentration range, (c) finding out relationship between concentration and time of exposure, (d) identification of the mechanism of cytotoxic effect, and (e) prediction of an effective drug for the treatment of cancer or other ailments.

Cells die following two mechanisms: necrosis—the unplanned death that occurs when cells are exposed to a severe physical or chemical insult or deprived of nutrients, and apoptosis—the programmed cell death that removes unwanted cells (Chapter 3). The test of cytotoxicity does not identify the cellular mechanisms by which cells die with respect to the specific compound. Rather, it simply shows the killing power of the compounds or a mediator cell (cytotoxic T cell), irrespective of the death mechanism followed. The most popular assays for determining cytotoxicity is based on alterations of plasma membrane permeability leading to the release of cellular components in the medium or the uptake of dyes that are normally excluded by the viable cells. A serious limitation of membrane permeability based assay is that in most cases the targets are intracellular components and not the plasma membrane. In these cases, a cell may be irreversibly damaged and finally die, but the plasma membrane remains unaffected. Thus, cell membrane based assays tend to underestimate cellular damage when compared to other methods. Despite of this limitation, permeability-based assays have been widely accepted to determine cytotoxic effect of drugs. Alternatively, cytotoxicity can be determined by MTT assay, in which the dead cells do not metabolize tetrazolium salts to form chocolate-brown metabolite, which is measured by a colorimeter. In case of apoptosis, cellular metabolism is continued and cells also metabolize MTT. Thus, like permeability assays as mentioned above, the colorimetric assays may also underestimate cellular damage. These assays detect cell death only at the later stages of apoptosis when the metabolic activity of the cells is reduced or stopped. Regardless of these disadvantages, the colorimetric based assays are popular for cytotoxicity assay.

Overall, the cytotoxicity assays are based on two basic principles, (1) measuring membrane integrity where cells are susceptible to the active agents, and in turn the plasma membrane is affected, and (2) measuring activity of cellular constituents, where membrane integrity is not much affected but the cellular processes. The alternate assays that are commonly used to determine cytotoxic effect of a drug is described below:

15.6.1 Assays based on integrity of plasma membrane

The techniques used to determine plasma membrane leakage are based on the uptake or exclusion of certain molecules with special light-absorbing or fluorescent properties. The viable (intact plasma membrane) and dead (damaged plasma membrane) cells are discriminated by differential staining of cellular constituents. Since fluorescent dyes stain the nucleus, test samples can be analyzed by microscopy or flow-cytometry. This kind of single cell based analysis is followed for quantitative determination of cytotoxicity or viability of cells. Different assays belonging to this category are mentioned below:

Nuclear staining—Propidium iodide (PI) is an analogue of ethidium bromide, which intercalates the nucleic acid helix with a resultant increase in fluorescence intensity. PI is excited at 488 nm and the emission maximum for the DNA-PI complex occurs at approximately 615 nm, and nuclei turn red color. PI stains the nuclei only when the plasma membrane is leaky. In other words, viable cells do not stain with PI. Thus, enumerating the number of PI^+ cells, one can screen drug sensitive cell lines. The number of PI^+ cells can be determined by flow cytometry or by direct counting using fluorescent microscope. Direct counting of PI^+ cells by fluorescent microscope is preferable for adhered cells, whereas, flow-cytometry-based assay is preferable for non-adhered cells. The procedure for nuclear staining with PI is as follows:

(i) Wash-off the drug
(ii) Incubate cells with PI solution (3 μg/ml final concentration) for 5-10 min
(iii) Count total and PI^+ cells using microscope or by flow-cytometry.

$$\text{Dead cells (\%)} = [(\text{PI}^+ \text{ cells})/(\text{Total cells})] \times 100 \qquad 15.16$$

Dye exclusion—The most commonly used vital dye to determine membrane integrity is trypan blue. The viable cells exclude the dye, whereas dead cells stain blue. Trypan blue dye exclusion assay is most suitable for suspension cultures; it can be used for adhered cells too. The major limitation of this assay is that serum proteins inhibit the binding of dye to the damaged membrane, thus increases false negative value. Percent dead cells can be determined using similar relationship, as mentioned above.

LDH release assay—Lactate dehydrogenase (LDH) is a stable cytoplasmic enzyme present in all animal cells. It is released into the cell culture supernatant, if the plasma membrane integrity is lost. LDH activity is measured by a single point assay, by the spectrophotometric measurement. The recently introduced CytoTox-ONE homogeneous membrane integrity assay (Promega) is a fluorometric method for estimating the number of nonviable cells. The assay measures the released LDH from cells with damaged membrane. LDH released into the culture medium is reacted with resazurin for 10-minute, which results in the formation of fluorescent compound, resorufin (Figure 15.6). The fluorescent signal is measured in a fluorometer with 560 nm excitation and 590 nm emission filter setting. The amount of fluorescence produced is proportional to the number of dead cells in the test plate.

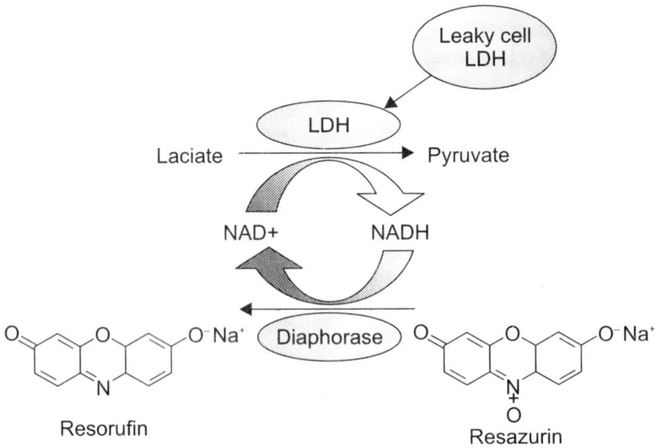

Figure 15.6 The released LDH from the dead cells is coupled with an enzymatic reaction that converts resazurin into the fluorescent compound, resorufin (*Source*: Promega, Madison, USA].

15.6.2 Assays based on cellular activity

Many drugs cause less damage to the plasma membrane, but they have anti-proliferating activity. Cell membrane integrity-based cytotoxicity assays are not recommended for this group of drugs. In such cases, the relative proliferation capacity of the test cells is determined in the presence of different concentrations of the drug(s). It is needless to mention that these assays are also suitable in cases where the drugs disintegrate plasma membrane.

Tetrazolium dye reduction—The most widely used cytotoxicity assay technique involves a tetrazolium salt, MTT (3-[4,5-Dimethylthiazol-2-yl]-2,5-diphenyl-tetrazolium bromide), in the living cells it is metabolized to chocolate-brown color formazan crystals by mitochondrial enzyme. It is a rapid colorimetric method, thus many samples can be tested in a short span of time. This method can be used for determination of antiproliferative/cytotoxic effect of drugs in cases of both suspended and adhered cells. There are few limitations in this assay—the absorption spectra of formazan depends on pH, exposure to light, cell density, and the type of cells. The procedure for MTT assay is explained below:

1. Culture the test cells in quadruplet wells in a flat bottom 96-well plate in a medium containing different concentrations of drug for 3-5 days.
2. Add MTT solution in each well so that the final concentration is 1 mg/ml. Normally, 25 μl MTT solution (5 mg/ml stock concentration in PBS) is added in 100 μl culture. Incubate the culture in a CO_2 incubator for 2 h. Care should be taken to avoid exposure of light to the culture plate, as MTT is light sensitive.
3. Add 100 μl of lysis buffer in each well (20% SDS in 50% dimethyformamide in water), and further incubate the plate for 6-8 h to lyse the cells and to dissolve the formazan crystals. Measure the optical density of the solution at 580-590 nm. Thus, the cytotoxic effect is calculated as,

$$\text{Cytotoxicity } (\%) = [(OD_C - OD_D)/OD_C] \times 100 \qquad 15.17$$

where OD_C is optical density of the sample (control) in the absence of drug (here cell viability is considered 100%), and OD_D is optical density of the test sample in the presence of drug.

Radioisotope incorporation—Another most popular method to determine antiproliferation/cytotoxic effect of drug is by measuring incorporation of [^{3}H]-thymidine by the test cells. Again, it is a relative test like MTT, where control. (drug untreated) cells incorporate highest level of [^{3}H]-thymidine into DNA, and its incorporation is reduced with an increase in cytotoxicity. This assay cannot be used in case of drugs like 5-fluorouracil and methotrexate, which inhibit pyrimidine biosynthesis due to an increased uptake of exogenous [^{3}H]-thymidine by salvage pathway. This assay can be used for both suspended and adhered cells. The procedure for [^{3}H]-thymidine incorporation assay is as follows:

1. Culture test cells in quadruplet wells in flat bottom 96-well plate in the presence of medium containing different concentrations of drug; the control sample should be cultured without the drug.
2. Add 0.5-1.0 μCi of [^{3}H]-thymidine per well about 12-18 h before completion of 3-5 days. At the end of pulse, cells are harvested on a glass wool filter paper, washed with water, and treated with scintillation fluid for 5-10 min before counting the incorporation of radioisotope in the individual sample with the help of a β-counter. The percent antiproliferation/cytotoxicty is calculated as,

$$\text{Antiproliferation/cytotoxicity } (\%) = [(CPM_C - CPM_D)/CPM_C] \times 100 \qquad 15.18$$

where, CPM_C is an average incorporation by the control cells in the absence of the drug (here cell viability is considered 100%), and CPM_D is an average incorporation in the presence of the drug.

ATP synthesis—The assay is based on the fact that ATP is rapidly degraded if cells die, and a viable cell contains about 2×10^{-15} moles of ATP. Thus, estimation of cell viability can be determined if it is assumed that the ATP content per viable cell remains fairly constant. ATP content may vary from cell line to cell line and with the culture media. Cellular ATP in viable cells is determined by the following biochemical reactions:

$$\text{Intracellular ATP} \longrightarrow \text{Free ATP}$$

$$\text{Free ATP} + \text{Luciferin} \longrightarrow \text{Adenyl-luciferin} + \text{PPi}$$

$$\text{Adenyl-luciferin} + O_2 \longrightarrow \text{Oxyluciferin} + \text{AMP} + CO_2 + \text{Light}$$

When ATP is the limiting reagent, the light emitted is proportional to the ATP present, which is, in turn, proportional to the number of viable cells in the sample. The amount of ATP in the sample may be calculated as follows:

$$ATP_S = [ATP_{IS} \times L_S] / [L_{S+IS} - L_S]$$

where, ATP_S is the ATP in the cell sample (moles), ATP_{IS} is the ATP in the added internal standard (moles), L_S is the light emitted by the cell sample, and L_{S+IS} is the light emitted by the cell sample plus the internal standard. The intensity of light is measured in relative unit using a luminometer. If the amount of ATP per cell is known (obtained experimentally or from the literature), the number of viable cells may be estimated as:

$$\text{Number of viable cells/ml} = [ATP_S \times DF]/\text{ATP per cell} \qquad 15.19$$

where, DF is the dilution fraction of sample.

There are many experimental variables that may dictats the success of a cytotoxicity assay. These are (i) storage of drug, its solubility in aqueous medium, and working dilution, and (ii) test cells, its density and drug exposure time. The interpretation of data is an another important factor for the success of the assay. Tetrazolium dye reduction and radioisotope incorporation assays are also used to determine proliferation of cells in the presence of mitogen.

15.7 SUMMARY

1. Quantitative analysis and interpretation of results are two most important exercises in any experiment.
2. Cell culture process control would be possible on the basis of informations on state variables on real time basis.
3. The efficacy of a therapeutic product is decided on the basis of the biological activities, which are tested *in vitro* or *in vivo*.
4. The effect of all cytotoxic drugs/agents cannot be determined utilizing all the available methods.

ABBREVIATIONS

AP Alkaline phosphatase
AHA Aminohexanoic acid
ABTS 2,2′-azo-bis(3-ethylbenzthiazoline-6-sulfonic acid

ATP	Adenosine triphosphate
DAB	3,3′- Diaminobenzidine
ELISA	Enzyme-linked immnunosorbent assay
G6P	Glucose-6-phosphate
G6PDH	Glucose-6-phosphate dehydrogenase
HK	Hexokinase
HRP	Horseradish peroxidase
LDH	Lactate dehydrogenase
MTT	3-[4,5-Dimethylthiazol-2-yl]-2,5-diphenyl-tetrazolium bromide
NAD	Necotinamide adenine dinucleotide
OTR	Oxygen transfer rate
OUR	Oxygen uptake rate
OPD	o-phenylenediamine
PITC	Phenyl isothiocyanate
PAGE	Polyacrylamide gel electrophoresis
pNPP	Para-nitrophenylphosphate
PI	Propidium iodide
RIA	Radio immuno assay
PBS	Phosphate buffered saline
SDS	Sodium dodicyl sulphate
GPC	Gel permiation chromatography
HPLC	High pressure liquid chromatography
TEA	Triethylamine
TMB	3,3′5,5′- tetramethylbenzidine

REFERENCES

1. Mather JP, Roberts PE (1998) Introduction to Cell and Tissue Culture: Theory and Technique. Plenum Press. New York.
2. Ozturk SS, Meyerhoff ME, Palsson BO (1989) *Biotechnol Letters* 3, 217.
3. Smiley AL, Hu WS, Wang DIC (1989) *Biotechnol. Bioeng*. 33, 1182.
4. Taguchi H and Humphery AE (1996) J Ferment. Technol., (Japan) 44, 881.
5. van Wezel AL (1973) In: Tissue culture: *Methods and application* (PF Kruse, MK Patterson, eds,) and p. 372. Academic Press, New York.

SUGGESTION FOR FURTHER READING

1. Lundin A. (1982). In: *Clinical and Biochemical Luminescence* (LJ Kricke and TJN carter, eds.), p. 43. Dekker, New York.
2. Lundin A, *et al*. (1986) *Methods Enzymol* 133, 27.
3. Sanford KK, Earle WR, Evans VJ, Waltz HK, Shannon JE (9151) *J. Natl. Cancer Inst*. 11, 773.
4. Strechler BL (1974). In: Methods of Enzymatic Analyisis (HU Bergmeyer, ed.), vol. 4, p. 2112. Academic Press.
5. Wilson AP (1992). In: *Animal Cell Culture* (RI Freshney, ed.), p. 263. IRL 263. IRL Press, Oxford.

16

Establishing Primary Cultures and Cryopreservation of Cells/Tissues

Primary tissues obtained from *neonatal* or young mammals can be cultured for about 5 passages to generate large number of cells. Although these cells do not have long life *in vitro*, they are useful for isolation and production of viruses. Again adult mice or rats-derived cells beyond 4 to 6 months of age do not proliferate much, it is extremely important to select the animals before isolating cells. Furthermore, every tissue might require certain specific culture conditions for their optimum growth that are supposed to be maintained while culturing them. Some of the common steps for initiating primary cultures are given below.

1. Sterilize the site of dissection with 70% ethanol to avoid bacterial/fungal contamination. Isolation procedure should be carried out under sterile condition using a Class IIA biosafety cabinet. All surgical instruments should be autoclaved
2. Dissect-out the tissue and cut finely to expose more surfaces for enzyme action.
3. Treat the chopped tissue with suitable enzyme(s) and immediately remove enzyme solution after disaggregation of tissue.
4. Initiate primary culture on a suitable tissue culture plate using higher cell number with rich media.

Derivations of few important primary cultures and general techniques for cryopreservation of cells/tissues are described below.

16.1 ESTABLISHING PRIMARY CULTURES

16.1.1 Isolation of mouse bone marrow cells

Bone marrow cells are used for investigations on hematopoiesis in vitro/in vivo and immunological properties of the cells. Various steps are followed to isolate mouse mononuclear bone marrow cells, which are described below:

1. Euthanize a 6 to 8 weeks old C57BL/6J mouse by CO_2 asphyxiation, thoroughly spray 70% ethanol in all the external areas for disinfection.
2. Make an incision in each hind leg using scissors; gently pull down the skin to expose the muscles. Cut the hind leg just above the pelvic/hip joint using scissors and also remove the lower portion of the hind leg. Cut the hind leg at just below the knee-joint through ligaments to remove tibia. Remove surrounding muscle from femur and tibia with the help of a scalpel.
3. Transfer bones in a petri plate containing IMDM supplemented with 3% FBS and 1x antibiotics and antimycotic mixture (Sigma).
4. Trim one end of the femur or tibia carefully using a sharp scissor to expose the interior marrow shaft, flush the contents of marrow with 2 ml of same medium from other side of the bone using a 1-ml insulin syringe with a 26G needle. Collect the contents into a 50-ml centrifuge tube.
5. The firmly adhered cells of the endosteal region are then recovered in the second stage by digesting the crushed bones with 0.03% collagenase type IV for 15 min at 37°C.
6. Pool the cell suspensions, centrifuge at 1000 rpm for 5 min. After centrifugation, remove the supernatant, dislodge the pellet by tapping and add 5 ml of Gay's solution to lyse erythrocytes. After treatment with gay's solution for 90 sec, add 45 ml PBS in the tube and further centrifuge at 1200 rpm for 5 min.
7. Remove the supernatant and resuspend cell pellet in 5 ml of medium for washing to remove traces of Gey's solution. Finally cells are resuspended in 5 ml of medium and enumerate of cell number is performed with the help of a haemocytometer.
8. Suspend cells (two femurs and 2 tibiae of a C57Bl/6J mouse can yield 70 to 80×10^6 cells) in 10 ml of IMDM supplemented with 10% FBS and culture in a 100 mm petri-plate by incubating at 37°C in a 5% CO_2 incubator for 5 to 10 days.

16.1.2 Mouse embryonic fibroblast (MEF)

MEF cells are most important in culturing mouse and human ES cells. It has been shown that the *pluripotency* of ES cells is maintained up to 200 generations or more if cultured in the presence of MEF as a feeder layer. MEF produce many growth factors, of which leukemia inhibitory factor (LIF) is most important for maintainance of ES cells. The detailed procedure for establishing MEF culture is mentioned below:

1. Select pregnant mice from 12-14 day post coitus (dpc) from healthy mouse colony, sacrifice by CO_2 asphyxiation, and swab the abdomen portion with 70% ethanol.
2. Make a median line incision in the abdominal wall tear the ventral skin transversely over the diaphragm, and expose the ventral surface of the abdominal wall.

3. Take a pair of sterile scissors and forceps cut the diaphragm longitudinally. At this stage the uteri filled with embryos (6-8 in case of C57BL6/J mouse) will be visible. Transfer the intact uteri to a 10 cm Petri plate containing 10 ml of sterile Hank's balanced salt solution (HBSS) and antibiotic mixture.
4. Dissect out the embryos by tearing uterus with a pair of sterile forceps. At this point all the embryos will be released from the membranes, wash them thoroughly with sterile HBSS. One such embryo, attached with placenta, is shown in Figure 16.1A.
5. Cut and remove the head, the limbs, the tail, the internal organs (Figure 16.1B), and thoroughly wash the tissue with HBSS.
6. Transfer selected portions of the embryonic tissue in a new Petri plate and cut finely with the help of a scalpel.

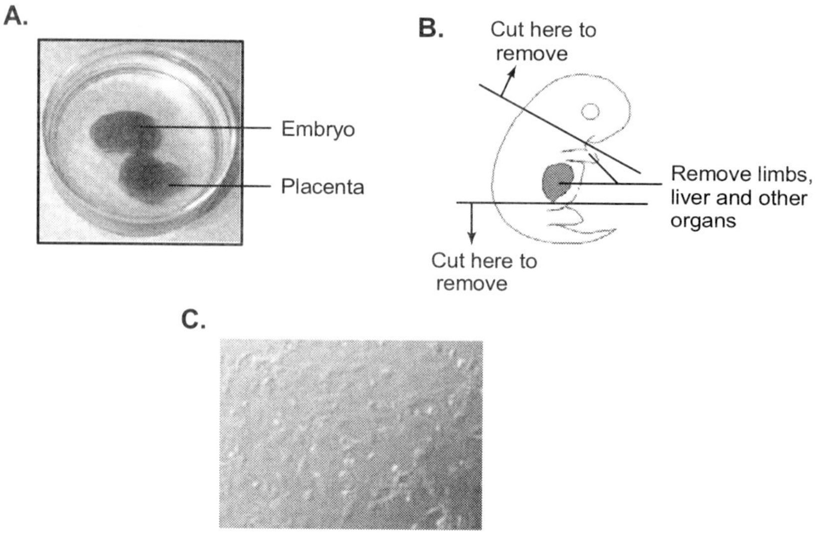

Figure 16.1 Mouse embryonic fibroblast cell isolation. **A.** Mouse embryo with placenta **B.** Schematic diagram to show the portion to be removed from the embryo; **C.** MEF cells growing in culture.

7. Transfer tissue fragments to a trypsinizing flask (250 ml) containing a teflon-coated magnetic bar, add 50 ml of 0.125% (w/v) trypsin (crude-Difco 1:250) in Ca^{+2}/Mg^{+2}-free PBS, stir at 50 rpm for 30 min in 37°C water bath.
8. Collect the supernatant cell suspension (40 ml) in a 50 ml conical tube, add 10 ml of medium containing 10% FBS to stop the action of trypsin, centrifuge at 500g for 5 min, resuspend cell pellet in 3 ml complete medium, and store in ice.
9. Add 40 ml of fresh trypsin solution to the remaining tissue, repeat steps 7 and 8 until complete dissociation occurs.
10. Pool cell suspension, enumerate viable cell number by trypan blue dye exclusion method, dilute cell suspension to 1×10^6 per ml in same medium,

and seed a tissue culture plate at a density of about 2×10^5 cells/sqcm. The yield of viable cells from 6 embryos will be approximately 5×10^7. The cells are cultured at 37°C in a 5% CO_2 incubator. Observe the growing cells in an inverted microscope. A complete monolayer is formed within 3-5 days of culture (Figure 16.1C).

16.1.3 Monkey kidney cells

Monkey kidney cells are essential substrates for the propagation of live polio-viruses. Handling live monkey tissues could be potentially hazardous, as many pathogenic viruses (e.g., herpes B, Marburg) may be present in the latent state. So adequate care should be taken while separating monkey kidney and process-ing the tissue. Following steps may be adopted to establish primary kidney cell culture.

1. Select a healthy young monkey, anaesthetize, and open the abdomen. Fix clamps on the renal artery and renal vein, 3 cm away from the kidney. Drain-off entire blood from the kidney by puncturing vein near the clamp.
2. Wash the kidney several times with HBSS to remove blood and connective tissues. Cut the tissue pieces from the cortical portion of the kidney with the help of a pair of curved scissors.
3. Tissue pieces are further chopped into small fragments of about 2 mm^3. Transfer tissue fragments to a trypsinizing flask containing a teflon-coated magnetic bar, add sufficient volume (200 ml) of prewarmed trypsin 0.25% (w/v) in Ca^{+2}/Mg^{+2}-free PBS at 37°C, stir at low speed for 15 min and discard the supernatant.
4. Pour fresh trypsin solution and continue trypsinization for 30 min. Withdraw the cell suspension, filter through a sterile muslin cloth/nylon filter to remove larger fragments. Collect the filtrate, stop the action of trypsin by adding 20 ml medium containing 10% FBS, centrifuge at 500g for 5 min, resuspend cell pellet in 10 ml complete medium, and store in ice.
5. Continue trypsinization as in step 4 for remant tissue, pool cell suspension, and count the number of viable cells recovered. The yield of viable cells from a kidney of young monkey will be about 5×10^9 to 1×10^{10}. Culture kidney cells in roller bottles at 37°C in a 5% CO_2 incubator, and observe everyday in an inverted microscope. Complete monolayer is formed within 4-6 days of culture. Replace the medium in between, if needed. Photomicrograph showing Vero cells (continues kidney cell line) growing in culture (Figure 16.2).

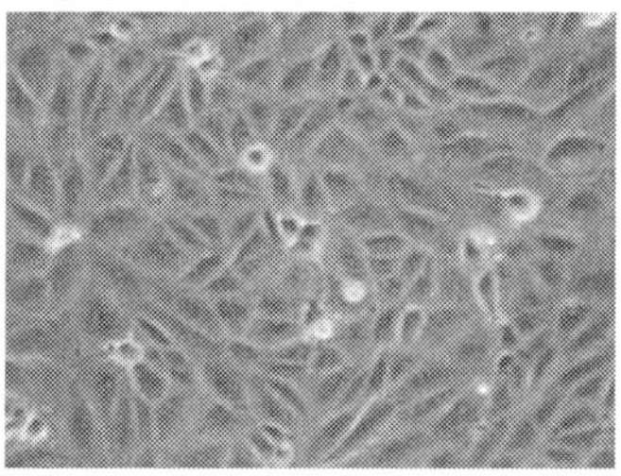

Figure 16.2 Monkey kidney epithelial cells (Vero) in culture.

Modified methods have been recommended to improve both yield and quality of the viable kidney cells, these are:

1. Cold trypsin diffusion method– Trypsin is ineffective at 4°C, but can diffuse through the tissue as effectively as at 37°C. Whole kidney is immersed in cold trypsin at 4°C for 20–24 h. Due to long exposure, trypsin diffuses inside the tissue and is rapidly activated upon warming at 37°C. Thus the overall exposure time of trypsin to the tissue at 37°C will be low, resulting in increase of viable cells.
2. *In vivo* perfusion– Alternately, warm trypsin is passed through the kidney of an anaesthetized animal with the help of a catheter inserted in the aorta, and the out-flow is collected through an incision at vena cava (Griffith, 1985). This procedure improves yield of viable cells, but requires expertise to carry out perfusion.

16.1.4 Rat hepatocyte

About 70% cells in liver are hepatocytes; remaining are biliary epithelial, endothelial and other non-parenchyma cells. Two-step perfusion technique is generally followed to isolate 90–95% viable hepatocytes (Bahjat *et al.*, 2000). In the first step, liver is perfused with Krebs Ringer buffer containing EDTA to separate desmosomal cell contacts. In the next step, type IV collagenase solution is pumped into the liver to dissolve the extracellular matrix components. It is important to note that hepatocytes are more fragile than any other animal cells, therefore vigorous shaking or pipetting may lead to the damage of the cells. The details of the isolation steps of rat hepatocytes are given below.

1. Before starting perfusion, rat is given intravenous injection of heparin (1000 units), anaesthetized with the help of intraperitonealinjection of ketamine (100 mg/kg b.w.) and xylazine (10 mg/kg b.w.), and an incision is made through the abdominal skin on the ventral midline.
2. The viscera are displaced to reveal vena cava and portal vein. A ligature is placed around the vena cava after the introduction of a 24-gauge catheter.
3. The liver is perfused with warm buffer (2 ml/min) at 37°C (Krebs Ringer buffer containing 20 mM glucose and 0.2 mM EGTA, pH7.4). The portal vein is then cut off, and the anterior vena cava is clamped between the diaphragm and the heart (Figure 16.3A). The perfusion rate is increased gradually to 5 ml/min for a total volume of 100 ml. If perfusion is successful, the liver turns from red to light brown color.
4. The perfusion is continued with digestion buffer (20 mM glucose, 1.37 mM $CaCl_2$, 0.72% BSA fraction V and 0.03% type IV collagenase at pH 7.4) of 50 ml total volume.
5. Transfer the perfused liver on a large Petri plate containing 10 ml of ice cold Ham F12/DMEM. Hold the liver firmly with a forcep and scrap out the cells gently with a spatula. If perfusion is successful, all hepatocytes will come out, leaving scaffold and arterial tissues.
6. Filter-remove the aggregated tissue through a 100 μm nylon filter; purify the hepatocytes in 2–3 steps low speed (50 g) centrifugation. Finally, hepatocytes are resuspended in above medium containing 10% FBS, insulin

and glucagon (10 μg/ml) and hydrocortisone (1 μM), and cultured on collagen coated plate. The yield and purity of hepatocytes generally varies between 30–40% and 80–95%, respectively. The attachment of cell on culture plate occurs within 3 h, but hepatocytes do not proliferate, though liver specific enzymes are produced for several weeks. Instead of collagen, Matrigel-coated plates can be used to maintain better viability of hepatocytes. The photomic-rograph of rat hepatocytes is shown in Figure 16.3B.

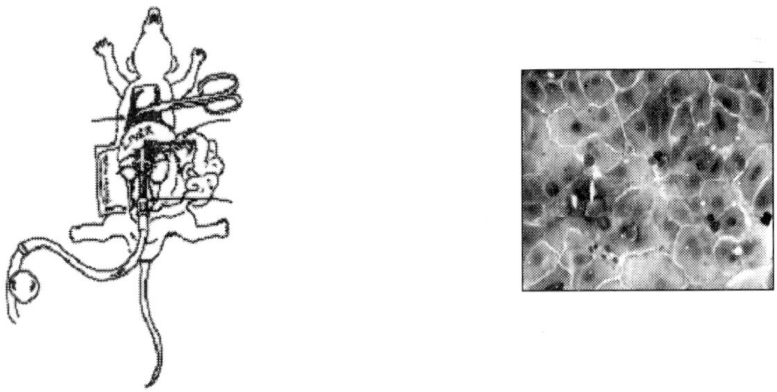

Figure 16.3 Isolation of rat hepatocyte. **A.** *In-situ* perfusion of rat liver. **B.** Hepatocyte growing in culture.

16.1.5 Human umbilical vein endothelial cells

Endothelial cells (EC) provide the internal lining of arteries, veins, lymphatic vessels, and capillaries. Besides many aspects of vascular biology, EC are involved in exchange of metabolites (nutrients) between the blood and the tissues. The main function of EC is to provide a nonthrombogenic surface and to provide many regulatory substances in heart and large vessels. The natural source of EC for cell biology research is umbilical vein. The method for isolation of human umbilical vein EC and their primary culture are shown below:

1. Collect the umbilical cord after separating cord blood (within 10–15 h of delivery); cut 1–2 cm of cord at each end to ensure sterility.
2. Cannulate the vein at both the ends with blunt needles, and clump the opening with arty-forceps.
3. Perfuse the vein with sterile HBSS until the exit solution is colorless. Remove trapped solution by quizzing the cord.
4. Fill the cord with 0.25% pre-warm trypsin solution in HBSS, plug both the ends and incubate at 37°C for 15–20 min.
5. The cord is massaged with fingers to release loosely adhered cells in trypsin solution. Flush the vein with additional 2-cord volume of HBSS and discard the cord.

6. Collect entire amount of trypsin-digested cell suspension in a 50 ml falcon tube. Inactivate trypsin by adding DMEM containing 10% FCS. Centrifuge the cell suspension at 450g for 10 min.
7. Wash the cell pellet twice in complete medium, and set up the culture in the same medium.
8. Seed the cells on a fibronectin- or gelatin-coated culture dish of flask, and incubate in a 5% CO_2 incubator. Wash the adhered cells with fresh medium after two days of initiation of the culture and continue incubation till confluent.
9. Once the culture become confluent, cells are trypsinized and subcultured in a new flask (fibronection/geletin coated) at a split ratio of 4:1.

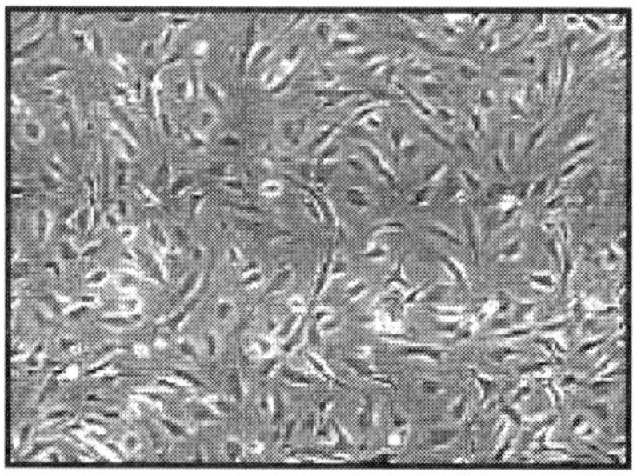

Figure 16.4 Human umbilical vein endothelial cells (HUVEC) growing in culture.

Umbilical vein endothelial cells can be serially propagated many times. However, after 5–7 passages, the cells lose specific properties, grow slowly, and undergo hypertrophy. The photomicrograph of human umbilical vein endothelial cells is shown in Figure 16.4. Endothelial cells form cobblestone morphology when confluent, and prolonged maintenance of confluent culture results in the formation of sprouting phenotype.

16.1.6 Mammary epithelial cells

The mammary gland is developed from the ectoderm. During development the gland generates a series of branching ducts that end in sac-like lobules. The mammary gland is structrally arraged with a continuous epithelium consisting of an outer basal layer of contractile myoepithelial cells and an inner layer of luminal cells. The isolation and culture steps of mouse mammary epithelial cells are shown below:

1. Dissect out mammary glands of mouse and transfer them in a sterile glass Petri dish, mince them like a pulp. Transfer the tissue in a tube containing enzyme solution (collagenase and hyaluronidase) in PBS containing 2% FBS, and incubate 6–8 h at 37°C with occasional stirring.
2. After incubation, pellet-down the cells at 450g, wash the pellet with a 1:4 mixture of cold HBSS supplemented with 2% FBS and ammonium chloride

(8%) and further centrifuge at 450g for 5 minutes. The resultant pellet contains epithelial cell organoids, well as stromal and immune cells.

3. Add 3 ml of pre-warmed Trypsin-EDTA to the above cell pellet and gently resuspend the pellet. The cell suspension may be highly viscous due to release of DNA from dead cells. To this, add 10 ml of cold HSBB and spin at 450*g* for 5 minutes.
4. Remove supernatant, add 2 ml of pre-warmed dispase and 200 μl DNASeI (1 mg/ml) and break the pellet. Add 10 ml of cold HBSS and filter the cell suspension through a 40 μm cell strainer and collect in a centrifuge tube.
5. Mammary epithelial cells are seeded in a tissue culture plate at a density of 1–5 × 10^4 cells/cm^2 in complete medium (50:50 Ham F12 and DMEM supplemented with human transferrin, insulin, bovine serum albumin, EGF, isoproterenol, ethanolamine, 3′,3′,5-triiodo-L-thyronine) containing 5% FBS.
6. Twenty-four hours later the medium is replaced with complete culture medium as above lacking serum.
7. Mammary epithelial cells are sub-cultured by washing with HBSS followed by detached with pre-warmed Trypsin-EDTA solution. Once the cells are detached from the culture Plats, equal volume of cold HBSS is added and the cells are centrifuged at 450g and resuspend in fresh medium. The photomicrograph of growing mammary epithelial cells is shown in Figure 16.5.

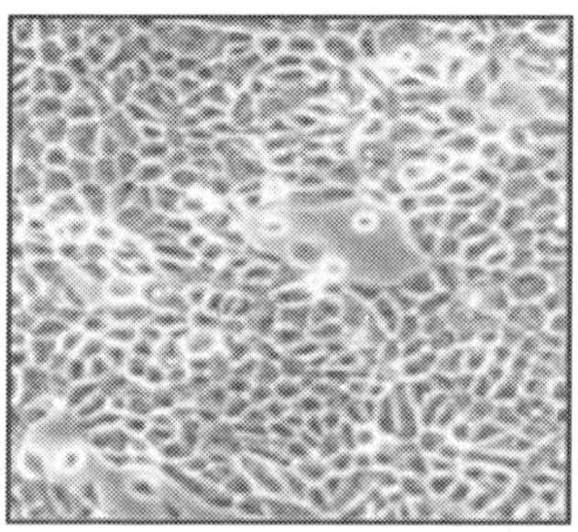

Figure 16.5 Mammary epithelial cells growing in culture.

16.2 FREEZING AND RECONSTITUTING CELLS AND TISSUES

The general procedure for freezing and reconstituting cells, tissues, and embryos are described below:

16.2.1 Cultured cells

1. Select the culture at the end of exponential phase, examine that cells are free from microbial contamination.
2. Harvest the cells by trypsinization (in case of anchorage dependent cells), or simply collect the cells by centrifugation (in case of anchorage

independent cells). Suspend the cell pellet at densities of 2×10^6 to 2×10^7 cells/ml in culture medium containing 20% serum and keep the cell suspension on ice. Higher concentrations of serum/albumin (>20%) may be used.

3. Prepare solution of cryo-protectant in serum free culture medium at double the final concentration to be used. The most commonly used cryo-protectant is DMSO at a final concentration of 10%. However, DMSO induces differentiation in many cells (e.g., HL60 and hepatocyps), where glycerol is the choice of cryo-protectant. Properly diluted cryo-protectant solution should be cooled on ice before use.
4. Add equal volume of cryo-protectant solution drop wise on cell suspension with gentle mixing the content. Dispense 1.0 to 1.5 ml mixture in the properly labeled 2.0 ml freezing vials, and freeze cell suspension in a programmable freezer at a specified freezing rate (normally 1° to 3°C/min)
5. Rapid thawing is advisable for reconstituting frozen cells. Thaw the frozen cells by incubating vial(s) in a 37°C water bath. Once thawed, the cell suspension is immediately diluted 10 to 15 times its original volume with cold medium. Spin-down and wash the cells to remove cryo-protectant. Finally, the cells are resuspended in a complete medium prior to planning for the culture.

16.2.2 Human embryos (4-cells stage)

Slow-cooling method is adopted for cryo-preservation of the early cleavage stage of embryos (Loong et al., 1991). The reagents to be used are: Solution A- 12% (v/v) 1,2-propanediol in isotonic medium; Solution B- solution A containing 0.1 M sucrose; Solution C- 10% (v/v) 1,2-propanediol in isotonic medium containing 0.2 M sucrose; Solution D- 0.2 M sucrose in isotonic medium. The steps are as follows:

1. Incubate embryos at room temperature in solution A for 15 min.
2. Transfer embryos to the equal volume of solution B, load them into plastic straws (capacity 0.25 ml), and seal the straw properly.
3. Place the straw in the cooling chamber of the cell-freezer and wait until the total time spent in 12% 1,2-propanediol reaches 30 min.
4. Carry out the freezing program as follows: Cool straw to 20°C at rate of 1°C/min, hold at that temperature for 5-9 min. Cool to –7°C at rate of 2°C/min, hold at that temperature for 15 min. Seed with cold forceps and wait 5 min for ice formation, hold for 10 min. Cool to -30°C at rate of 0.3°C/min. Then cool to -150°C at rates of 10°C/min. Transfer straws to LN_2 container.
5. For reconstitution, remove straw from LN_2 storage tank and warm at room temperature for 40 sec, then in water bath at 25°C.
6. Transfer embryos to solution C and wait for 5 min, and then to solution D for 5 min.

7. Wash embryos in two changes of isotonic medium, and slowly warm to 37°C.
8. Transfer embryos to a complete culture medium, and incubated in a CO_2 incubator.

16.2.3 Tissue

In case of tissues, the freezing and thawing processes are complicated by the presence of different cell types (each cell may respond quite differently to the same freezing and thawing conditions). Nevertheless, the presence of extracellular matrix components further complicates the freezing process, because cellular responses of matrix-attached cells may change. Again, penetration of cryo-protectant is a problem in case of freezing of tissues. The water content in most of the tissues is about 80%, which must be taken into consideration when designing a cryo-preservation protocol. The addition of a cryo-protectant to the tissue is more difficult by this extra volume, as it should penetrate, and equilibrate with the cells to provide the maximal protective effect. This mean that the microenvironment surrounding each cell must first equilibrate with the cryo-protective agent, and then the cell. For small tissues, this is usually accomplished by immersing the tissue in a solution containing the appropriate concentration of cryo-protectant. For example, the equilibrated DMSO concentration in the tissue will be closer to 7.5%, if 1 ml of tissue is placed in 1 ml of 15% DMSO solution. Thus the volume of tissue must be taken into account when designing these procedures. Since this equilibration is a slow diffusion dependent process, prolong contact with cryoprotectant may lead to toxic effects to the cells. Therefore, the process optimization for freezing tissues is more complicated than free cells or embryos. Due to the porous nature of the tissue and the presence of the vasculature, there is possibility of mechanical damage in the tissue and in the arteries as a result of the formation of ice crystals. Similarly, removal of the cryo-protectant from thawed tissue is carried out at slow rate and which is absolutely essential before transplantation of the tissue. Currently, tissues banking is a routine activity in hospitals. Bone, cartilage, heart valves, bone marrow, blood vessels, skin, corneas, islets of langerhans, etc. are cryopreserved in LN_2.

The cryopreservation procedures that are currently in use for these tissues can vary greatly from cell bank to cell bank. Some of the current practices are protected by intellectual patent protection law, and others are known to result in low recovery of cells. Freezing and thawing of some important tissues are presented below, which may not be the optimized conditioned for storing.

1. Heart valves: Place the valve in Media 199 containing 10% DMSO, further supplement with 0.1 mg/ml cephapirin, 0.2 mg/ml azlocilin, 0.08 mg/ml tobramycin, and 0.1 mg/ml amphotericin B. Cool the mixture at rate of 1°C/min, and store in LN_2.

2. Small elastic arteries: Soak the arteries in corneal-potassium-TES (CPTES) buffer containing 15% w/w DMSO until equilibrium is established. Freeze the suspension at rate of 1°C/min, and store in LN_2. During thawing, the warming rate is maintained below 50°C/min up to temperature – 120°C. Once above temperature is attained, tissue is thawed in a 37°C water bath. Uncontrolled thawing leads damage to the arteries.

3. *Aricular cartilage:* Soak the tissue in MEM containing 1M DMSO for 30 minutes before cooling at a rate of 1°C/min to -60°C, later plunge the tissue in LN_2. Thawing of the tissue is carried out in a 37°C water bath.

ABBREVIATIONS

BSA	Bovine serum albumin
DMEM	Dulbecco's modified Eagle medium
ES	Embryonic stem
EGTA	Ethylene glycol-bis (2-aminoethyl ether) N,N,N',N'-tetra acetic acid
EGF	Epithelial growth factor
EDTA	Ethylene diamine tetra acetic acid
EC	Endothelial cell
FBS	Fetal bovine serum
HBSS	Hank's balanced salt solution
LIF	Leukemia inhibitory factor
MEF	Mouse embryonic fibroblast
PBS	Phosphate buffered saline

REFERENCES

1. Freshney RI (1987) *Culture of animal cells. A manual of basic technique.* 2nd edition, p. 207. Alan R. Liss, Inc. New York.
2. Shaw AJ (1996) *Epithelial cell culture. The Practical Approach series*, IRL Press, Oxford University Press.
3. Loong EPL, Tam PPL, Chiu TTY, Panesar NS and Heines CJ (1991). *J Hong Kong Med Assoc* 43, 102.
4. Renard JP and Babinet C (1984). *J Experiment. Zoology* 230, 443.

SUGGESTION FOR FURTHER READING

1. Celis JE (1994) *Cell Biology- A Laboratory Handbook,* vol. 1. Academic Press, San Diego.
2. Alberts B, Bray D, Lewis J, *et al.* (1989) *Molecular Biology of The Cell*, 2nd edition, Garland Publishing Co, New York.
3. Spector DL, Goldman RD, Leinwand LA (1998) *Cells—A Laboratory Manual*, vol 1. Cold Spring Harbor Laboratory Press, Cold Spring.

Glossary

Active immunity: It refers to immunity produced by the body following exposure to antigens.

Allele: One of a pair, or series, of variant forms of a gene that occur at a given locus in a chromosome. In a normal diploid cell there are two alleles of each gene (one from each parent), which occupy the same relative position (locus) on homologous chromosomes.

Antibody: Antibody is defined as an immunoglobulin capable of specific combination with the antigen that caused its production in a susceptible animal.

Antigen: Any foreign substance that elicits an immune response (e.g., the production of specific antibodies) when injected to a susceptible animal and is capable of combining with the specific antibodies formed.

Apoptosis: Genetically programmed cell death.

Apoptosomes: Activation of the caspases that initiate apoptosis typically requires cognate scaffold proteins, including Apaf-1 in mammals. Each scaffold protein oligomerizes procaspases into a complex called the apoptosome.

Articular Cartilage: It is a complex, living tissue that lines the bony surface of joints. It's function is to provide a low friction surface enabling the joint to withstand imparted load due to normal and sports activities.

Asymmetric membrane: In asymmetric membrane the pores are gradually decrease in size from the upstream to the downstream side. The larger pores on the upstream side act as a prefilter, while the absolute rated downstream side pores act as absolute cut-off layer.

Asymmetric division: By which a stem cell divides into two unequal daughter cells, only one of them resembles the parent stem cell and the other differentiated form.

Autocrine signalling: It is a form of signalling in which a cell secretes a chemical messenger (called the autocrine agent or growth factor) that signals the same cell.

Basement membrane: Basement membrane is a structure that supports an overlying epithelium or endothelium. It consists of an electron dense membrane called basal lamina, about 300-700 A° thickness, and an underlying network of reticular collagen fibrils that average 300 A° in diameter. This network is 0.1-2 μm in thickness.

Biosafety: Prevention of large-scale loss of potential hazardous biological substances from the working place, which might affect on ecology and human health.

Biocompatibility: The properties of being biologically compatible, which do not produce any toxic, injurious compound, or elicit immunological response in living tissue.

Biodegradable: A biodegradable product has the ability to break down, safely and relatively quickly, by biological means, into the raw materials by which the product has been made.

Blastocyst: A preimplanted embryo of 30-40 cells in the inner cell mass. Bastocyst consists of a sphere made up of an outer layer of cells (the trophectoderm), a fluid-filled cavity (the blastocoel), and a cluster of cells on the interior (the inner cell mass).

Blebbing: Bleb is an irregular bulge in the plasma membrane formed due to apoptosis of cells. Blebbing is the term used to describe the formation of blebs.

Bubble point test: The most widely used non-destructive integrity test of an sterile filter is the bubble point test. In this test, the liquid is held in the pores of the filter by surface tension and capillary forces. The minimum pressure required to force liquid out of the pores is a measure of the pore diameter.

Catabolic pathway: A pathway by which an organic molecule is degraded in order to release energy for growth and other cellular processes.

CD molecules (cluster of differentiation): Any surface antigen(s) that is associated with a specific sub-population of cell. There are various molecules express on stem cells, e.g., CD34 is present on the surface of hematopoietic stem cells.

cDNA (complementary DNA): The double-stranded DNA complement of an mRNA sequence; synthesized *in vitro* from a mature RNA template using reverse transcriptase (to create a single strand of DNA from the RNA template) and DNA polymerase (to create the double-stranded DNA).

cDNA library: A collection of cDNA clones that were generated *in vitro* from the mRNA sequences isolated from an organism or a specific tissue or cell type or population of an organism.

Centromere: The portion of the chromosome to which the spindle fibers attach during mitotic and meiotic division. It appears as a constriction when chromosomes contract during cell division.

Chromatin: Substance of which eukaryotic chromosomes are composed. It consists of primarily DNA with proteins (mainly histones), and small amounts of RNA. Chromatin is readily stained with chromaticity dye.

Cloning vector: A small self-replicating DNA molecule, a plasmid or a viral DNA chromosome, in which foreign DNA is inserted in the process of cloning genes. It can carry inserted DNA, which perpetuate in a host cell.

Chaperones: Large group of unrelated protein families whose role is to stabilize unfolded proteins, unfold them for translocation across membranes or for degradation, and/ or to assist in their correct folding and assembly.

Clonality: A line of cells that is genetically identical to the originating cells.

Continuous cell line: A permanently established cell that will proliferate indefinitely provided that it has the appropriate medium and space.

Compensatory hyperplasia: Compensatory hyperplasia is the proliferation of differentiated cells to regenerate new tissue. The differentiated cells are divide while maintaining all or some of their differentiated functions.

Complementarity determining regions (CDR): The N-terminal 110 amino acid sequences of the H and L chains constitute the variable regions, which show remarkable diversity and which is the site of antigen-binding of an immunoglobulin.

Containment area: Protection from hazardous fumes and pathogens in area by a system of baffles, bypassing inlets, and filtering the harmful substances to maximize work safety.

Cryoprotectant: A cryoprotectant is a chemical that is used to protect biological tissue from freezing damage.

Cytotoxicity: It is the quality of being toxic to cells by a chemical substance or immune cells.

Cytokinesis: It is the process whereby the cytoplasm of a single cell is divided to spawn two daughter cells, begins during the late stages of mitosis.

Dedifferentiation: The process by which cells lose their specialization and proliferate by cell division to form a mass of cells which, in response to appropriate stimuli, may later differentiate again to form either the same cell type or a different one. De-differentiation occurs in response to wounding and in tissue cultures.

Differentiation: The process whereby an unspecialized cell acquires the features of specialized cells (e.g. hematopoietic stem cells into myeloid or lymphoid cells).

Dilution rate (D): The ratio of the inflow volume flow rate (dV/dt) and the culture volume (V). In continuous fermentation, a measure of a rate at which the existing medium is replaced with fresh medium, D is the reciprocal of the hydraulic retention time.

Diploid cells: A cell of tissue having two chromosome sets, as opposed to the haploid situation of gametes, which have only one chromosome set.

Ectoderm: The outer most germ layer, which differentiates to give rise to many important tissues and structures including the outer layer of the skin and its appendages (the sweat glands, hair, and nails), the teeth, the lens of the eye, parts of the inner ear, the nerves, brain, and spinal cord.

Embryogenesis: It is the process by which the embryo is formed and develops; it starts with the fertilization of egg. The fertilized egg, called zygote undergoes rapid mitotic divisions, the formation of two exact genetic replicates of the original cell, with no significant growth, a process known as cleavage and differentiation, leading to development of an embryo.

Embryoid bodies (EBs): Clumps of cellular structures that arise when ES cells are culture without MEF cells or LIF. EBs contain tissue from all three of the germ layers: endoderm, ectoderm, and mesoderm.

Embryonic stem (ES) cells: Primitive, undifferentiated cells from the embryo that have the potential to differentiate in to various tissues *in vitro*.

Embryo splitting: The splitting of young embryos into several sections, each of which develops into an animal. A form of animal cloning, i.e., of producing animals that are genetically identical. In practice, the number of identical organisms that can be produced from a single embryo is less than 10.

Endotoxin: It is the part of the outer membrane of the cell wall of Gram-negative bacteria, occasionally used to refer to any cell-associated bacterial toxin. It is mostly referred to the lipopolysaccharide (LPS) complex associated with the outer membrane of Gram-negative bacteria.

Endoderm: Lower layer of a group of cells derived from the inner cell mass of the blastocyst; it later becomes the lungs and digestive organs.

Endocytosis: It is a process by which cells absorb material from the outside by engulfing it with their cell membrane. It is used by all cells of the body because most substances important to them are big molecules, which cannot pass through the plasma membrane.

Epiblast: A sheet of cells appears at the end of the first week of developing embryo in humans. A portion of epiblast cells differentiate into amnioblasts that contribute to development of the amniotic membrane and extraembryonic mesoderm. During the third week of developing embryo, primitive streak is formed within the epiblast and then epiblast cells ingress first to displace the hypoblast, forming the definitive endoderm. Additional epiblast cells migrate into the space between the epiblast and definitive endoderm to form the intraembryonic mesoderm. At the conclusion of this process of gastrulation, the epiblast is termed as definitive ectoderm.

Epitope: The small site on an antigen to which a complementary antibody may specifically bind.

Expression vector: A cloning vector that has been constructed in such a way that, after insertion of a DNA molecule, its coding sequence is properly transcribed and the RNA is translated. The cloned gene is put under the control of a promoter sequence for the initiation of transcription, and often also has a transcription termination sequence at its end. Such promoters are termed high level; examples include P1 (the leftward promoter of phage l) and the promoter of the yeast PGK (phosphoglycerate kinase) gene.

Explant culture: It is a technique used for the isolation of cells from a piece of tissue. Explant culture can also refer to the culturing of the tissue, where cells are left in their surrounding ECM to more accurately mimic the *in vivo* environment.

Extracellular matrix: The microenvironment next to a cell that allows for structural support, orientation, and connections for cell-to-cell interactions. They are protein and glycoprotein in nature.

Euchromatin: Genetic material that is stained less intensely by certain dyes during interphase, and that comprises many different kinds of genes.

Feed batch culture: Culture of cells where nutrients are added periodically to the bioreactor.

Feedback inhibition: The process by which the accumulated end product of a biochemical pathway stops synthesis of that product. A late metabolite of a synthetic pathway regulates synthesis at an earlier step of the pathway.

Gastrulation: The period of development of an organism from fertilization of the ovum until birth.

Gap junctions: Communicating cell-cell junctions that allow ions and small molecules to pass from the cytoplasm of one cell to the cytoplasm of another cell.

Gene splicing: A stage in the processing of mRNA, occurring only in eukaryote cells, in which intervening sequences (introns) are removed from the primary RNA transcript, and the coding exons are joined together to form the mature mRNA molecule.

Glycosylation: The covalent addition of sugar or sugar-related molecules to proteins or polynucleotides.

Glycolysis: Sequence of reactions that converts glucose into pyruvate, with the concomitant production of ATP.

Good manufacturing practice: It is a set of regulations, codes, and guidelines for the manufacture of drug substances, drug products, medical devices, *in vivo* and *in vitro* diagnostic products, and foods.

gp120: A membrane protein of HIV that interacts with membrane proteins of its target cell.

Haploid: Refers to the gamete having one chromosome set, as opposed to the diploid situation of cells or tissues, where there are two chromosome sets.

Heteroploid: Term given to a cell culture when the cells comprising the culture possess nuclei containing chromosome numbers other than the diploid number.

Heterochromatin: Regions of chromosomes that stain darkly even during interphase; thought to be for the most part genetically inactive.

HEPA filter: A filter capable of screening out particles larger than 0.3 μm. HEPA filters are used in laminar air flow cabinets (hoods) for sterile transfer work.

Hematopoiesis: It is the development of blood cells, which initially occurs in the yolk sack, then liver, and eventually the bone marrow. In normal adults it occurs in marrow and lymphatic tissues.

Hypertrophy: Enlargement or overgrowth of an organ or part of the body due to the increased size of the constituent cells. Hypertrophy occurs in the biceps and heart because of increased work.

Immunogenic: The ability of an antigen to provoke an immune response.

Immunocompromised: In medicine, immunodeficiency is a state in which the immune system's ability to fight infectious diseases is compromised or entirely absent. Most cases of immunodeficiency are either congenital or acquired. Living animal who has an immunodeficiency is said to be immunocompromised.

Knock-out: An animal resulting from an embryonic stem cell in which a normal functional gene has been replaced by a non-functional form of the gene. This technique is used extensively in mice – much can be learned about the function of a gene by studying the phenotype of animals that lack the peptide product of that gene.

Lysosome: A membrane-bound sac within the cytoplasm of animal cells that contains enzymes responsible for the digestion of material in food vacuoles, the dissolution of foreign particles entering the cell and, on the death of the cell, the breaking down of all cell structures. This is the digestive system of the cell.

LTR: Long terminal repeats, flank the ends of the HIV genome and contain a Psi-sequence near the 5′ end of the RNA-genome

Major histocompatibility (MHC) antigen: A cell-surface macromolecule that allows the immune system to distinguish foreign or "non-self" from "self". A better term is histoglobulin. These are the antigens that must be matched between donors and recipients during organ and tissue transplants to prevent rejection.

Master cell bank: A culture of fully characterized cells distributed into containers in a single operation, processed together in such a manner as to ensure uniformity and stored in such a manner as to ensure stability. A master cell bank is usually stored at –70°C or lower.

Memory cells: Long-lived B and T cells that mediate rapid secondary immune responses to a previously encountered antigen.

Metabolic quotients: Specific consumption and production rates of nutrients and by-products, respectively, by cells.

Mesoderm: The middle layer of the embryonic disk, which consists of a group of cells derived from the ICM of the blastocyst. Mesoderm is the precursor to bone, muscle, connective tissues, etc.

Meiosis: A process where two successive cells divide and produce cells, eggs, or sperm that contain half the number of chromosomes in the somatic cells.

Miochondria: Small cytoplasmic organelle that carries out aerobic respiration. Oxidative phosphorylation takes place to produce ATP.

Multiplicity of infection (MOI): Number of virus particles used per cell for infection.

Multipotent stem cells: Stem cells potential to differentiate into different terminal cells, at the same time able to self-renew (e.g. hematopoietic and neuronal stem cells).

Multilineage: Formation of different cell types having various functional properties, depending upon the type of cells (e.g. T cell, B cell, erythrocytes).

Mutation: A sudden, heritable change appearing in an individual as the result of a change in the structure of a gene (gene mutation), or changes in the structure of chromosomes (chromosome mutation), or in the number of chromosomes (genome mutation).

Nano-structures: A nanostructure is an intermediate size between molecule and microscopic structures. Nanotextured surfaces have one dimension on the nanoscale from 0.1 to 100 nm. Nanotubes have two dimensions on the nanoscale, i.e., the diameter of the tube is between 0.1 and 100 nm; its length could be much greater. Finally, spherical nanoparticles have three dimensions on the nanoscale, i.e., the particle is between 0.1 and 100 nm in each spatial dimension.

Non-growth associated product: The synthesis of biochemical by an organism is not associated with its growth. Non-growth associated products are maximally synthesized when the culture reaches to the stationary growth phase.

Niche: Niches are subsets of tissue cells, extracellular substrates, and growth factors that house stem cells, and controls self-renewal and progeny production.

Oncogenes: A gene that causes cells to grow in an uncontrolled manner, i.e., that causes cancer. Oncogenes are mutant forms of normal functional genes (called proto-oncogenes) that have a role in normal cell proliferation. Oncogenes are found in tumours and in retroviruses; in the latter case, having been picked up along with retroviral genes during retroviral replication in the host. Once incorporated as part of the retroviral genome, they are activated at inappropriate times and places by retroviral promoters, thereby becoming oncogenes.

Ontogeny: Developmental life history of an organism.

Oxygen transfer rate: The amount of oxygen transferred from gas phase to the liquid (medium) phase in unit time.

Passage: A round of cell growth and proliferation in culture.

Pasteur effect: The inhibition effect of oxygen on the process of glycolysis or fermentation.

Paracrine signaling: The signaling effects of the target cell causes by the cells of close proximity by releasing chemical factor(s).

Passive transport: It means moving of biochemicals, and atomic substances, across membranes without involving chemical enegry. It dependents on the permeability of the cell membrane, which, in turn, is dependent on the organization and characteristics of the membrane lipids and proteins.

Passive immunity: 1. Natural acquisition of antibodies by the foetus or neonate (newborn) from the mother. 2. The artificial introduction of specific antibodies by the injection of serum from an immune animal. In both cases, temporary protection is conferred on the recipient.

Perfusion bioreactor: In a bioreactor, cells are perfused with fresh medium without allowing them to leave the bioreactor.

Plating efficiency: Percentage of inoculated cells give rise to colonies.

Plasticity: The ability of stem cells from one adult tissue to generate the differentiated types of another tissue.

Pluripotent stem cells: Pluripotent stem cells can eventually specialize in any bodily tissue, but they cannot themselves develop into a human being.

Population doubling: A doubling in the number of cells when grown in culture.

Primordial germ cells: These are the precursors of *gametes* (sperm and oocyte) in adult mammals.

Primary culture: A culture started from cells, tissues or organs taken directly from organisms. A primary culture may be regarded as such until it is sub-cultured for the first time. It then becomes a cell line.

Promoter: A nucleotide sequence of DNA to which RNA polymerase binds and initiates transcription. It usually lies upstream of a coding sequence. A promoter sequence aligns the RNA polymerase so that transcription will initiate at a specific site.

Primary containment: It is the technique by which initially the hazardous materials are contained within a defined space. Biological Safety Cabinets are among the most effective and the most commonly used primary containment devices in laboratories working with infectious agents.

Proto-oncogenes: A normal gene that could develop into one that causes a transformation of normal cells into cancerous tumor cells, especially a viral gene that transforms a host cell into a tumor cell.

Protein engineering: Generating proteins with modified structures that confer properties such as higher catalytic specificity or thermal stability.

Quiescent cell: Cells are at resting state, but not necessarily dormant, and having potential to resume metabolic activity.

Reporter gene: Inserted into a genome along with a new gene to show the position and existence of the new gene.

Reverse transcription: The process of converting RNA to DNA.

Ribosome: RNA- and protein-rich cytoplasmic organelles that are sites of protein synthesis.

Scaffolds: It is a temporary framework used to support the growth of cells for tissue engineering applications.

Scale up: Conversion of a process, such as fermentation of a microorganism, from a small scale to a larger scale.

Selection pressure: The intensity of selection acting on a population of organisms or cells in culture. Its effectiveness is measured in terms of differential survival and reproduction, and consequently in change in the frequency of alleles in a population.

Senescence: The last stage in the post-embryonic development of multicellular organisms, during which loss of functions and degradation of biological components occur. In physiological ageing process, cells and tissues deteriorate and finally die.

Signal transduction: The biochemical events that conduct the signal of a hormone or growth factor from the cell exterior, through the cell membrane, and into the cytoplasm. This involves a number of molecules, including receptors, ligands and messengers.

Site-specific mutagenesis: A technique to change one or more specific nucleotides within a cloned gene in order to create an altered form of a protein with one or more specific amino acid changes.

Somatic cells: Any cell of a plant or animal other than a germ cell or precursor of it.

Stromal cell: A non-blood cell that is derived from mesenchymal stem cells, which is capable of supporting growth of hematopoietic stem/precursor cells.

Somatic stem cells: Cells from the body that compose the tissues, organs, and parts of that individual other than the germ (sex) cells. Each somatic cell contains a full set of chromosomes, whereas sex cells only contain half.

Somatic cell nuclear transfer: A process by which a nucleus from a single cell (for example a skin cell) is transferred into an unfertilized egg, from which the nucleus (the genetic contents) has been removed. The resulting reconstructed embryo is then allowed to develop to the blastocyst stage.

Splicing: During the maturation of eukaryotic mRNA, the process that eliminates intervening intron sequences and covalently joins exon sequences of RNA.

Synchronous culture: A culture in which the majority of cells is dividing at the same time or is at a specific phase of the cell cycle.

Sterility: It is the absence of living organisms.

Stemness: It is the property of stem cells by which they divide extensively to maintain their pool and differentiated progeny, without compromising with their inherent properties.

Shuttle vector (bifunctional vector): A plasmid capable of replicating in two different host organisms because it carries two different origins of replication and can therefore be used to 'shuttle' genes from one to the other.

Telomerase: An enzyme that is composed of a catalytic protein component and a RNA template and that synthesizes DNA at the ends of chromosomes and confers replicative immortality to cells.

Teratocarcinoma: A tumor that occurs mostly in the testis.

Totipotent stem cells: Stem cells which are capable of forming every type of body cell. Each totipotent cell could replicate and differentiate and become a human being. All cells within the early embryo are totipotent until the 16-cell stage or so.

Transgene: A gene that has been incorporated from one cell or organism and passed on to successive generations.

Transactivation: It is an increased rate of gene expression triggered either by endogenous cellular or viral proteins.

Ultrastucture: It is the detailed structure of a cell or tissue that can be observed by transmission electron miscopy.

Umbilical cord: In mammals the umbilical cord is a tube that connects a developing embryo to the placenta.

Unipotent stem cells: The precursor stem cell that gives rise to one type of tissue specific cell after maturation.

Vaccine: A preparation of dead or weakened pathogens or of derived antigenic determinants that is used to induce formation of antibodies or immunity against the pathogen.

Validation: Action of proving, in accordance with the principles of GMP, that any procedure, process, equipment, material, activity or system actually leads to the expected results.

Wash-out: The loss of the slower growing microorganism when two organisms are being grown together in a chemstate.

Wild type: An organism as found in nature; the dominant allele usually found in nature, and from which mutations produce other dominants or recessives alleles.

Xenotransplantation: The transplantation of tissue from one species to another species, typically from non-human mammals to humans. This technology has become very important because of a worldwide shortage of organs for humans requiring a new organ. The most popular non-human species involved in this technology is the pig.

Zygote: An egg cell that comes from fertilization. It contains the complete set of chromosomes received from the union of the male (sperm) and female (egg) sex cells. The zygote develops into the organisms' adult body.

INDEX

A

Active immunization 168
Adeno-associated viral vector 310
Adenoviral vector 309
Adhesive junctions 6
Adult human stem cells 221, 240, 241, 265
    plasticity in 249
Adult skin 270
Adult stem cells therapy 318
Aerobic glycolysis 13
Air, gas, and vent filtration 99
Airlift bioreactor 192, 209, 210, 212
Amino acids 67
    estimation of 342
Amplifiable vector 163
Anaphase 26
Anchorage-dependent cell culture process 118
Animal cell culture facility 85, 86
Animal cell culture processes, limitations of 205
Animal cell culture, role of serum in 69
Animal cell culture-based processes 113
Animal cell culture-based products 181
Animal cell cytoskeleton 7
Animal cell, shear effects on 116
    cryopreservation of 59
Animal tissues, cryopreservation of 59
Antibiotics 68
Antibodies 165
    production of 171
    response to vaccination 166
    therapeutic use of 168
Antioxidants 75
Apoptosis 39
    detection of 43
    factor for 41
    mechanisms of 42
Apoptosome 43
Apoptotic cells, identification of 43
Appropriate translational initiation sequence 158
Aseptic operation 89
Aseptic process 88
Asn-glycosylation 18
Asn-linked oligosaccharide, biosynthesis pathway of 20
Assays, based on cellular activity 352
Asymmetric membranes 103
ATP synthesis 353
Attachment 76
Autohydrolysis 272
Autonomous regeneration 264
    limitation of 266
Autophagic cell death 41

B

Backbone shuttle sequences 158
Batch airlift bioreactor 194
Batch bioreactor 131
Bioartificial liver 294
Bioartificial organs 294, 295
Bioartificial pancreas 295
Bioassay 349
Bioreactor scale-up 209
Biosafety cabinet 93, 97
Biosafety levels 92
Blastocyst 222
Blood cells, densities of 144
Bone marrow microenvironment 242
Bone marrow, transplantation of 319
Bubble point 104
Bubble-free aeration 128

C

Cage aeration 128
Capsule filter 101
Carbohydrates 67
Carbon source concentration 115
Cartridge filter 101

Catabolic control 78
CDR-grafting 174, 175
Cell adhesion 117
Cell and tissue therapy, in regenerative medicine 267
Cell attachment, on microcarrier beads 121
Cell banks, 55
Cell culture 87
    common contaminants in 88
    related equipment 87, 107
Cell culture bioreactor 130
Cell culture medi
    oxygen supply in 126
    physicall/chemical properties of 72
    selection of 65
Cell culture process, problems in 206
Cell culture related products 331
Cell culture,
    aspects of 329
    technology and engineering
    ultra-pure water for 105
Cell cycle 24
    biochemical studies on 26
Cell cycle analysis 28
Cell densities 126
Cell growth 23
Cell junctions 6
Cell membrane, Annexin V staining of 43
Cell number, enumeration of 337
Cell retention devices 137, 138
Cell signaling 35
Cell spreading factors 76
Cell strain 50
    characteristic features of 51
Cell surface 5
Cell theory 3
Cell therapy 264
    approaches of 301
Cell-freezing 60
Cells and tissues, freezing and reconstituting of 363
Cells, characterization of 320
    classification of 49
    for manufacturing biological 52
    sources of 49
    identification of 34
    recovery of 59
    separation of 143, 144, 148
Cellular contaminants 53
Cellular therapy 300
Centrifugal elutriation 145
Centrifuge 107
Channeling problem 216
Check point 31
Chemotherapy 320
Chimeric antibodies 173
Clean air systems, applications of 98
Clean zone, clean room, specification for 91
Clinical grade lentiviral vector 314
Clone, expansion of 171
    stability of 183, 184
    and productivity 184
Cloning 301
    risk of 306
Cloning efficiency 70
c-Myc 252
CO_2 incubator 108
Collagens 278
Compensatory hyperplasia 265, 266
Competitive ELISA 347, 348
Complementarity determining regions 166
Containment 92
Continuous cell lines, characteristic features of 51
Continuous flow bioreactor 133
Continuous perfusion bioreactor 135
Continuous perfusion stirred tank bioreactor 200
Coulter particle counter 339
Counting chamber 337
Culture environment 113, 115
Culture medium, analysis of 340
Culture plate 109
Culture vessels 109
Cultured cells 363
Cumalative population doubling 34
Cytodex microcariers,
    physical properties of 119
    surface chemistry of 119
Cytoplasm 7
Cytosol 7
Cytotoxicity assay 350

D

Dedifferentiation 265
Density dependent inhibition 30
Density gradient centrifugation 143
Density scale-up 206
Depth filter 100
Design of 3-D scaffold 272
Diffusional flow 104

Dihydrofolate reductase (DHFR) amplification 162
Diploid cells 24
Direct steam sterilization 102
Disc filter 101
Dissolved oxygen concentration 115
DNA delivery, and its chromosomal integration 159
DNA fragmentation by agarose gel electrophoresis 44
Dye exclusion 351

E

ECM components 278
    biological role of 280
Embryonic cells 221
    embryonic tissues,
Endocytosis 5
Endoplasmic reticulum 8, 9
Envelop vector plasmid 313
Enzyme-linked immunosorbent assay (ELISA) 345
Equilibrium zones 145
Equipment decontamination 90
ES cell line 234
ES cells 230
    characteristic features of 231
    maintenance of 235
    targeted differentiation of 238
Euchromatin 11
Expanded bed bioreactor 215
Expression vector 164
Extracellular matrix (ECM) 276
Extracellular signaling 36
Extraembryonic and of 224

F

Facilitated diffusion 5
Fatty acids 68
Fed-batch bioreactor 132
Fermentation processes, scale-up criteria for 205
Fetal thymus culture 289
Fiber bonding 274
Filter screw 109
Flow cytometer 27
Fluid mosaic plasma membrane model 6
Fluidized bed bioreactor 194, 196, 215
Fluorescence-activated cell sorter (FACS) 150
Freezing equipment 61
Freezing medium 60

G

Gap junctions 7
Gas foaming technique 274
Gastrulation 225
Gene amplification 161
Gene delivery system 309
Gene therapy 308
    strategy for 315
Germ cell tumors 230
Germline gene therapy 317, 318
Glucose metabolism, regulation of 15
Glucose, estimation of 340
Glycosylation 18
Golgi apparatus 8, 9
Growth factor receptors 36
Growth regulation 29
GS-based amplification 163

H

Haploid cells 24
Hayflick limit 34
Hayflick's phenomenon 33, 34
Hematopoiesis 241
    in mammals 242
Hematopoietic stem 318
Hematopoietic stem cells 241, 244
    ex vivo culture of 247
Hemispherical-bottomed animal cell culture bioreactor 213
Hemocytometer 338
Herpes simplex viral vector 310
Heterogeneously distributed perfusion 138
Heterologous proteins 161
High aerobic glycolysis 15
Hollow-fiber bioreactor 139, 187, 188, 190, 191
Hollow-fiber cartridges 189, 192
Hormones 69, 75
    growth factors 75
Host cell engineering 183
Human blastocyst 222
Human ES cell line 233
Human Mab, development of 172
Human umbilical vein endothelial cells 361
Human viral vaccines, HDCs and 53
Humanized antibody 172
Human-mouse chimera, making of 173
Human-mouse chimeric antibody, expression vectors for 174
Hybridoma, cloning and selection of 170

Hydrodynamic cell death 124, 125
Hypervariable regions 166

I

Immunized spleen cells, fusion with myeloma cell 170
Immunoglobulins 165
Immunomagnetic separation 148
In situ DNA fragmentation 44
In vitro culture method, advantages of 185
Indirect and direct immunomagnetic separation of cells 148
Induced pluripotent stem (iPS) cells 252, 253
Intercellular signaling 36
Intron 158
Inverted microscope 107
iPS cells, applications of 254
iPS technology, applications of 325
 limitations of 254

K

Kidney, decellularization of 292
Klf4 253

L

Lactate, estimation of 342
Large-scale animal cell culture process 179
LDH release assay 352
Lentiviral vector 310, 312, 313
 construction of 312
 for DNA transfer, 311
Life, beginning of 221
Limiting dilution technique 171
Lipids 68, 75
Liver, decellularization of 291
 recellularization of 293
Lungs, decellularization of 292
Lysosomes 10

M

Mab, commercial production of 186
Magnetic cell sorting technique 146, 147
Magnetically labeled cells, isolation of 147
Maintenance regeneration 264
Major mammalian ECM components 281
Mammalian cell culture medium, safe oxygenation of 128
Mammalian cell derived products 332
Mammalian cells,
 folding of proteins in 16
 growth of 29, 30
 organization of 3
 physical properties of 11
 post-translational modification of 18
 secretion of proteins in 16
 used for manufacturing vaccines 54
 pathway for recombinant protein production 157
Mammalian expression vector, 157
 structural components 159
Mammalian heat-shock proteins 16
Mammalian promoter/enhancer 159
Mammary epithelial cells 362
Manufacturers' working cell bank 58
Marker gene 159
Master cell bank 56
Mechanical force, and tissue remodeling 281
Media filtration 102
Media preparation
 and testing 79
 precautionary measures 79
Media, basic constituents of 66
Medium supplements 68
MEF feeder layer 236
Membrane filter 101, 103
Mesenchymal stem cells 241, 248
Metabolic quotients, determination of 345
Metabolism 78
Metaphase 26
Microbial cells, physical properties of 11
Microbial contamination 58
Microcarrier beads, high-density cell culture on 199
Microcarrier culture, benefits of 198
Microcarrier density 123
Microcarrier particles, growth of animal cells on 120
Minerals 67
Mitochondria 8
Mitosis 25
Mitotic index 27
Molecular chaperones 17
Monkey kidney cells 359
Monoclonal antibodies (Mab) 165, 171
 production of Rabbies vaccine 180, 185
Morphology 56
Mouse ascites method, advantages of 185
Mouse bone marrow cells, isolation of 356
Mouse embryonic fibroblast (MEF) 357
Mouse ES cell line 231
Mouse ES cells, targeted differentiation of 239
Mouse, immunization of 170

Mouse-mouse hybridoma 169
mRNA stability 160
MSCs 320
Multilineage group 247
Multipotent stem cells 227
Multitray flask 109

N

Nanotechnology, and fabrication of scaffold 287
Necrotic death 39
Negative selection 149
Neurons 238
Niche, structure of 258
    function of 258
N-linked glycans 19
Non-rotating bioreactors 285, 286
Normal cells, glucose metabolism in 13
Normal fertilization 304
Nuclear staining 351
Nucleic acid precursors 68
Nucleoplasm 10
Nucleus 10

O

Oct3/4 253
O-linked glycosylation 21
Optimum base media 76
Organ culture 288
    alternate techniques of 289
    decellularized matrices 290
Organ, decellularization of 290
Osmolality 73
Oxygen 69
Oxygenation
    by direct sparging of air, 127
    by increasing partial pressure 129
    by medium perfusion 129
    by surface aeration, 127

P

Packaging plasmid 312
Packed glass beads perfusion system 140
Pancreatic islets 238
Panning method 146
Parameters 115
Passive immunization 168
Passive transport 5
Pasteur effect 13
Perfusion bioreactor, cell retention systems in 136
Peroxisomes 10
pH 115
Phagocytosis 5
Phase separation and emulsification 274
Pinocytosis 5
Plasma membrane 5
Plasma membrane, integrity of, 351
Plasmid copy number 58
Plating efficiency 70, 349
Pluripotent stem cells 226, 228
Pluripotent stem cells therapy 323
Polyadenylation signal 159
Population doublings 50
Porous matrix from collagen 274
Porous microcarriers 195
Positive selection 149
Potential biocompatible materials 273
Potentiating group 247
Pre-implanted blastocyst in humans 223
Primary cells, characteristic features of 51
Primary cultures 356
Primordial germ (PG) cells 226
Process-specific filter validation 104
Progenitor cells 318
Proliferation index 27
Proliferation, molecular basis of 30
Prophase 25
Proteins 76
Proteins/antigens, assay of 345
Proteoglycans 279

Q

Quiescent cell 30

R

Radio immuno assay (RIA) 348
Radioisotope incorporation 353
Rat hepatocyte 360
Receptor tyrosine kinase, activation of 38
Recombinant proteins 54
Regenerative medicine 264, 300
Replacement assay 256
Replicative senescence 34
Reproductive cloning 303, 304
    ethical issues in 307
Respiration rate 12
Restriction point 31
Retroviral vector 309
Risk assessment 98
Roller apparatus 109
Roller bottle 109
Rotating bioreactors 285, 286
Rough endoplasmic reticulum 8

S

Sandwich ELISA 346, 347
Scaffold, recellularization of 293
Sedimentation velocity 145
Senescent cells 34
Separation efficiency 145
Serum free media,
  basic supplements of 74
  development of 74
  merits and demerits 74
Serum lot, selection of 70
Serum, constituents of 70
Serum-supplemented culture media 71
SFM, adaptation of cells to 76
Shear sensitive cells 139
Side population cells 245
Signal transduction 35
Solid microcarriers 118
Solvent casting/particulate leaching 274
Somatic cell gene therapy 316
Somatic cell nuclear transfer 302
  applications of 303
Somatic cells 240
Somatic nuclear cloning 305
Specific glucose uptake 12
Specific oxygen uptake rate 343
Spinner apparatus 109
Spinner flask 109
Steam-in-pace sterilization 102
Stem cell niche 255
  classification of 256
  composition of 259
  identification of 256
Stem cells 226, 240
  cryopreservation of 319
  harvesting of 319
Stem cells infusion 320
Sterile grade filtration 99
Stirred bioreactor 212
Stirred tank bioreactor 197
Storage container 61
Stress-induced cell death 39
Structural glycoproteins 279
Sub-G0/G1 fraction of cells 44
Surface tension 73
Suspension versus adhered cell culture 116, 117

T

T cell mediated antibody response 168
T flask 109
Telophase 26
Temperature 73
Tetrazolium dye reduction 352
Therapeutic antibodies 169
Therapeutic antibody production systems 187
Therapeutic cloning 303, 305
Three-stage batch air-lift bioreactor 193
Tight junctions 7
Tissue engineering 264, 268
  bioreactors in 282
  current approaches in 287
  scaffolds for 275
Tissue level regeneration, mechanism of 265
Tissue of origin 57
Tissue plasminogen activator (t-PA), production of 208
Tissue,
  composition of 269
  three-dimensional growth of 271
  biomaterials for 3-D growth of 272
Totipotent stem cells 226
Trace elements 67
Transfer vector plasmid 312
Tubular biological film bioreactor 140
Tumor cells, glucose metabolism in 13

U

Ultra-low temperature cell storage methods 62
Ultra-pure water generating system 107
Unfolded protein response 39
Unilineage group 248
Unipotent stem cells 227
Unwoven meshes 274

V

Verax microsphere 215
Viable cells 143
Vials, storing of 61
Viral vaccine, production of 199
Viscosity 73
Vitamins 67
Volume scale-up 206

W

Waste metabolite concentration 116